Fatigue Testing and Analysis

(Theory and Practice)

FATIGUE TESTING AND ANALYSIS

(Theory and Practice)

YUNG-LI LEE
DaimlerChrysler

JWO PAN
University of Michigan

RICHARD B. HATHAWAY
Western Michigan University

MARK E. BARKEY
University of Alabama

ELSEVIER
BUTTERWORTH
HEINEMANN

AMSTERDAM • BOSTON • HEIDELBERG • LONDON • NEW YORK
OXFORD • PARIS • SAN DIEGO • SAN FRANCISCO • SINGAPORE
SYDNEY • TOKYO

Elsevier Butterworth–Heinemann
200 Wheeler Road, Burlington, MA 01803, USA
Linacre House, Jordan Hill, Oxford OX2 8DP, UK

Library of Congress Cataloging-in-Publication Data

ISBN-13: 978-0-7506-7719-6
ISBN-10: 0-7506-7719-8

British Library Cataloguing-in-Publication Data
A catalogue record for this book is available from the British Library.

(Application submitted)

For information on all Butterworth–Heinemann publications
visit our Web site at www.bh.com

Transferred to Digital Printing 2011

Thank our families for their support and patience.

To my parents and my wife Pai-Jen.
- Yung-Li Lee

To my Mom and my wife Michelle.
- Jwo Pan

To my wife Barbara.
- Richard B. Hathaway

To my wife, Tammy, and our daughters Lauren and Anna.
- Mark E. Barkey

PREFACE

Over the past 20 years there has been a heightened interest in improving quality, productivity, and reliability of manufactured products in the ground vehicle industry due to global competition and higher customer demands for safety, durability and reliability of the products. As a result, these products must be designed and tested for sufficient fatigue resistance over a large range of product populations so that the scatters of the product strength and loading have to be quantified for any reliability analysis.

There have been continuing efforts in developing the analysis techniques for those who are responsible for product reliability and product design. The purpose of this book is to present the latest, proven techniques for fatigue data acquisition, data analysis, test planning and practice. More specifically, it covers the most comprehensive methods to capture the component load, to characterize the scatter of product fatigue resistance and loading, to perform the fatigue damage assessment of a product, and to develop an accelerated life test plan for reliability target demonstration.

The authors have designed this book to be a useful guideline and reference to the practicing professional engineers as well as to students in universities who are working in fatigue testing and design projects. We have placed a primary focus on an extensive coverage of statistical data analyses, concepts, methods, practices, and interpretation.

The material in this book is based on our interaction with engineers and statisticians in the industry as well as based on the courses on fatigue testing and analysis that were taught at Oakland University, University of Michigan, Western Michigan University and University of Alabama. Five major contributors from several companies and universities were also invited to help us enhance the completeness of this book. The name and affiliation of the authors are identified at the beginning of each chapter.

There are ten chapters in this book. A brief description of these chapters is given in the following.

Chapter 1 (Transducers and Data Acquisition) is first presented to address the importance of sufficient knowledge of service loads/stresses and how to measure these loads/stresses. The service loads have significant effects on the results of fatigue analyses and therefore accurate measurements of the actual service loads are necessary. A large portion of the chapter is focused on the strain gage as a transducer of the accurate measurement of the strain/stress, which is the most significant predictor of fatigue life analyses. A variety of methods to identify the high stress areas and hence the strain gage placement in the test part are also presented. Measurement for temperature, number of temperature cycles per unit time, and rate of temperature rise is included. The inclusion is to draw the attention to the fact that fatigue life prediction is based on both the number of cycles at a given stress level during the service life and the service environments. The basic data acquisition and analysis techniques are also presented.

In Chapter 2 (Fatigue Damage Theories), we describe physical fatigue mechanisms of products under cyclic mechanical loading conditions, models to describe the mechanical fatigue damages, and postulations and practical implementations of these commonly used damage rules. The relations of crack initiation and crack propagation to final fracture are discussed in this chapter.

In Chapter 3 (Cycle Counting Techniques), we cover various cycle counting methods used to reduce a complicated loading time history into a series of simple constant amplitude loads that can be associated with fatigue damage. Moreover the technique to reconstruct a load time history with the equivalent damage from a given cycle counting matrix is introduced in this chapter.

In Chapter 4 (Stress-Based Fatigue Analysis and Design), we review methods of determining statistical fatigue properties and methods of estimating the fatigue resistance curve based on the definition of nominal stress amplitude. These methods have been widely used in the high cycle fatigue regime for decades and have shown their applicability in predicting fatigue life of notched shafts and tubular components. The emphasis of this chapter is on the applications of these methods to the fatigue design processes in the ground vehicle industry.

In Chapter 5 (Strain-Based Fatigue Analysis and Design), we introduce the deterministic and statistical methods for determining the fatigue resistance parameters based on a definition of local strain amplitude. Other accompanied techniques such as the local stress-strain simulation and notch analysis are also covered. This method has been recommended by the SAE Fatigue Design & Evaluation Committee for the last two decades for its applicability in the low and high cycle fatigue regimes. It appears of great value in the application of notched plate components.

In Chapter 6 (Fracture Mechanics and Fatigue Crack Propagation), the text is written in a manner to emphasize the basic concepts of stress concentration factor, stress intensity factor and asymptotic crack-tip field for linear elastic materials. Stress intensity factor solutions for practical cracked geometries under simple loading conditions are given. Plastic zones and requirements of linear elastic fracture mechanics are then discussed. Finally, fatigue crack propagation laws based on linear elastic fracture mechanics are presented.

In Chapter 7 (Fatigue of Spot Welds), we address sources of variability in the fatigue life of spot welded structures and to describe techniques for calculating the fatigue life of spot-welded structures. The load-life approach, structural stress approach, and fracture mechanics approach are discussed in details.

In Chapter 8 (Development of Accelerated Life Test Criteria), we provide methods to account for the scatter of loading spectra for fatigue design and testing. Obtaining the actual long term loading histories via real time measurements appears difficult due to technical and economical reasons. As a consequence, it is important that the field data contain all possible loading events and the results of measurement be properly extrapolated. Rainflow cycle counting matrices have been recently, predominately used for assessing loading variability and cycle extrapolation. The following three main features are covered: (1) cycle extrapolation from short term measurement to longer time spans, (2) quantile cycle extrapolation from median loading spectra to extreme loading, and (3) applications of the extrapolation techniques to accelerated life test criteria.

In Chapter 9 (Reliability Demonstration Testing), we present various statistical-based test plans for meeting reliability target requirements in the accelerated life test laboratories. A few fatigue tests under the test load spectra should be carried out to ensure that the product would pass life test criteria.

The statistical procedures for the choice of a test plan including sample size and life test target are the subject of our discussion.

In Chapter 10 (Fatigue Analysis in the Frequency Domain), we introduce the fundamentals of random vibrations and existing methods for predicting fatigue damage from a power spectral density (PSD) plot of stress response. This type of fatigue analysis in the frequency domain is particularly useful for the use of the PSD technique in structural dynamics analyses.

The authors greatly thank to our colleagues who cheerfully undertook the task of checking portions or all of the manuscripts. They are Thomas Cordes (John Deere), Benda Yan (ISPAT Inland), Steve Tipton (University of Tulsa), Justin Wu (Applied Research Associates), Gary Halford (NASA-Glenn Research Center), Zissimos Mourelatos (Oakland University), Keyu Li (Oakland University), Daqing Zhang (Breed Tech.), Cliff Chen (Boeing), Philip Kittredge (ArvinMeritor), Yue Chen (Defiance), Hongtae Kang (University of Michigan-Dearborn), Yen-Kai Wang (ArvinMeritor), Paul Lubinski (ArvinMeritor), and Tana Tjhung (DaimlerChrysler).

Finally, we would like to thank our wives and children for their love, patience, and understanding during the past years when we worked most of evenings and weekends to complete this project.

Yung-Li Lee, DaimlerChrysler
Jwo Pan, University of Michigan
Richard B. Hathaway, Western Michigan University
Mark E. Barkey, University of Alabama

TABLE OF CONTENTS

1 TRANSDUCERS AND DATA ACQUISITION 1
Richard B. Hathaway, Western Michigan University
Kah Wah Long, DaimlerChrysler

2 FATIGUE DAMAGE THEORIES 57
Yung-Li Lee, DaimlerChrysler

3 CYCLE COUNTING TECHNIQUES 77
Yung-Li Lee, DaimlerChrysler
Darryl Taylor, DaimlerChyrysler

4 STRESS-BASED FATIGUE ANALYSIS AND DESIGN 103
Yung-Li Lee, DaimlerChrysler
Darryl Taylor, DaimlerChyrysler

5 STRAIN-BASED FATIGUE ANALYSIS AND DESIGN 181
Yung-Li Lee, DaimlerChrysler
Darryl Taylor, DaimlerChyrysler

6 FRACTURE MECHANICS AND FATIGUE CRACK PROPAGATION 237

Jwo Pan, University of Michigan
Shih-Huang Lin, University of Michigan

7 FATIGUE OF SPOT WELDS 285

Mark E. Barkey, University of Alabama
Shicheng Zhang, DaimlerChrysler AG

8 DEVELOPMENT OF ACCELERATED LIFE TEST CRITERIA 313

Yung-Li Lee, DaimlerChrysler
Mark E. Barkey, University of Alabama

9 RELIABILITY DEMONSTRATION TESTING 337

Ming-Wei Lu, DaimlerChrysler

10 FATIGUE ANALYSIS IN THE FREQUENCY DOMAIN 369

Yung-Li Lee, DaimlerChrysler

About the Authors

Dr. Yung-Li Lee is a senior member of the technical staff of the Stress Lab & Durability Development at DaimlerChrysler, where he has conducted research in multiaxial fatigue, plasticity theories, durability testing for automotive components, fatigue of spot welds, and probabilistic fatigue and fracture design. He is also an adjunct faculty in Department of Mechanical Engineering at Oakland University, Rochester, Michigan.

Dr. Jwo Pan is a Professor in Department of Mechanical Engineering of University of Michigan, Ann Arbor, Michigan. He has worked in the area of yielding and fracture of plastics and rubber, sheet metal forming, weld residual stress and failure, fracture, fatigue, plasticity theories and material modeling for crash simulations. He has served as Director of Center for Automotive Structural Durability Simulation funded by Ford Motor Company and Director for Center for Advanced Polymer Engineering Research at University of Michigan. He is a Fellow of American Society of Mechanical Engineers (ASME). He is on the editorial boards of International Journal of Fatigue and International Journal of Damage Mechanics.

Dr. Richard B. Hathaway is a professor of Mechanical and Aeronautical engineering and Director of the Applied Optics Laboratory at Western Michigan University. His research involves applications of optical measurement techniques to engineering problems including automotive structures and powertrains. His teaching involves Automotive structures, vehicle suspension, and instrumentation. He is a 30-year member of the Society of

Automotive Engineers (SAE) and a member of the Society of Photo-Instrumentation Engineers (SPIE).

Dr. Mark E. Barkey is an Associate Professor in the Aerospace Engineering and Mechanics Department at the University of Alabama. He has conducted research in the areas of spot weld fatigue testing and analysis, multiaxial fatigue and cyclic plasticity of metals, and multiaxial notch analysis. Prior to his current position, he was a Senior Engineering in the Fatigue Synthesis and Analysis group at General Motors Mid-Size Car Division.

1

Transducers and Data Acquisition

Richard Hathaway
Western Michigan University
Kah Wah Long
DaimlerChrysler

1.1 INTRODUCTION

This chapter addresses the sensors, sensing methods, measurement systems, data acquisition, and data interpretation used in the experimental work that leads to fatigue life prediction. A large portion of the chapter is focused on the strain gage as a transducer. Accurate measurement of strain, from which the stress can be determined, is one of the most significant predictors of fatigue life. Prediction of fatigue life often requires the experimental measurement of localized loads, the frequency of the load occurrence, the statistical variability of the load, and the number of cycles a part will experience at any given load. A variety of methods may be used to predict the fatigue life by applying either a linear or weighted response to the measured parameters.

Experimental measurements are made to determine the minimum and maximum values of the load over a time period adequate to establish the repetition rate. If the part is of complex shape, such that the strain levels cannot be easily or accurately predicted from the loads, strain gages will need to be applied to the component in critical areas. Measurements for temperature, number of temperature cycles per unit time, and rate of temperature rise may be included. Fatigue life prediction is based on knowledge of both the number of cycles the part will experience at any given stress level during that life cycle and other influential environmental and use factors. Section 1.2 begins with a review of surface strain measurement, which can be used to predict stresses and ultimately lead to accurate fatigue life prediction. One of the most commonly accepted methods of measuring strain is the resistive strain gage.

1.2 STRAIN GAGE FUNDAMENTALS

Modern strain gages are resistive devices that experimentally evaluate the load or the strain an object experiences. In any resistance transducer, the resistance (R) measured in ohms is material and geometry dependent. Resistivity of the material (ρ) is expressed as resistance per unit length × area, with cross-sectional area (A) along the length of the material (L) making up the geometry. Resistance increases with length and decreases with cross-sectional area for a material of constant resistivity. Some sample resistivities (μohms-cm^2/cm) at 20°C are as follows:

Aluminum: 2.828
Copper: 1.724
Constantan: 4.9

In Figure 1.1, a simple wire of a given length (L), resistivity (ρ), and cross-sectional area (A) has a resistance (R) as shown in Equation 1.2.1:

$$R = \rho\left(\frac{L}{\frac{\pi}{4}D^2}\right) = \rho\left(\frac{L}{A}\right)$$

If the wire experiences a mechanical load (P) along its length, as shown in Figure 1.2, all three parameters (L, ρ, A) change, and, as a result, the end-to-end resistance of the wire changes:

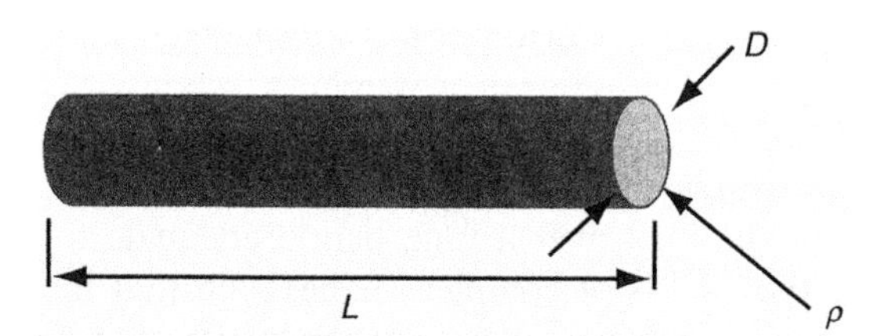

R = Resistance (ohms) L = Length ρ = Resistivity [(ohms × area) / length)]
A = Cross-sectional Area:
Area is sometimes presented in circular mils, which is the area of a 0.001-inch diameter.

FIGURE 1.1 A simple resistance wire.

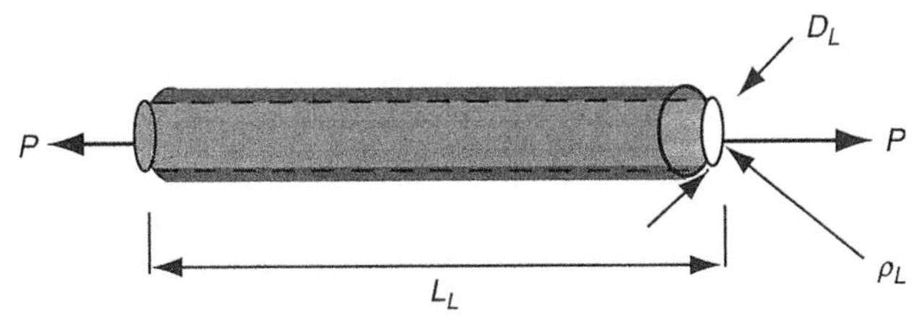

FIGURE 1.2 A resistance wire under mechanical load.

$$\Delta R = \left(\rho_L \times \frac{L_L}{\frac{\pi}{4} D_L^2} \right) - \left(\rho \times \frac{L}{\frac{\pi}{4} D^2} \right) \tag{1.2.2}$$

The resistance change that occurs in a wire under mechanical load makes it possible to use a wire to measure small dimensional changes that occur because of a change in component loading. The concept of strain (ε), as it relates to the mechanical behavior of loaded components, is the change in length (ΔL) the component experiences divided by the original component length (L), as shown in Figure 1.3:

$$\varepsilon = \text{strain} = \frac{\Delta L}{L} \tag{1.2.3}$$

It is possible, with proper bonding of a wire to a structure, to accurately measure the change in length that occurs in the bonded length of the wire. This is the underlying principle of the strain gage. In a strain gage, as shown in Figure 1.4, the gage grid physically changes length when the material to which it is bonded changes length. In a strain gage, the change in resistance occurs when the conductor is stretched or compressed. The change in resistance (ΔR) is due to the change in length of the conductor, the change in cross-sectional area of the conductor, and the change in resistivity ($\Delta \rho$) due to mechanical strain. If the unstrained resistivity of the material is defined as ρ_{us} and the resistivity of the strained material is ρ_s, then $\rho_{us} - \rho_s = \Delta \rho$.

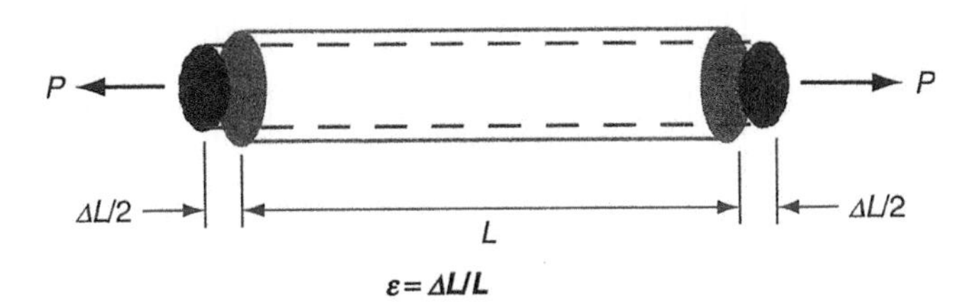

FIGURE 1.3 A simple wire as a strain sensor.

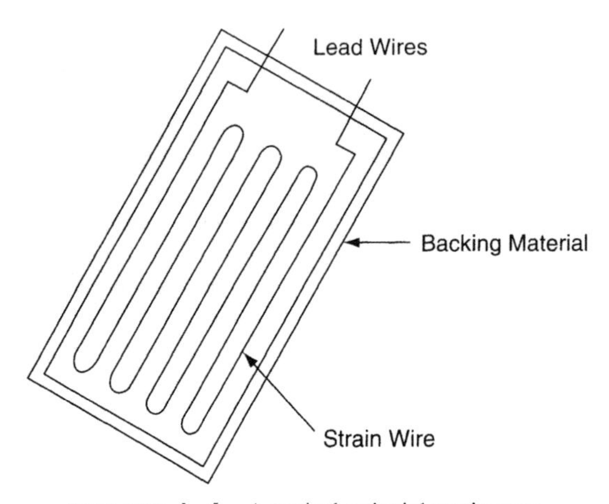

FIGURE 1.4 A typical uniaxial strain gage.

$$R = \frac{\rho L}{A} \qquad \frac{\Delta R}{R} \approx \frac{\Delta L}{L} + \frac{\Delta \rho}{\rho} - \frac{\Delta A}{A} \tag{1.2.4}$$

The resistance strain gage is convenient because the change in resistance that occurs is directly proportional to the change in length per unit length that the transducer undergoes. Two fundamental types of strain gages are available, the wire gage and the etched foil gage, as shown in Figure 1.5. Both gages have similar basic designs; however, the etched foil gage introduces some additional flexibility in the gage design process, providing additional control, such as temperature compensation. The etched foil gage can typically be produced at lower cost.

The product of gage width and length defines the active gage area, as shown in Figure 1.6. The active gage area characterizes the measurement surface and the power dissipation of the gage. The backing length and width define the required mounting space. The gage backing material is designed such that high

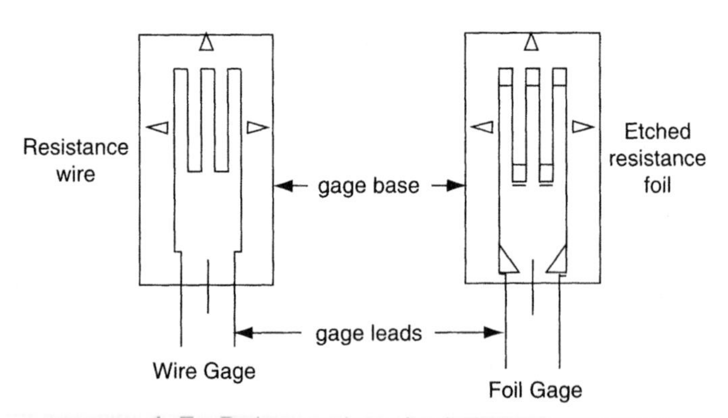

FIGURE 1.5 Resistance wire and etched resistance foil gages.

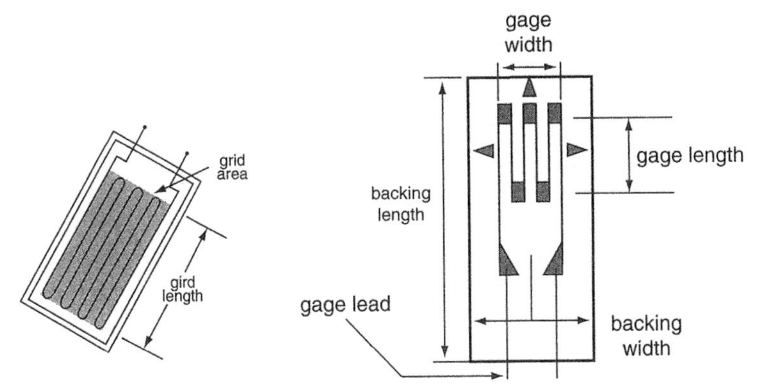

FIGURE 1.6 Gage dimensional nomenclature.

transfer efficiency is obtained between the test material and the gage, allowing the gage to accurately indicate the component loading conditions.

1.2.1 GAGE RESISTANCE AND EXCITATION VOLTAGE

Nominal gage resistance is most commonly either 120 or 350 ohms. Higher-resistance gages are available if the application requires either a higher excitation voltage or the material to which it is attached has low heat conductivity. Increasing the gage resistance (R) allows increased excitation levels (V) with an equivalent power dissipation (P) requirement as shown in Equation 1.2.5.

Testing in high electrical noise environments necessitates the need for higher excitation voltages (V). With analog-to-digital (A–D) conversion for processing in computers, a commonly used excitation voltage is 10 volts. At 10 volts of excitation, each gage of the bridge would have a voltage drop of approximately 5 V. The power to be dissipated in a 350-ohm gage is thereby approximately 71 mW and that in a 120-ohm gage is approximately 208 mW:

$$P_{350} = \frac{V^2}{R} = \frac{5^2}{350} = 0.071\text{W} \quad P_{120} = \frac{V^2}{R} = \frac{5^2}{120} = 0.208\text{W} \tag{1.2.5}$$

At a 15-volt excitation with the 350-ohm gage, the power to be dissipated in each arm goes up to 161 mW. High excitation voltage leads to higher signal-to-noise ratios and increases the power dissipation requirement. Excessively high excitation voltages, especially on smaller grid sizes, can lead to drift due to grid heating.

1.2.2 GAGE LENGTH

The gage averages the strain field over the length (L) of the grid. If the gage is mounted on a nonuniform stress field the average strain to which the active

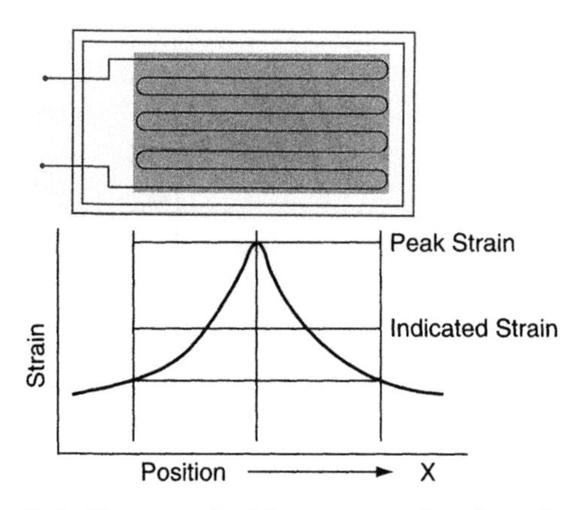

A gage length that is one-tenth of the corresponding dimension of any stress raiser where the measurement is made is usually acceptable.

FIGURE 1.7 Peak and indicated strain comparisons.

gage area is exposed is proportional to the resistance change. If a strain field is known to be nonuniform, proper location of the smallest gage is frequently the best option as shown in Figure 1.7.

1.2.3 GAGE MATERIAL

Gage material from which the grid is made is usually constantan. The material used depends on the application, the material to which it is bonded, and the control required. If the gage material is perfectly matched to the mechanical characteristics of the material to which it is bonded, the gage can have pseudo temperature compensation with the gage dimensional changes offsetting the temperature-related component changes. The gage itself will be temperature compensated if the gage material selected has a thermal coefficient of resistivity of zero over the temperature range anticipated. If the gage has both mechanical and thermal compensation, the system will not produce apparent strain as a result of ambient temperature variations in the testing environment. Selection of the proper gage material that has minimal temperature-dependant resistivity and some temperature-dependent mechanical characteristics can result in a gage system with minimum sensitivity to temperature changes in the test environment. Strain gage manufacturers broadly group their foil gages based on their application to either aluminum or steel, which then provides acceptable temperature compensation for ambient temperature variations.

The major function of the strain gage is to produce a resistance change proportional to the mechanical strain (ε) the object to which it is mounted experiences. The gage proportionality factor, commonly called the gage factor (GF), which makes the two equations of 1.2.6 equivalent, is defined in Equation 1.2.7. Most common strain gages have a nominal gage factor of 2, although special gages are available with higher gage factors.

$$\varepsilon = \text{Strain} = \frac{\Delta L}{L} \qquad \varepsilon \propto \frac{\Delta R}{R} \tag{1.2.6}$$

$$GF = \frac{\left(\frac{\Delta R}{R}\right)}{\varepsilon} = \frac{\left(\frac{\Delta R}{R}\right)}{\left(\frac{\Delta L}{L}\right)} \quad \text{therefore} \quad \Delta R = GF \times \varepsilon \times R \tag{1.2.7}$$

The gage factor results from the mechanical deformation of the gage grid and the change in resistivity of the material (ρ) due to the mechanical strain. Deformation is the change in length of the gage material and the change in cross-sectional area due to Poisson's ratio. The change in the resistivity, called piezoresistance, occurs at a molecular level and is dependent on gage material.

In fatigue life prediction, cyclic loads may only be a fraction of the loads required to cause yielding. The measured output from the instrumentation will depend on the gage resistance change, which is proportional to the strain. If the loads are relatively low, Equation 1.2.7 indicates the highest output and the highest signal-to-noise ratio is obtained with high-resistance gages and a high gage factor.

Example 1.1. A 350-ohm gage is to be used in measuring the strain magnitude of an automotive component under load. The strain gage has a gage factor of 2. If the component is subjected to a strain field of 200 microstrain, what is the change in resistance in the gage? If a high gage factor 120-ohm strain gage is used instead of the 350-ohm gage, what is the gage factor if the change in resistance is 0.096 ohms?

Solution. By using Equation 1.2.7, the change in resistance that occurs with the 350-ohm gage is calculated as

$$\Delta R = GF \times \varepsilon \times R = 2 \times 200 \times 10^{-6} \times 350 = 0.14\,\text{ohms}$$

By using Equation 1.2.7, the gage factor of the 120-ohms gage is

$$GF = \frac{\frac{\Delta R}{R}}{\varepsilon} = \frac{\frac{0.096}{120}}{200 \times 10^{-6}} = 4$$

1.2.4 STRAIN GAGE ARRANGEMENTS

Strain gages may be purchased in a variety of arrangements to make application easier, measurement more precise, and the information gained more comprehensive. A common arrangement is the 90° rosette, as shown in Figure 1.8. This arrangement is popular if the direction of loading is unknown or varies. This gage arrangement provides all the information required for a Mohr's circle strain analysis for identification of principle strains. Determination of the principle strains is straightforward when a three-element 90° rosette is used, as shown in Figure 1.9.

Mohr's circle for strain would indicate that with two gages at 90° to each other and the third bisecting the angle at 45°, the principle strains can be identified as given in Equation 1.2.8. The orientation angle (ϕ) of principle strain (ε_1), with respect to the X-axis is as shown in Equation 1.2.9, with the shear strain (γ_{xy}) as given in Equation 1.2.10:

$$\varepsilon_{1,2} = \frac{(\varepsilon_x + \varepsilon_y)}{2} +/- \frac{\sqrt{(\varepsilon_x - \varepsilon_y)^2 + \gamma_{xy}^2}}{2} \quad (1.2.8)$$

$$\tan 2\phi = \frac{\gamma_{xy}}{\varepsilon_x - \varepsilon_y} \quad (1.2.9)$$

$$\gamma_{xy} = 2\varepsilon_{45} - \varepsilon_x - \varepsilon_y \quad (1.2.10)$$

The principle strains are then given by Equations 1.2.11 and 1.2.12:

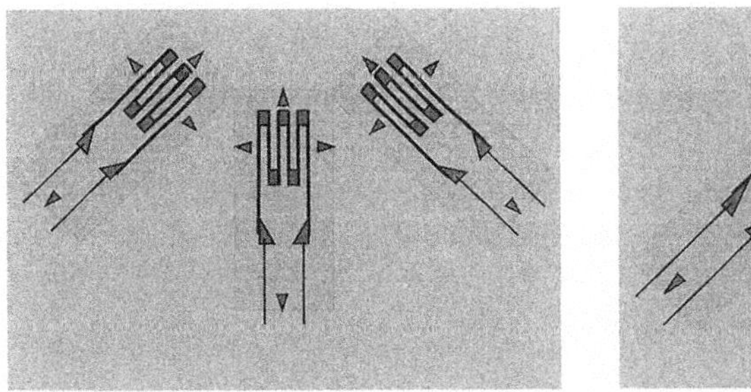
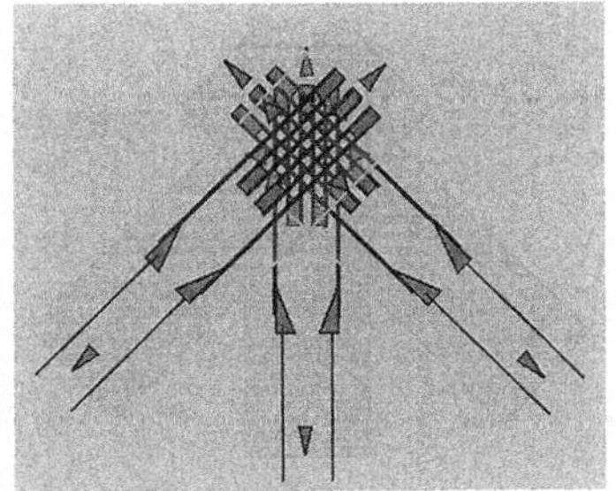

FIGURE 1.8 Three-element rectangular and stacked rectangular strain rosettes.

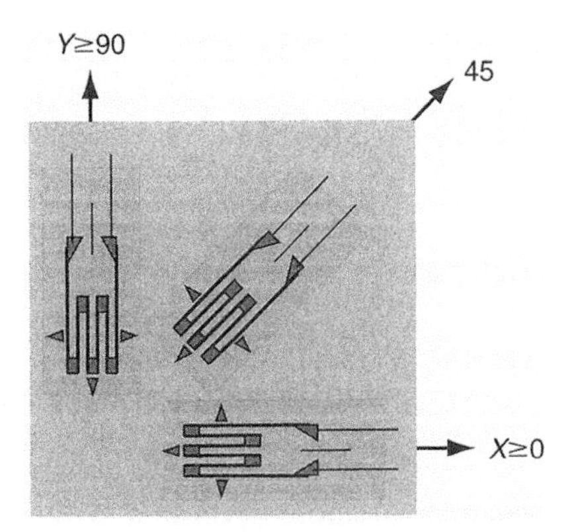

FIGURE 1.9 Rectangular three-element strain rosette.

$$\varepsilon_1 = \frac{(\varepsilon_x + \varepsilon_y)}{2} + \frac{\sqrt{(\varepsilon_x - \varepsilon_y)^2 + (2\varepsilon_{45} - \varepsilon_x - \varepsilon_y)^2}}{2} \qquad (1.2.11)$$

$$\varepsilon_2 = \frac{(\varepsilon_x + \varepsilon_y)}{2} - \frac{\sqrt{(\varepsilon_x - \varepsilon_y)^2 + (2\varepsilon_{45} - \varepsilon_x - \varepsilon_y)^2}}{2} \qquad (1.2.12)$$

Correspondingly, the principle angle (ϕ) is as shown in Equation 1.2.13:

$$\tan 2\phi = \frac{2\varepsilon_{45} - \varepsilon_x - \varepsilon_y}{\varepsilon_x - \varepsilon_y} \qquad (1.2.13)$$

With principle strains and principle angles known, principle stresses can be obtained from stress–strain relationships. Linear stress–strain relationships are given in Equations 1.2.14–1.2.25. In high-strain environments, these linear equations may not hold true.

The linear stress–strain relationships in a three-dimensional state of stress are shown in Equations 1.2.14–1.2.16 for the normal stresses. The stresses and strains are related through the elastic modulus (E) and Poisson's ratio (μ):

$$\varepsilon_x = \frac{1}{E}\left[\sigma_x - \mu(\sigma_y + \sigma_z)\right] \qquad (1.2.14)$$

$$\varepsilon_y = \frac{1}{E}\left[\sigma_y - \mu(\sigma_x + \sigma_z)\right] \qquad (1.2.15)$$

$$\varepsilon_z = \frac{1}{E}\left[\sigma_z - \mu(\sigma_x + \sigma_y)\right] \qquad (1.2.16)$$

The relationship between shear strains and shear stresses are given in Equation 1.2.17. Shear strains and shear stresses are related through the shear modulus (G):

$$[\gamma_{xy}] = \frac{1}{G}\sigma_{xy} \quad [\gamma_{xz}] = \frac{1}{G}\sigma_{xz} \quad [\gamma_{yz}] = \frac{1}{G}\sigma_{yz} \qquad (1.2.17)$$

Equations 1.2.18–1.2.20 can be used to obtain the normal stresses given the normal strains, with a three-dimensional linear strain field:

$$\sigma_x = \frac{E}{(1+\mu)(1-2\mu)} = \left[(1-\mu)\varepsilon_x + \mu(\varepsilon_y + \varepsilon_z)\right] \qquad (1.2.18)$$

$$\sigma_y = \frac{E}{(1+\mu)(1-2\mu)} = \left[(1-\mu)\varepsilon_y + \mu(\varepsilon_x + \varepsilon_z)\right] \qquad (1.2.19)$$

$$\sigma_z = \frac{E}{(1+\mu)(1-2\mu)} = \left[(1-\mu)\varepsilon_z + \mu(\varepsilon_x + \varepsilon_y)\right] \qquad (1.2.20)$$

Shear stresses are directly obtained from shear strains as shown in Equation 1.2.21:

$$\sigma_{xy} = G\lfloor\gamma_{xy}\rfloor \quad \sigma_{xz} = G\lfloor\gamma_{xz}\rfloor \quad \sigma_{yz} = G\lfloor\gamma_{yz}\rfloor \qquad (1.2.21)$$

Equations 1.2.22 and 1.2.23 can be used to obtain principle stresses from principle strains:

$$\sigma_1 = \frac{E}{(1+\mu^2)}(\varepsilon_1 + \mu\varepsilon_2) \tag{1.2.22}$$

$$\sigma_2 = \frac{E}{(1+\mu^2)}(\varepsilon_2 + \mu\varepsilon_1) \tag{1.2.23}$$

Principle stresses for the three-element rectangular rosette can also be obtained directly from the measured strains, as shown in Equations 1.2.24 and 1.2.25:

$$\sigma_1 = \frac{E}{2}\left[\frac{(\varepsilon_x + \varepsilon_y)}{(1-\mu)} + \frac{1}{(1+\mu)}\sqrt{(\varepsilon_x - \varepsilon_y)^2 + (2\varepsilon_{45} - \varepsilon_x - \varepsilon_y)^2}\right] \tag{1.2.24}$$

$$\sigma_2 = \frac{E}{2}\left[\frac{(\varepsilon_x + \varepsilon_y)}{(1-\mu)} - \frac{1}{(1+\mu)}\sqrt{(\varepsilon_x - \varepsilon_y)^2 + (2\varepsilon_{45} - \varepsilon_x - \varepsilon_y)^2}\right] \tag{1.2.25}$$

1.3 UNDERSTANDING WHEATSTONE BRIDGES

The change in resistance that occurs in a typical strain gage is quite small, as indicated in Example 1.1. Because resistance change is not easily measured, voltage change as a result of resistance change is always preferred. A Wheatstone bridge is used to provide the voltage output due to a resistance change at the gage. The strain gage bridge is simply a Wheatstone bridge with the added requirement that either gages of equal resistance or precision resistors be in each arm of the bridge, as shown in Figure 1.10.

1.3.1 THE BALANCED BRIDGE

The bridge circuit can be viewed as a voltage divider circuit, as shown in Figure 1.11. As a voltage divider, each leg of the circuit is exposed to the same

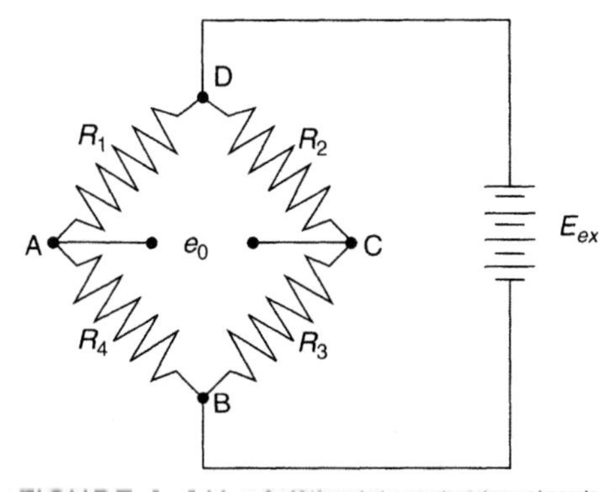

FIGURE 1.10 A Wheatstone bridge circuit.

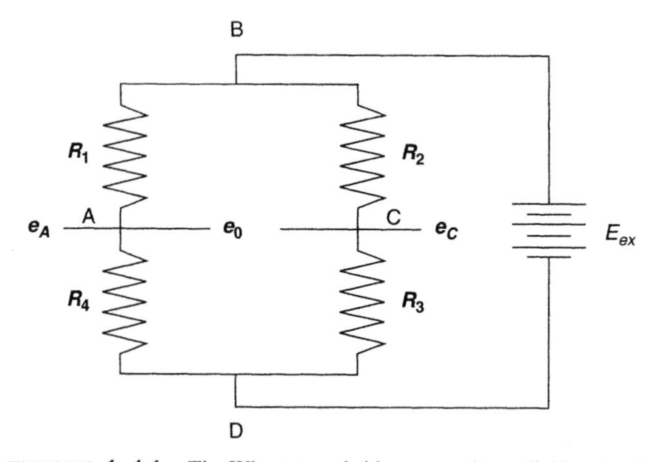

FIGURE 1.11 The Wheatstone bridge as a voltage divider circuit.

excitation voltage (E_{ex}). The current that flows through each leg of the circuit is the excitation voltage divided by the sum of the resistances in the leg, as shown in Equation 1.3.1. If the resistance value of all resistors is equal ($R_1 = R_2 = R_3 = R_4 = R$), the current flow from the source is the excitation voltage (E_{ex}) divided by R, as shown in Equation 1.3.2:

$$I_A = \frac{E_{ex}}{R_1 + R_4} \quad I_C = \frac{E_{ex}}{R_2 + R_3} \tag{1.3.1}$$

$$I_A = \frac{E_{ex}}{2R} \quad I_C = \frac{E_{ex}}{2R} \quad I_{ex} = I_A + I_C = \frac{E_{ex}}{R} \tag{1.3.2}$$

As a voltage divider circuit, the voltages measured between points A and D and between C and D, at the midpoint, are as shown in Equation 1.3.3:

$$e_A = \left(\frac{R_4}{R_1 + R_4}\right) E_{ex} \quad e_C = \left(\frac{R_3}{R_2 + R_3}\right) E_{ex} \tag{1.3.3}$$

The strain gage bridge uses the differential voltage measured between points A and C in Figure 1.11 to determine the output of the bridge resulting from any imbalance, as indicated in Equation 1.3.4:

$$e_o = e_A - e_C = E_{ex}\left[\frac{R_4}{R_1 + R_4} - \frac{R_3}{R_2 + R_3}\right] \tag{1.3.4}$$

If the bridge is initially balanced, points A and C are of equal potential, which implies that the differential e_0 equals zero, as shown in Equation 1.3.5:

$$e_o = 0 = E_{ex}\left[\frac{R_4}{R_1 + R_4} - \frac{R_3}{R_2 + R_3}\right] \tag{1.3.5}$$

With the bridge output zero, $e_A - e_C = 0$, Equation 1.3.6 results for a balanced bridge. Note that the bridge can be balanced without all resistances being equal (Figure 1.12). If $R_2 = R_3$, then $R_1 \equiv R_4$, or if $R_3 = R_4$, then $R_1 \equiv R_2$ for a balanced bridge resulting in zero differential output:

$$\left(\frac{R_4}{R_1 + R_4}\right) = \left(\frac{R_3}{R_2 + R_3}\right) \text{ therefore } R_1 = \frac{R_2 R_4}{R_3} \tag{1.3.6}$$

1.3.2 CONSTANT-CURRENT WHEATSTONE BRIDGE

The constant-current Wheatstone bridge (Figure 1.13) employs a current source for excitation of the bridge. The nonlinearity of this circuit is less than that of the constant-voltage Wheatstone bridge (Dally and Riley, 1991). The constant-current bridge circuit is mainly used with semiconductor strain gages. The voltage drop across each arm of the bridge and the output voltage are as shown in Equations 1.3.7 and 1.3.8:

$$V_{AD} = I_1 R_1 \quad V_{DC} = I_2 R_2 \tag{1.3.7}$$

$$e_o = V_{AD} - V_{DC} = I_1 R_1 - I_2 R_2 \tag{1.3.8}$$

If the constant-current bridge is balanced, then the voltage drop V_{AD} is equal to the voltage drop V_{DC}. The voltage drop across the total bridge (V_{BD}) is then as shown in Equation 1.3.9:

Bridge Excitation		5 V				
Example #	Bridge Type	Resistance Values		Bridge Out/Excitation (V/V)	Bridge Out/Excitation (mV/V)	Bridge Output at 5(mV)
1	balanced bridge	R_1	120	0	0.0	0.0
		R_2	120			
		R_3	120			
		R_4	120			
2	one leg unbalanced	R_1	120	0.002075	2.075	10.37
		R_2	120			
		R_3	120			
		R_4	121			
3	two legs unbalanced and opposite	R_1	119	0.004167	4.167	20.83
		R_2	120			
		R_3	120			
		R_4	121			
4	three legs unbalanced	R_1	119	0.008333	8.333	41.67
		R_2	121			
		R_3	119			
		R_4	121			
5	four legs unbalanced, improperly wired	R_1	119	0	0.0	0.0
		R_2	119			
		R_3	121			
		R_4	121			

FIGURE 1.12 Classical bridge output comparisons.

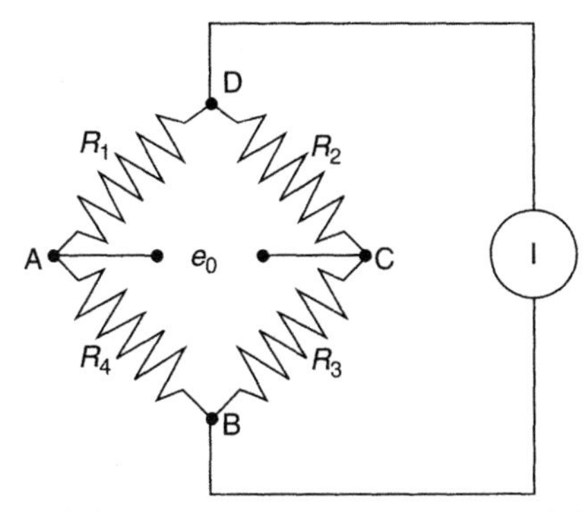

FIGURE 1.13 A constant-current Wheatstone bridge circuit.

$$e_o = V_{AD} - V_{DC} = I_1R_1 - I_2R_2 = 0 \tag{1.3.9}$$

Because the bridge is representative of a parallel circuit, the voltage drop on each leg is equivalent and equal to the circuit voltage drop, as shown in Equation 1.3.10:

$$V_{BD} = I_1(R_1 + R_4) = I_2(R_2 + R_3) \tag{1.3.10}$$

By transposing Equation 1.3.10 and adding the current in each leg to obtain the source current (I), Equation 1.3.11 is obtained:

$$I = I_2\left(\frac{R_1 + R_2 + R_3 + R_4}{R_1 + R_4}\right) \tag{1.3.11}$$

The individual currents in each leg (I_1, I_2) in terms of total circuit current (I) can then be obtained by applying Equations 1.3.12 and 1.3.13:

$$I_1 = I\left(\frac{R_2 + R_3}{R_1 + R_2 + R_3 + R_4}\right) \tag{1.3.12}$$

$$I_2 = I\left(\frac{R_1 + R_4}{R_1 + R_2 + R_3 + R_4}\right) \tag{1.3.13}$$

Substitution of Equations 1.3.12 and 1.3.13 into Equation 1.3.9 results in Equation 1.3.14 for the constant-current bridge:

$$\begin{aligned} e_{OCI} &= I\left[\left(\frac{R_2 + R_3}{R_1 + R_2 + R_3 + R_4}\right)R_1 - \left(\frac{R_1 + R_4}{R_1 + R_2 + R_3 + R_4}\right)R_2\right] \\ e_{OCI} &= \left[\frac{I}{R_1 + R_2 + R_3 + R_4}\right](R_1R_3 - R_2R_4) \end{aligned} \tag{1.3.14}$$

Equation 1.3.14 clearly indicates that for bridge balance, $R_1 \times R_3 = R_2 \times R_4$.

If the circuit is designed as a quarter bridge with R_1 being the transducer ($R_1 = R + \Delta R_1$), R_2 is equal in resistance to $R_1 (R_2 = R)$, and R_3 and

R_4 are multiples of $R(R_3 = R_4 = kR$, where k is any constant multiplier), a balanced, more flexible output circuit can be designed as shown in Figure 1.14.

By substituting the resistance values in the circuit of Figure 1.14, Equation 1.3.15 is obtained:

$$e_{oCI} = \left[\frac{I}{(R+\Delta R_1) + R + kR + kR}\right][(R+\Delta R)(kR) - R(kR)]$$
$$e_{oCI} = \left[\frac{I}{(\Delta R_1) + R(2+2k)}\right][(\Delta R_1)(kR)] \quad (1.3.15)$$

By further reducing Equation 1.3.15, Equation 1.3.16 results for a constant-current bridge. Equation 1.3.16 provides insight into the sensitivity and linearity of the constant-current bridge:

$$e_{oCI} = \left[\frac{I\Delta R_1}{\frac{(\Delta R_1)}{2R(1+k)} + 1}\right]\left(\frac{k}{1+k}\right) \quad (1.3.16)$$

1.3.3 CONSTANT-VOLTAGE WHEATSTONE BRIDGE

The constant-voltage Wheatstone bridge employs a voltage source for excitation of the bridge. The output voltage is as shown in Equation 1.3.17:

$$e_{oCV} = E_{ex}\left[\frac{R_4}{R_1 + R_4} - \frac{R_3}{R_2 + R_3}\right] \quad (1.3.17)$$

If the circuit is designed as a quarter bridge with R_1 being the transducer ($R_1 = R + \Delta R_1$), R_2 is equal in resistance to R_1, ($R_2 = R$), and R_3 and R_4 are

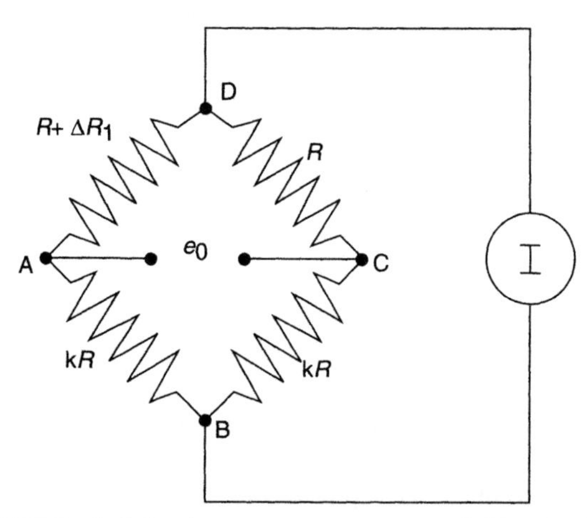

FIGURE 1.14 A constant-current quarter-bridge circuit.

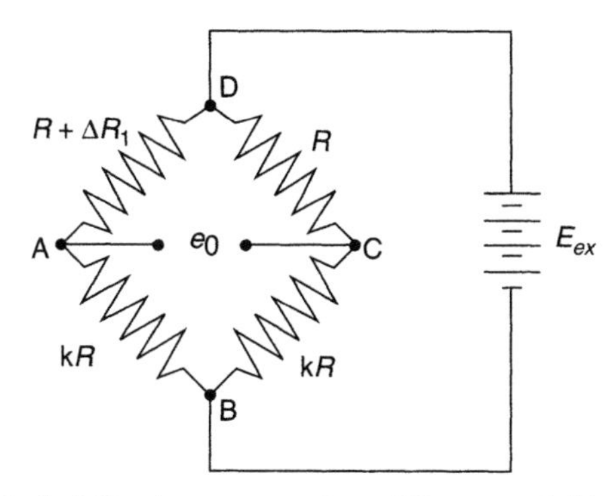

FIGURE 1.15 A constant-voltage Wheatstone bridge circuit.

multiples of $R(R_3 = R_4 = kR)$, a balanced, more flexible output circuit can be designed, as shown in Figure 1.15.

By substituting the resistance values of Figure 1.15 into Equation 1.3.18, the sensitivity and linearity of the constant voltage Wheatstone bridge can be calculated:

$$e_{o_{CV}} = E_{ex}\left[\frac{kR}{(R+\Delta R_1)+kR} - \frac{kR}{R+kR}\right]$$

$$e_{o_{CV}} = \frac{E_{ex}}{R}\left[\frac{\Delta R_1}{\left(\frac{\Delta R_1}{R(1+k)}+1\right)}\right]\left(\frac{k}{(1+k)^2}\right) \tag{1.3.18}$$

1.4 CONSTANT–VOLTAGE STRAIN GAGE BRIDGE OUTPUT

Depending on the configuration, either one gage, called a quarter bridge; two gages, called a half bridge; or four gages, called a full bridge, are used. In a quarter bridge, one of the resistors is replaced with a strain gage, whereas the other three arms employ high-precision resistors with a nominal value the same as that of the gage. If a half bridge is used, the gages are usually positioned in the bridge such that the greatest unbalance of the bridge is achieved when the gages are exposed to the strain of the part; the remaining two arms receive precision resistors. If a full bridge is used, all four resistors in the bridge are replaced with active strain gages.

With strain gages installed in the bridge arms, the bridge output is easily determined. Figure 1.16 shows the previously examined bridge with a strain gage replacing one of the bridge resistors (R_4). This bridge is referred to as a quarter bridge as only one arm has been equipped with a strain-sensing device. The anticipated bridge output is important, because most output is

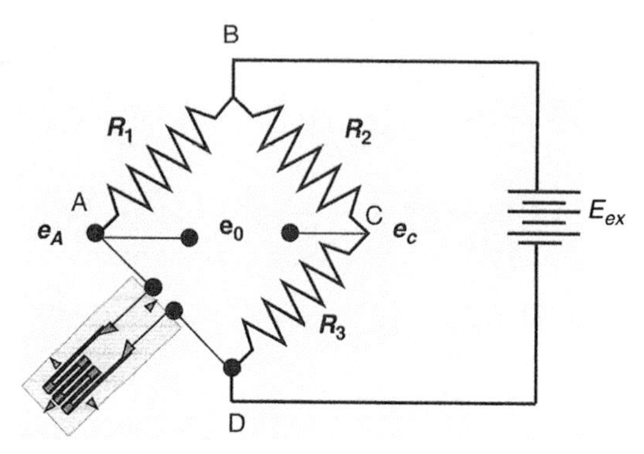

FIGURE 1.16 A constant-voltage quarter-bridge circuit.

digitized through an A–D conversion process and then analyzed by a computer. Combining Equations 1.2.7 and 1.3.18 and letting $R_1 = R_2 = R_3 = R_{gage} = R$ results in Equation 1.4.1:

$$\frac{e_{OCV}}{E_{ex}} = \left[\frac{(R_4 + \Delta R)}{R_1 + (R_4 + \Delta R)} - \frac{R_3}{R_2 + R_3}\right] = \left[\frac{(R_4 + GF \times \varepsilon \times R)}{R_1 + (R_4 + GF \times \varepsilon \times R)} - \frac{R_3}{R_2 + R_3}\right]$$
$$= \left[\frac{(1 + GF \times \varepsilon)}{2 + (GF \times \varepsilon)} - \frac{1}{2}\right] = \left[\frac{GF \times \varepsilon}{4 + (2 \times GF \times \varepsilon)}\right] \approx \frac{GF \times \varepsilon}{4} \tag{1.4.1}$$

Example 1.2. A component is manufactured from material that has an elastic modulus (E) of 210 GPa and a tensile yield strength (σ_{yp}) of 290 Mpa. Predict the strain and the bridge output if the component is expected to be loaded to approximately 70% of its yield strength and the single gage has a gage factor of 2.

Solution. The tensile strain at yield would be approximately 1380×10^{-6} mm/mm or 1380 microstrain.

The strain to be expected is $70\% \times 1380 \times 10^{-6} \approx 1000 \times 10^{-6}$ or 1000 microstrain.

If the gage factor is 2 and the material is at 1000 microstrain, the bridge output would be approximately 0.50 mV/V:

$$\frac{e_{OCV}}{E_{ex}} = \left[\frac{GF \times \varepsilon}{4 + (2 \times GF \times \varepsilon)}\right] = \frac{2 \times 0.0010}{4 + 4 \times 0.0010}$$
$$= 0.0005\,\text{V/V } or \text{ } 0.50\,\text{mV/V}$$

If the bridge excitation is 10 V, the bridge output with the material loaded to 70% of yield is only 5.0 mV, or approximately 0.50 mV per 100 microstrain.

If the bridge is configured with additional strain gages, the output of the bridge can be enhanced, provided the gages are properly positioned in the bridge. If a component under test experiences a bending load, it may be advantageous to mount the gages such that one gage experiences a tensile strain while the second gage experiences a compressive strain. The two gages when wired into the bridge result in a half-bridge configuration, as shown in Figure 1.17. For the half bridge, the bridge output is predicted by using Equation 1.4.2:

$$\begin{aligned}\frac{e_{OCV}}{E_{ex}} &= \left[\frac{(R_4+\Delta R)}{(R_1-\Delta R)+(R_4+\Delta R)} - \frac{R_3}{R_2+R_3}\right]\\ &= \left[\frac{(R_4+GF\times\varepsilon\times R)}{(R_1-GF\times\varepsilon\times R)+(R_4+GF\times\varepsilon\times R)} - \frac{R_3}{R_2+R_3}\right]\\ &= \left[\frac{(R_4-R_1+2GF\times\varepsilon)}{2R_1+2R_4}\right]\\ \frac{e_{OCV}}{E_{ex}} &= \left[\frac{2\times GF\times\varepsilon}{4}\right]\end{aligned} \tag{1.4.2}$$

The component of Example 1.2 is instrumented with two strain gages and wired into the bridge to provide maximum output. The bridge output can be predicted by using Equation 1.4.2.

In the analysis, let the initial values of the bridge be as follows, with $R_{gage1} = R_2 = R_3 = R_{gage2} = R$.

For the half bridge shown in Figure 1.17, the output, with two active gages is approximately 1.0 mV/V at 1000 microstrain. If the bridge excitation is 10 V, the bridge output with the material loaded to about 70% of the yield is 10.0 mV, or approximately 1.0 mV per 100 microstrain.

If the bridge is configured with additional strain gages, the output of the bridge can be enhanced, provided the gages are properly positioned in the bridge. If a component under test experiences a bending or torsional load, it may be advantageous to mount the gages such that two gages experience a tensile strain while the other two gages experience a compressive strain. The

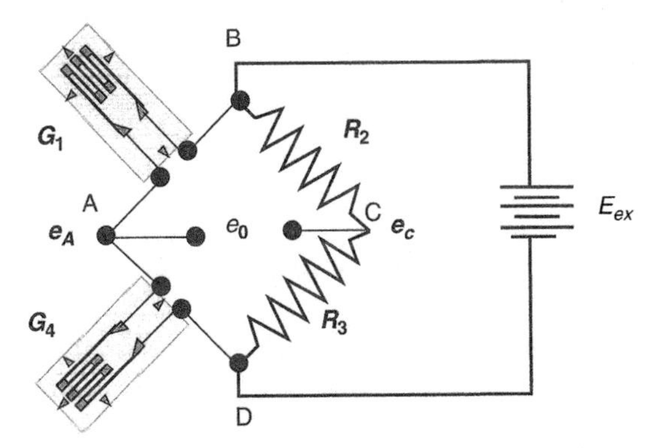

FIGURE 1.17 A constant-voltage half-bridge circuit.

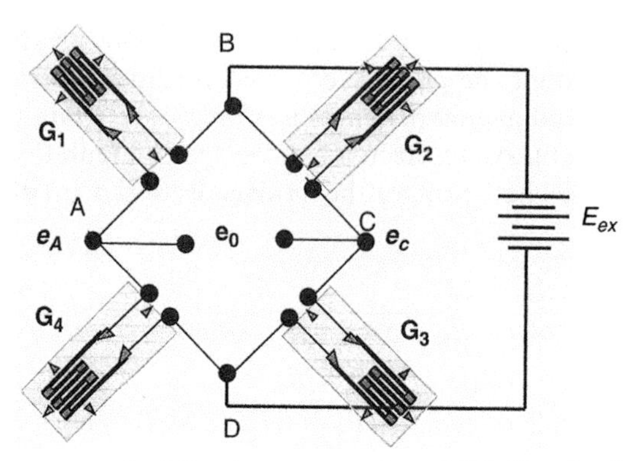

FIGURE 1.18 A constant-voltage full-bridge circuit.

four gages, when wired into the bridge, complete what is referred to as a full bridge, as shown in Figure 1.18.

$$\frac{e_{OCV}}{E_{ex}} = \left[\frac{2 \times GF \times \varepsilon}{4}\right] = \frac{4 \times 0.0010}{4} = 0.001\,\text{V/V}\ \textit{or}\ 1.0\,\text{mV/V}$$

Assume that the gages G_4 and G_2 are in tension and G_1 and G_3 in compression. This arrangement will provide maximum bridge output:

$$\begin{aligned} \frac{e_{OCV}}{E_{ex}} &= \left[\frac{(R_4 + \Delta R)}{(R_1 - \Delta R) + (R_4 + \Delta R)} - \frac{R_3 - \Delta R}{(R_2 + \Delta R) + (R_3 - \Delta R)}\right] \\ &= \left[\frac{(R_4 + (GF \times \varepsilon \times R))}{R_1 + R_4} - \frac{(R_3 - (GF \times \varepsilon \times R))}{R_1 + R_4}\right] \\ \frac{e_{OCV}}{E_{ex}} &= \left[\frac{(2R \times GF \times \varepsilon + 2 \times R \times GF \times \varepsilon)}{4R^2}\right] = [GF \times \varepsilon] \end{aligned} \tag{1.4.3}$$

For the constant, voltage strain gage bridge, the ratio of the output to excitation voltages is given by Equation 1.4.4, where N is the number of active gages, GF the gage factor, ε the strain, e_{OCV} the bridge output, and E_{ex} the bridge excitation:

$$\frac{e_{OCV}}{E_{ex}} = \left[\frac{N \times GF \times \varepsilon}{4}\right] \tag{1.4.4}$$

1.5 GAGE AND LEAD-WIRE COMPENSATION

Lead wires and lead-wire temperature changes affect the balance and the sensitivity of the bridge. If temperature changes occur during measurement, drift will occur. Dummy gages are used to compensate for component temperature variations that occur during the test.

1.5.1 TWO-LEAD-WIRE CONFIGURATION

A gage that has two lead wires between the gage and the bridge may introduce an error if the lead wires are subjected to any temperature change or temperature variation (Figure 1.19). If long lead wires are used in a two-wire system, the measurement system may demonstrate a loss of balance control, a loss of temperature compensation, or a change in signal attenuation. For balance, without temperature change, Equation 1.5.1 applies:

$$\frac{R_2}{R_3} = \frac{R_g + R_{L_1} + R_{L_2}}{R_4} \tag{1.5.1}$$

For balance with temperature change occurring along the lead length, Equation 1.5.2 prevails. If a temperature change occurs on the leads, the resistance of the leads changes and error is introduced during the test.

$$\frac{R_2}{R_3} = \frac{R_g + R_{L_1} + R_{L_2} + \left(\frac{\Delta R_{L_1}}{\Delta T}\right)\Delta T + \left(\frac{\Delta R_{L_2}}{\Delta T}\right)\Delta T}{R_4} \tag{1.5.2}$$

1.5.2 THREE-LEAD-WIRE CONFIGURATION

A three-lead configuration can be used to eliminate lead temperature errors (Figure 1.20). This can be important when long lead wires are used between the gage and the instrument system.

$$\frac{R_2}{R_3} = \frac{R_g + R_{L_1} + R_{L_2} + \left(\frac{\Delta R_{L_1}}{\Delta T}\right)\Delta T + \left(\frac{\Delta R_{L_2}}{\Delta T}\right)\Delta T}{R_4 + R_{L_2} + R_{L_3} + \left(\frac{\Delta R_{L_2}}{\Delta T}\right)\Delta T + \left(\frac{\Delta R_{L_3}}{\Delta T}\right)\Delta T} \tag{1.5.3}$$

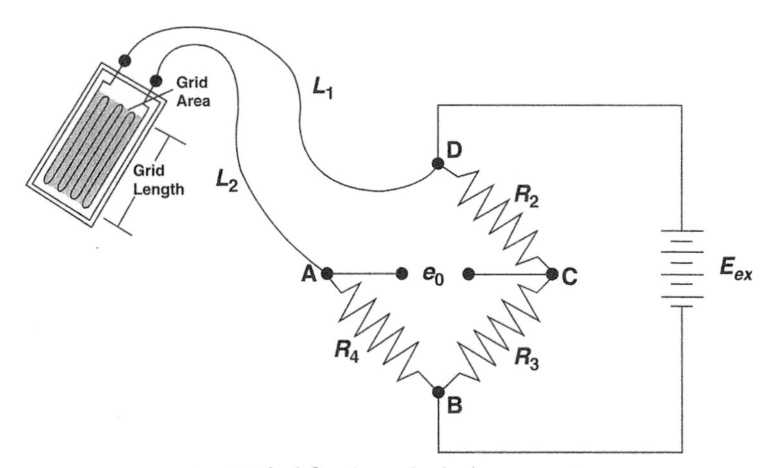

FIGURE 1.19 A two-lead-wire gage system.

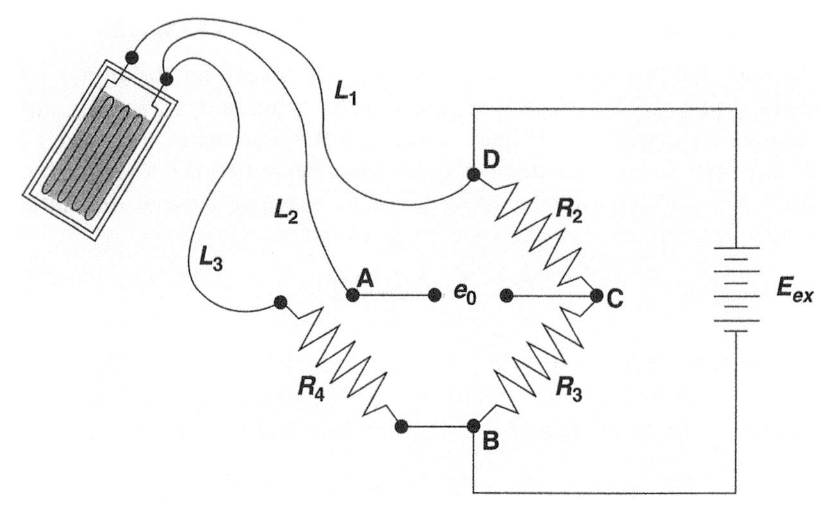

FIGURE 1.20 A three-lead-wire gage system.

If all lead wires are of equal length, subject to the same temperature, and of the same material, then Equation 1.5.4 is the governing equation. A three-lead system compensates for lead-wire resistance and resistance change due to ΔT along the leads.

$$R_{L_1} = R_{L_2} = R_{L_3} \quad \frac{\Delta R_{L_1}}{\Delta T} = \frac{\Delta R_{L_2}}{\Delta T} = \frac{\Delta R_{L_3}}{\Delta T} \tag{1.5.4}$$

$$\frac{R_g}{R_4} = \frac{R_2}{R_3} \tag{1.5.5}$$

1.5.3 LEAD-WIRE SIGNAL ATTENUATION

If long leads are used, the resistance in the leads will tend to reduce the gage factor of the system, reducing sensitivity, and, potentially, introducing error. The gage factor is based on gage resistance and gage resistance change, $k = (\Delta R/R)/\varepsilon$, as defined previously. If long leads are introduced and the resistance for each lead is R_L then Equation 1.5.6 can be applied:

$$k = \frac{\dfrac{\Delta R_g}{R_g + 2R_L}}{\varepsilon} = \frac{\Delta R_g/R_g}{\varepsilon[1 + (2R_L/R_g)]} \tag{1.5.6}$$

Equation 1.5.6 is used to quantify the reduced gage factor and the resulting signal loss associated with lead-wire resistance.

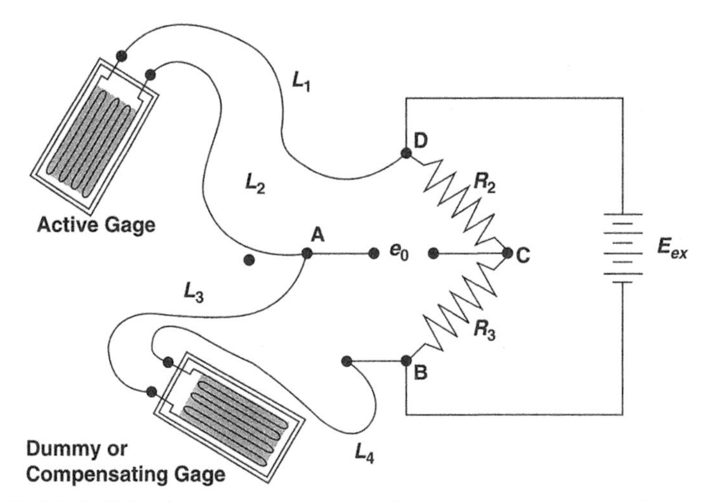

FIGURE 1.21 A gage system incorporating a dummy or compensating gage.

1.5.4 DUMMY GAGES

Dummy gages can be used as temperature compensation devices, as shown in Figure 1.21. The dummy gage and its lead wires provide compensation for temperature changes of the test specimen, lead wires, and lead-wire resistance changes if properly designed. The dummy gage is ideally mounted on the same type of material with the same mass as the test component or mounted on the test component itself in an unstressed area. The dummy gage must be exposed to identical conditions, excluding the load, as the test component.

1.6 STRAIN GAGE BRIDGE CALIBRATION

System calibration is a very important part of any measuring system (Figure 1.22). Calibration is performed by shunting a high-calibration resistor across one arm of the bridge circuit. When possible, the shunt is placed across one of the active arms. With the calibration resistor in position, the bridge is unbalanced and a known output is produced. The output voltage as a function of the excitation voltage for this bridge is given quite accurately by Equation 1.6.1 and exactly by Equations 1.3.17 and 1.4.4:

$$\frac{\Delta E}{V_{IN}} = \frac{R_g}{4 \times (R_g + R_C)} \qquad (1.6.1)$$

Example 1.3. A 60-kohm shunt resistor is placed across the active arm of a 120-ohm bridge.

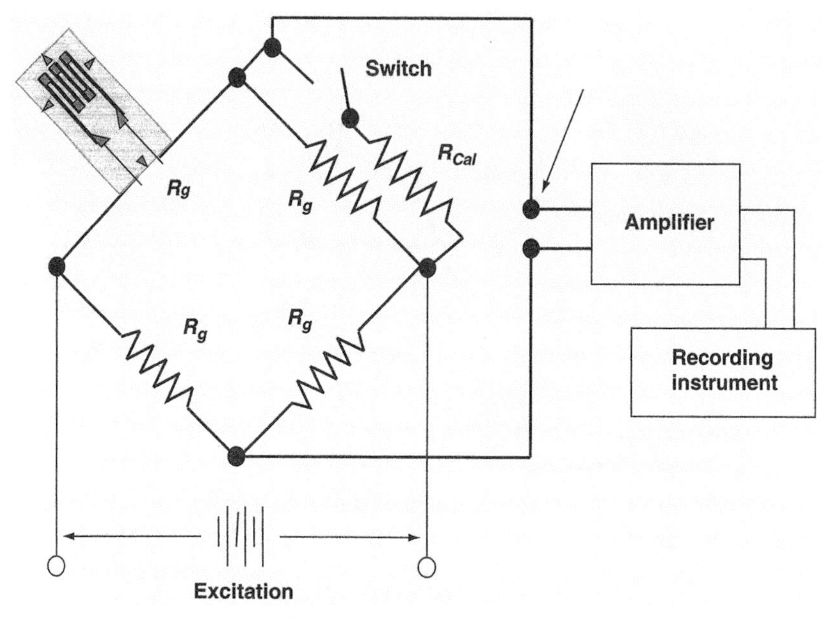

FIGURE 1.22 Bridge calibration arrangement.

Determine the equivalent resistance of the shunted arm and determine the bridge output.

If 120-ohm gages with a gage factor (GF) of 2 are used, determine the indicated strain.

Solution. The equivalent resistance in the shunted arm (R_1) is predicted from equations for a parallel resistance circuit as follows:

$$R_l = \frac{R_s \times R_g}{R_s + R_g} = \frac{120 \times 60000}{60120} = 119.76\,\text{ohms}$$

The bridge output in terms of the excitation voltage, as derived in Equation 1.3.17, is as follows:

$$\frac{e_o}{E_{ex}} = \left[\frac{120}{119.76 + 120} - \frac{120}{120 + 120}\right] \times 1000 = 0.50\,\text{mV/V}$$

For a previously balanced bridge, with a 60-kohm shunt resistor in one arm, a GF of 2, using a 120-ohm gage, the indicated strain is then given as previously shown in Equation 1.4.4.

$$\varepsilon_{ind} = \left(\frac{e_o}{E_{ex}}\right) \times \frac{4}{GF} = 0.00050 \times \frac{4}{2} = 1000 \text{ microstrains}$$

1.7 STRAIN GAGE TRANSDUCER CONFIGURATION

Strain gages are commonly used to construct transducers for measuring loads. This section reviews some of the more common types of load-sensing configurations. In most cases, the load cell is configured to maximize the intended quantity and minimize the influence of the other indirect quantities. Isolation of the intended and secondary quantities is best performed with the transducer or system configuration; however, it may have to be performed at the data analysis stage. The interdependency of the desired quantity and any unintended load-related quantities produces a cross-talk between the measured data.

1.7.1 CANTILEVER BEAM IN BENDING

One of the simplest types of strain gage sensors is a cantilever beam in bending, as shown in Figure 1.23. For a cantilever beam, with the load applied between the end of the beam and the gage, the flexural outer fiber stresses and the strain on the top surface of the beam at the gage location are as given in Equation 1.7.1:

$$\sigma = \frac{Mc}{I} = E\varepsilon \tag{1.7.1}$$

where σ is the stress, E the elastic modulus, M the applied moment, I the moment of inertia, ε the strain, and c the distance from the neutral axis. This load system is sensitive to axial loading and is sensitive to temperature variations of the beam.

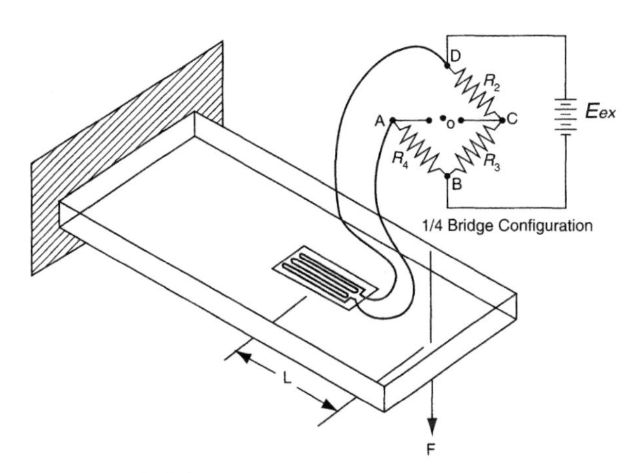

FIGURE 1.23 A single-gage cantilever beam in bending.

1.7.2 CANTILEVER BEAM IN BENDING WITH AXIAL LOAD COMPENSATION

The cantilever beam shown in Figure 1.24 has gages on both surfaces. Because of the configuration of the bridge, the gages compensate for any applied axial load and compensate for uniform temperature changes while having increased sensitivity over the system shown in Figure 1.24. For the cantilever beam, Equation 1.7.2 is used, which is derived from Equation 1.4.4:

$$\frac{e_{OCV}}{E_{ex}} = \left[\frac{(2 \times R \times GF \times \varepsilon + 2 \times R \times GF \times \varepsilon)}{4R}\right] = [GF \times \varepsilon] \qquad (1.7.2)$$

1.7.3 TENSION LOAD CELL WITH BENDING COMPENSATION

The bridge shown in Figure 1.25 has two gages in the axial loaded direction and two gages in the Poisson orientation. This cell has both bending and torsion compensation and is therefore sensitive to tensile and compressive loads only. The resulting bridge output is as given in Equation 1.7.3:

$$\frac{\Delta E}{E_{ex}} = \frac{GF}{2} \times (1 + \nu)\varepsilon \qquad (1.7.3)$$

1.7.4 SHEAR FORCE LOAD CELL

The shear force load cell is designed with high sensitivity to shear force while canceling the bending and torsion load output. The bridge output of Figure 1.26, as a function of the excitation voltage (E_{ex}), is as shown in Equation 1.7.4:

$$\frac{\Delta E}{E_{ex}} = GF \times \varepsilon \qquad (1.7.4)$$

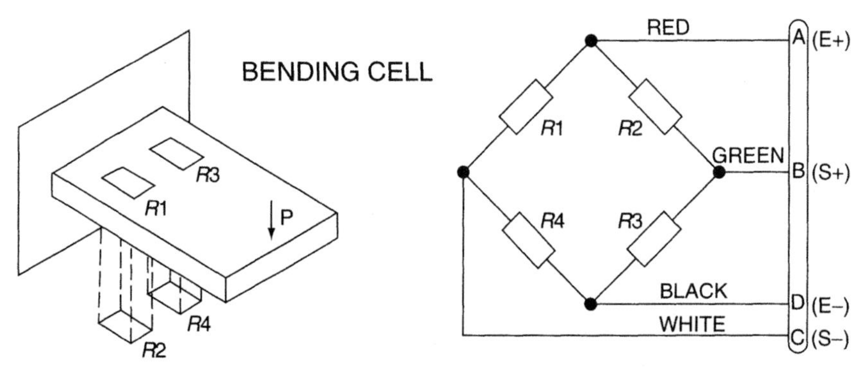

FIGURE 1.24 A multigage cantilever beam system.

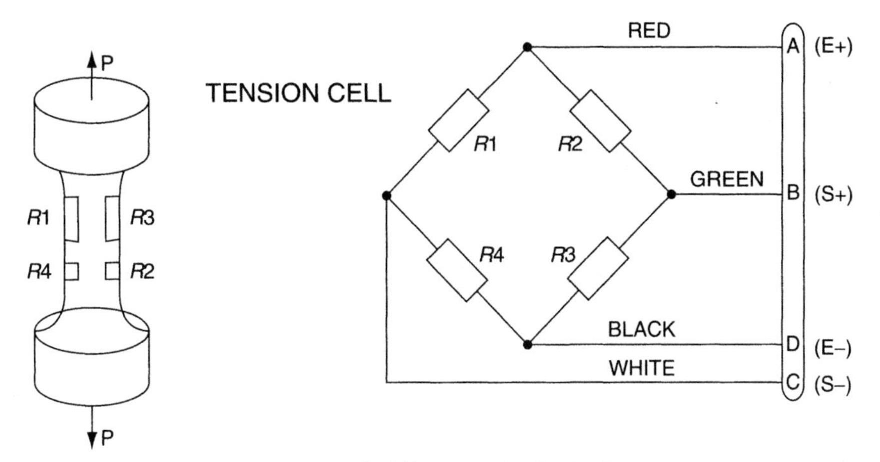

FIGURE 1.25 A tension load cell.

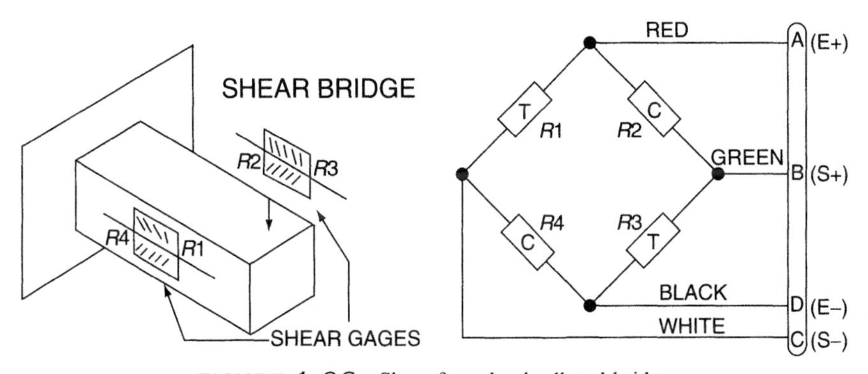

FIGURE 1.26 Shear force load cell and bridge.

1.7.5 TORSION LOAD CELL

The torsion cell shown in Figure 1.27 is highly sensitive to torsional loading and insensitive to bending loads. It is common to build this cell with a pair of 90° gages located exactly 180° around the shaft from each other.

1.7.6 COMMERCIAL S-TYPE LOAD CELL

A typical commercial-type load cell is shown in Figure 1.28. The commercially available S-type load cell is a high-output load-measuring device that offers temperature compensation and inherent damage protection for the strain gages. These units are available in a range of load capacities.

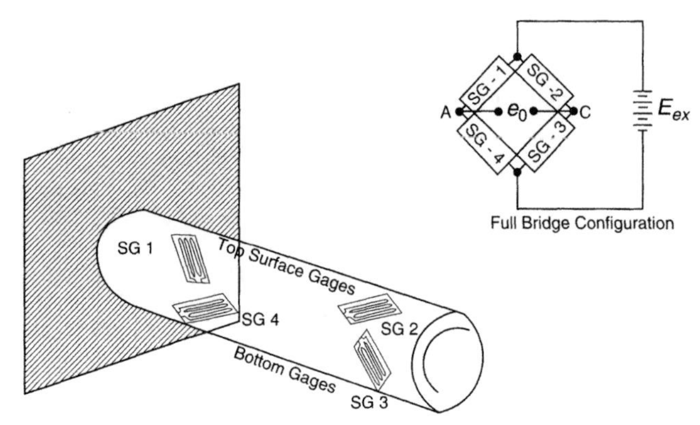

FIGURE 1.27 A torsion load cell.

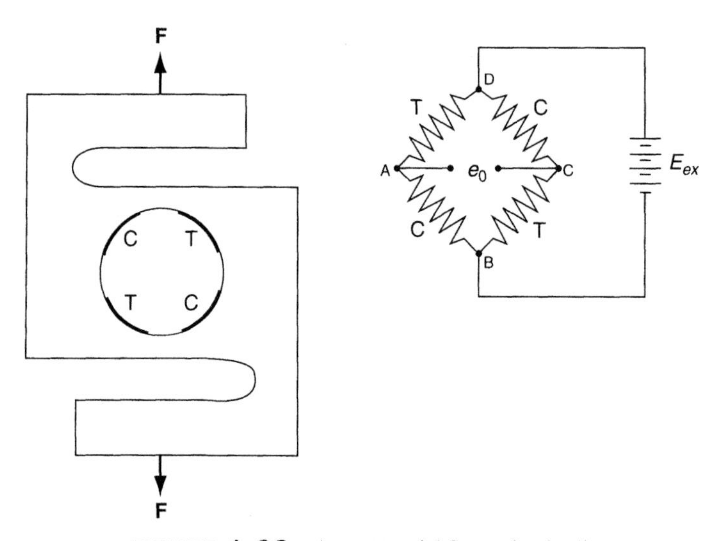

FIGURE 1.28 A commercial S-type load cell.

1.8 MATRIX-BASED LOAD TRANSDUCER DESIGN

When analyzing complex shapes, there is often a difficulty in separating the load–strain directional relationships. The interdependency of the load and strain fields produces a cross-talk between the measured data. Separation of the loads into orthogonal loading directions and the strain related to those orthogonal load components enable the experimenter to examine the direct and indirect relationships that exist between the applied loads and the strain.

The number of gages, gage locations, and orientations are optimized by maximizing the determinant of $|A^{T}A|$ in order to minimize the estimated load variance. This so-called D-optimal technique can be used to separate the loading directions and strains (Lee et al., 1997). The matrix $[A]$ is called the sensitivity coefficient matrix. A linear elastic relationship between the nth strain vector ($\{S\}_{n\times 1}$) and the pth applied load vector ($\{L\}_{p\times 1}$) is assumed for the analysis.

$$\{S\}_{n\times 1} = [A]_{n\times p}\{L\}_{p\times 1} \tag{1.8.1}$$

The n elements of the strain matrix can be replaced with the n bridge output elements (in mV/V) with similar results. With the given $\{S\}_{n\times 1}$ and $\{L\}_{p\times 1}$, the best estimated $\{L\}_{p\times 1}$ is as given in Equation 1.8.2:

$$\{\hat{L}\} = ([A]^T[A])^{-1}[A]^T\{S\} \tag{1.8.2}$$

The estimated load variance–covariance matrix is as given in Equation 1.8.3:

$$s^2(\hat{L}) = (A^T A)^{-1}\sigma^2 \tag{1.8.3}$$

where s is the strain measurement error, A the sensitivity coefficient matrix ($\hat{L}$), and σ the standard deviation.

1.8.1 SENSITIVITY COEFFICIENT MATRIX

The sensitivity coefficient matrix tabulates the measured strain from each gage when a specific load case is applied (Figure 1.29). If each load case is orthogonal to the other, the dependency of each strain output to the load case is established.

Example 1.4. The vertical and lateral loads on an automotive differential cover are to be determined. Because of the complex shape of the structure, a matrix-based transducer method is used. The transducer consists of two axial bridges mounted to the torque arm, as shown in Figure 1.30. The sensitivity of the transducer is given by the following matrix:

$$[A] = \begin{bmatrix} 2.1676\times 10^{-4} & -5.1492\times 10^{-4} \\ 8.6979\times 10^{-5} & 4.0802\times 10^{-4} \end{bmatrix} \text{ mV/V/lb}$$

Find the relationship between the orthogonal load and the bridge output.

	Measured strain per unit load (με/lb)		
	Load case 1	Load case 2	Load case 3
Gage #1	a_{11}	a_{12}	a_{13}
Gage #2	a_{21}	a_{22}	a_{23}
Gage #3	a_{31}	a_{33}	a_{33}
Gage #4	a_{41}	a_{42}	a_{43}
Gage #5	a_{51}	a_{52}	a_{53}

FIGURE 1.29 A matrix technique array.

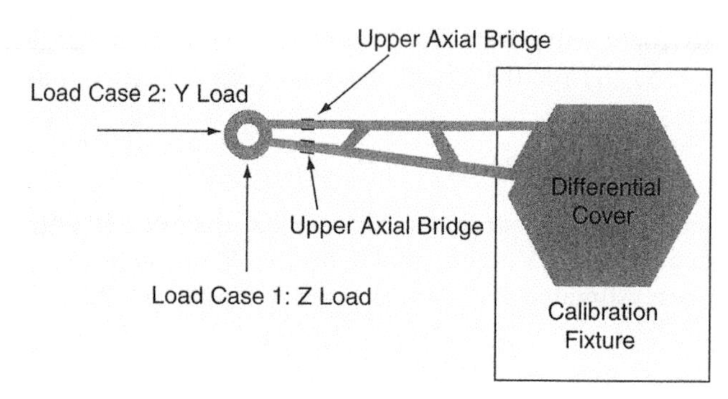

FIGURE 1.30 Loading example of an automotive differential torque arm.

Solution.

$$[A]^T[A] = \begin{bmatrix} 5.4552 \times 10^{-8} & -7.6127 \times 10^{-8} \\ -7.6127 \times 10^{-8} & 4.3162 \times 10^{-7} \end{bmatrix} \begin{matrix} \text{upper bridge, } S_u \\ \text{lower bridge, } S_l \end{matrix}$$

$$\begin{matrix} L_Y & L_Z \\ \text{mv/v/lb} & \text{mv/v/lb} \end{matrix}$$

$$[A]^T = \begin{bmatrix} 2.1676 \times 10^{-4} & 8.6979 \times 10^{-5} \\ -5.1492 \times 10^{-4} & 4.0802 \times 10^{-4} \end{bmatrix}$$

$$([A]^T[A])^{-1} = \begin{bmatrix} 24316060 & 4288682 \\ 4288682 & 3073234 \end{bmatrix}$$

$$([A]^T[A])^{-1}[A]^T = \begin{bmatrix} 3062.5 & 3864.9 \\ -652.8 & 1627.0 \end{bmatrix}$$

$$\{\hat{L}\} = \begin{bmatrix} 3062.5 & 3864.9 \\ -652.8 & 1627.0 \end{bmatrix} \{S\}$$

$$\hat{L}_Y = 3062.5 \times S_u + 3864.9 \times S_l$$

$$\hat{L}_Z = -652.8 \times S_u + 1627.0 \times S_l$$

where S_u and S_l are the upper and lower bridge outputs (mV/V) and L_Y and L_Z the applied loads (lb).

1.8.2 POTENTIAL SOURCES OF ERROR IN MATRIX TECHNIQUES

Many experimental and analytical techniques have inherent errors; the matrix techniques are no exception. Knowing the potential source of errors is paramount to understanding the quality of the obtained data. The sources of error in the matrix methods include the following: (1) assumption of a linear elastic relationship between strains and loads, (2) statistically based formula dependent on the calibrated sample size, (3) a calibration loading frequency that is different from that of the measured system, and (4) a calibrated boundary condition different from the actual boundary conditions that exist in the real application.

1.9 TRANSDUCER (GAGE) PLACEMENT AND IDENTIFICATION OF REGIONS OF INTEREST

Proper placement of strain gages in the strain field is critical. The fatigue life prediction is directly a function of the magnitude of strain to which the material is exposed. If the fatigue life prediction of a component that has a complex shape is necessary, it is imperative that the maximum strain be accurately measured. It is beneficial to have an understanding of where the high-stressed areas of a part are located or to be able to provide an initial full-field view of the strains or deformations before gages are applied. In this section, a limited number of full-field techniques are introduced to inform the reader of some of the techniques available.

A number of methods have been traditionally used to locate the area of maximum strain, including brittle lacquer treatments as well as a variety of optical methods including stress coatings. In most cases, techniques that require minimum surface preparation are preferred.

1.9.1 BRITTLE COATING TECHNIQUE

Brittle coating is a full-field method by which a component is sprayed with a special lacquer coating that cracks when exposed to certain strain levels. This technique is a relatively simple and direct method of obtaining strain levels and regions of high strain where high accuracy is not required.

In fatigue life prediction, this technique can be used to identify critical high-stress areas, which can allow the test engineer to locate strain gages in the most critical areas of the part.

Brittle coating can be used to identify the principle strain orientation as the cracks propagate in a direction normal to the principle strain direction, which can be used to orient the strain gage on the part. This technique is very sensitive to application methods as the sensitivity is a function of the coating thickness and the environmental conditions during the curing process.

1.9.2 PHOTO-STRESS COATING

Photostress coating is a full-field method by which a special coating is applied to a prototype or the actual component. The applied coating is double refractive, or more often called birefringent. The coating has a reflective base material with the birefringent coating on the surface. The area of interest is then illuminated with a monochromatic light source and viewed through a polariscope. A steady-state load can then be applied to the object under study and stress-related fringes can be viewed through the polariscope. The photostress coating method is an extension of the more classical photoelasticity, which requires a part model be made of special material that has a high stress-optic coefficient. The fringes form as a result of the change in material's refractive index when loaded because the refractive index is stress related.

Photostress coating requires very careful application of the coating material, which may be difficult on complex shapes. This technique can be very informative in identifying the magnitude and direction of the stress field if steady-state loads representative of the actual loads can be applied. This technique is appropriate if very high accuracy in the load strain relationship is not required. In fatigue life prediction, it can be used to identify strain directions and magnitudes, allowing the test engineer to locate strain gages in the critical areas of the part for further dynamic analysis.

The special coating material, when applied, relies on a perfect bond between the object and the base material, and the base material's elastic modulus must be low in comparison to the elastic modulus of the project. Photostress coatings can be used to identify the principle strain orientation and the difference in the principle stresses.

1.9.3 OPTICAL STRAIN MEASURING TECHNIQUES

An interferometer is a device used to determine differences in optical path lengths to which light from the same source might be exposed. These path length differences can be used to evaluate strain and strain fields. These methods allow full-field evaluation of strain-related displacement to be performed.

1.9.3.1 Holographic Interferometry

Holographic interferometry (Kobayashi, 1993) makes use of a change in path length distance, which creates interference fringes in a double-exposure hologram, a real-time hologram, or a sandwich hologram (Figure 1.31). The double exposure is a commonly used technique in holographic interferometry. The technique employed is to expose the hologram to the object and reference beams from a single, coherent monochromatic light source of wavelength λ while the object is in an unloaded state. The object is then loaded, as would normally occur, and a second exposure on the same holographic plate is made. The doubly exposed hologram, when reconstructed, shows the basic object with fringes superimposed on it. These fringes reflect phase differences due to changes in path length from the undeformed to the deformed state.

$$I = I_1 + I_2 + 2\sqrt{I_1 I_2}\,\mathrm{Cos}\,\Delta\phi = 2I_0[1 + \mathrm{Cos}\,\Delta\phi] \qquad (1.9.1)$$

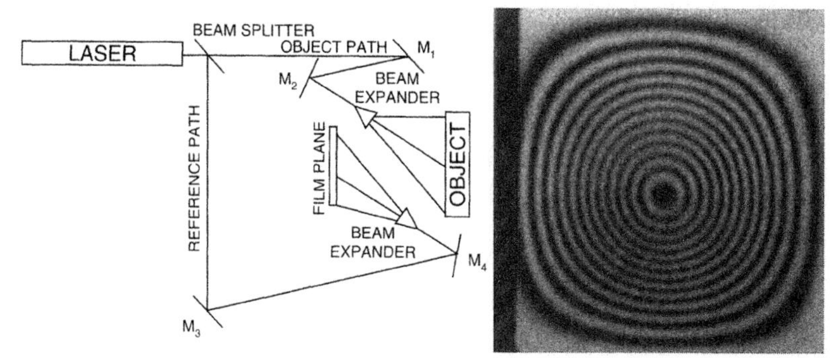

FIGURE 1.31 A holographic interferometry test system and a pattern formed from a displaced diaphragm.

$$\frac{(2n-1)\lambda}{2} = [(\text{Sin }\theta)\text{u} + (\text{Sin }\phi)\text{v} + (1 + \text{Cos }\theta)\text{w}] \qquad (1.9.2)$$

The angles θ and ϕ represent the sensitivity vectors to the three linear displacements u, v, and w. As an example, if v is the vertical displacement at a point and the illumination and sensing angles are at zero angle to the surface normal (horizontal plane), the interferometric pattern is insensitive to the displacement that occurs in the v direction. Three equations, resulting from using three different angles, can be used to separate the three unknowns in Equation 1.9.2.

1.9.3.2 Speckle-Shearing Interferometry

Speckle-shearing interferometry includes a variety of optical methods, allowing the measurement of in-plane and out-of-plane displacement derivatives (slope) of object surfaces. This discussion focuses on one method: shearography.

Shearography (Hung, 1982) has much in common with holographic interferometry and the more common speckle interferometric methods; however, there are many important differences. Shearography, speckle interferometry, and holographic interferometry require the use of a coherent light source. All these methods rely on interference phenomena in the recording.

Shearography is an optical method that results in the recording of fringe patterns in a transparent film or imaging system. The resulting speckle recording is a two-dimensional recording that contains information related to the displacement of the object surface. Shearography and speckle interferometry utilize the random variations on the object surface to provide constructive and destructive interference of the coherent light impinging on it. The random interference results in dark (destructive) and light (constructive) spots on the object surface called speckles.

Both shearography and speckle interferometry require an imaging lens because the speckle itself is being recorded. As a result, speckle interferometry results in a two-dimensional recording. The imaging lens must be of sufficient quality to focus the speckle onto the film plane. Shearography makes use of a special image-shearing camera to record the image. The camera consists of an imaging lens with a small-angle prism positioned either in front of the lens or in the iris plane of one half of the lens. The small angle prism deviates (bends) the light waves impinging on the portion of the lens covered by the prism. In both shearography and speckle interferometry, the speckle pattern, which is photographed during the first exposure, moves when the object is deformed and the surface displaces. The second exposure captures the speckles in the deformed position.

The image-shearing lens of shearography causes two images, one slightly displaced by a distance δx, to be displayed in the film plane. The second exposure results in a similar display of the deformed object. The interferometric pattern is then a fringe pattern resulting from the change between the two exposures. Each exposure has two sheared images, and, as a result, the fringes are a record of a change with respect to the double image spacing or the displacement derivative. The measurement of displacement derivatives also makes the method especially appealing for nondestructive testing where structural anomalies result in high strains detectable on the object surface.

The recording of the speckle in the photographic transparency is the recording of a complex waveform similar to that of holographic interferometry. The fringes viewed in a double-exposure hologram are the result of a total optical path length change in multiples of $\lambda/2$.

$$I = 2\, I_o[1 + \cos\delta] \quad I_o = A^2 \tag{1.9.3}$$

The hologram records a change in path length, whereas in speckle the transparency records relative phase changes between speckles (Figure 1.32).

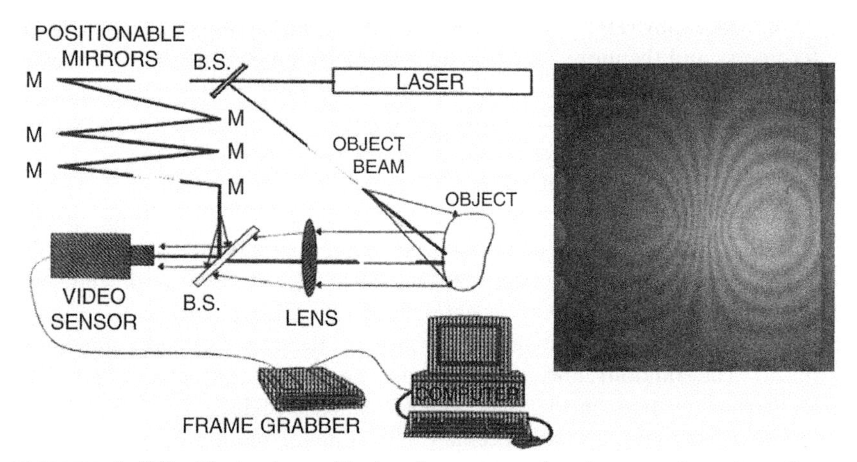

FIGURE 1.32 Electronic speckle interferometry and a shearography pattern from a deformed diaphragm.

The waveform recorded in the photographic transparency in shearography is due to a complex waveform that is completely random. This randomness occurs because of the scattering of the light rays when they impinge on the surface.

$$I = 4\, I_o \left[1 + \cos\left(\phi + \frac{\Delta}{2} \right) \times \cos\left(\frac{\Delta}{2} \right) \right] \tag{1.9.4}$$

$$\Delta = \frac{2\pi}{\lambda} \left[-\sin\theta \left(\frac{\partial u}{\partial x} \right) - (1 + \cos\theta) \left(\frac{\partial w}{\partial x} \right) \right] \delta x \tag{1.9.5}$$

1.10 INTRODUCTION TO TEMPERATURE MEASUREMENT

Temperature measurement may be an important part of a fatigue measurement system, as it may influence the material performance as well as the strain measurement system. Thermal cycling at elevated temperatures can also lead to fatigue-related failures. Although there are many different methods to measure temperature, this section focuses on the thermoelectric devices that could serve as sensors in the fatigue measurement system.

1.10.1 THERMOCOUPLE CHARACTERISTICS

A thermocouple is a measuring device that produces a voltage change because of relative temperature differences between the junction of two dissimilar metals and the output junction. Two dissimilar metals when joined form a junction that produces an electromotive force (EMF), which varies with the temperature to which the junction is exposed (Figure 1.33).

The junction of the two dissimilar materials forms the thermocouple junction. Materials for thermocouples are chosen for their temperature range and their sensitivity. The K-type thermocouple shown in Figure 1.34 is one of the most commonly used in an engineering environment. Thermocouples have a variety of different junction ends allowing the junction to be bolted, inserted into a cavity, or attached by using a heat-conducting material (Figure 1.35).

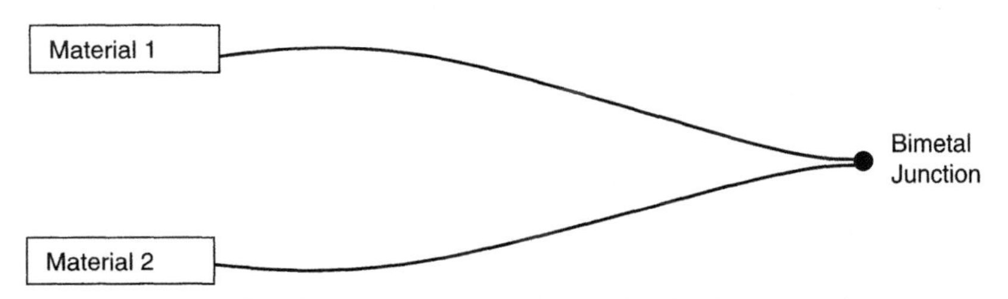

FIGURE 1.33 A typical thermocouple with the bimetal junction.

Base Material	Composition
Alumel	90% Nickel, 2% Aluminum + Silicon, Manganese
Chromel	10% Chromium, 90% Nickel
Constantan	60% Copper, 40% Nickel
Copper	Element
Iridium	Element
Iron	Element
Platinum	Element
Rhodium	Element

FIGURE 1.34 Common alloys used in thermocouples.

T-C Type	Thermocouple Materials	Temperature Range (Recom Range) {Maximum Range}	Features
More Common Types			
J	Iron–Constantan	0 to +500°C {0 to 750°C}	Low Cost Hi Output
K	Chromel–Alumel	0 to +1000°C {–200 to 1250°C}	Hi Output Durable, Ox. Resist.
T	Copper–Constantan	-200 to +300°C	LowCost Hi Output
Less Common Types			
E	Chromel–Constantan	-20 to +550°C {–200 to 900°C}	Hi Output Stable
R	Platinum– 13% Rhodium	700 to 1700°C {0 to 1450°C}	Hi Cost, Low Output
B	Platinum– 30% Rhodium	700 to 1700°C {0 to 1700°C}	Hi Cost, Low Output
N	Nickel–14.2% Chromium	-270 to 1300°C	Alternative to K-type

FIGURE 1.35 Thermocouple types.

1.10.2 THERMISTORS

Thermistors can be used as a temperature measurement device or very effectively as a monitoring and control device. Thermistors are made from semiconductor materials, typically metallic oxides, using cobalt, manganese, or nickel. The temperature–resistance relationship in a thermistor is non-linear and negative, as shown in Equation 1.10.1. The resistance change of the thermistor is used to produce a voltage change, which is easily read through the data acquisition system.

$$R = R_0 \times e^{\left[\beta\left(\frac{1}{T} - \frac{1}{T_0}\right)\right]} \quad (1.10.1)$$

In Equation 1.10.1, R is the resistance at temperature $T(^\circ K)$, R_0 the resistance at reference temperature T_0, (°K), e the base of natural log, and β a constant that varies from about 3400 to 4600, depending on thermistor composition. Thermistor resistance change is large and negative (usually 100 to 450,000 ohm-cm) and the practical operating range of thermistors is relatively low (−100 to +300°C).

1.10.3 ELECTRICAL RESISTANCE THERMOMETERS AND RESISTANCE-TEMPERATURE DETECTORS

The electrical resistance thermometer and resistance-temperature detectors (RTDs) are accurate methods of temperature measurement. The RTD relies on the change in resistance in the temperature-sensing material as an indicator of the thermal activity. Unlike thermistors, which are made of semiconductor materials and have a negative temperature–resistance relationship, the RTD has a positive temperature–resistance relationship, although the sensitivity is lower than that of a thermistor. RTD temperature–resistance characteristics may also be somewhat nonlinear. The RTD typically can be used over a higher temperature range than a thermistor, having temperature ranges of −250 to 1000°C. A constant-voltage bridge circuit, similar to that used with strain gages, is usually used for sensing the resistance change that occurs.

1.10.4 RADIATION MEASUREMENT

The temperature of a body can be determined by measuring the amount of thermal radiation emitted by the body. Thermal radiation is electromagnetic radiation emitted owing to the temperature of the body. Thermal radiation has a wavelength of approximately 100 to 100,000 nm. Ideal radiation is called blackbody radiation and is expressed in Equation 1.10.2:

$$E_b = \boldsymbol{\sigma} T^4 \qquad (1.10.2)$$

where E_b is the emissive power ($\mathrm{W/m^2}$) and T the absolute temperature (°K), σ the Steffan–Boltzmann constant $= 5.669 \times 10^{-8}\,\mathrm{W/(m^2 {}^\circ K^4)}$. The emissive power of a blackbody varies as a function of wavelength as governed by Planck's distribution.

$$E_{b\lambda} = \frac{(3.743 \times 10^8 \mathrm{W} \cdot \mu\mathrm{m}^4/\mathrm{m}^2) \times \lambda^{-5}}{e\left[\dfrac{(1.4387 \times 10^4 \mu\mathrm{m} \cdot K)}{\lambda T}\right] - 1} \qquad (1.10.3)$$

1.10.5 OPTICAL PYROMETERS

Optical pyrometers allow easy measurement of the temperature of objects that would normally be difficult to measure with contact devices. Examples are rotating components, hazardous materials, or high-electrical-field/high-voltage environments. The primary error sources in infrared pyrometry are in the field of view and the emissivity correction.

Optical pyrometers are designed with a field of view, basically defined by the instrument sensor and optics. The field of view is determined by the ratio of the object distance to spot diameter ratio, which is related to the divergence of the beam to the viewing area. It is important that if thermal gradients exist, proper distances be used so as to fill the viewing spot with the interest area. Optical pyrometers are designed for short or long viewing. With long viewing, a relatively small area can be examined at large distances (up to 20 m). For example, an instrument with a ratio of 120:1 would have a spot size of 5 cm at a 6-m distance. Infrared pyrometry is accomplished by measuring the thermal radiation emitted from the object. The emissivity (ε) of a body can range from 0 to 1.0, with 1.0 being an ideal emitter or blackbody. Generally, the higher the emissivity, the easier and more accurate is the infrared measurement. If reflectivity is high, correspondingly emissivity must be low. Emissivity is defined as the ratio of the emissive power of the object to that of a blackbody:

$$\varepsilon = \frac{E}{E_b} \tag{1.10.4}$$

where

$$E = \varepsilon \sigma T^4 \text{ or } T = \sqrt[4]{\frac{E}{\varepsilon \sigma}} \tag{1.10.5}$$

1.10.6 TEMPERATURE MEASUREMENT BY CHANGE OF STATE

If the only concern is the maximum temperature reached during the test, temperature devices that physically change state can be used. These are typically of the telltale variety and include temperature-sensing crayons, in which the substance melts, changes to a liquid, and flows if the temperature is reached. These typically have an operating range of 40 to 1000°C. Temperature tapes having discrete locations with thermally different characteristics in each location are available. These tapes behave similarly to a thermometer, with the temperature reached designated by the last location at which change in color occurred. Thermal paints are spray-on substances that liquefy at a specific temperature and change color. Similar to the crayons, stick-on pellets liquefy at a specified temperature.

1.11 THE GENERALIZED MEASUREMENT SYSTEM

The simplest form of measurement system might be considered a three-phase system (Figure 1.36 and 1.37). The system consists of the detector,

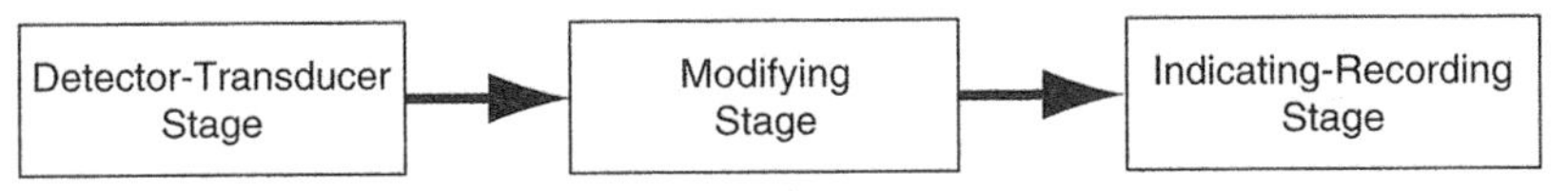

FIGURE 1.36 A three-phase measurement system.

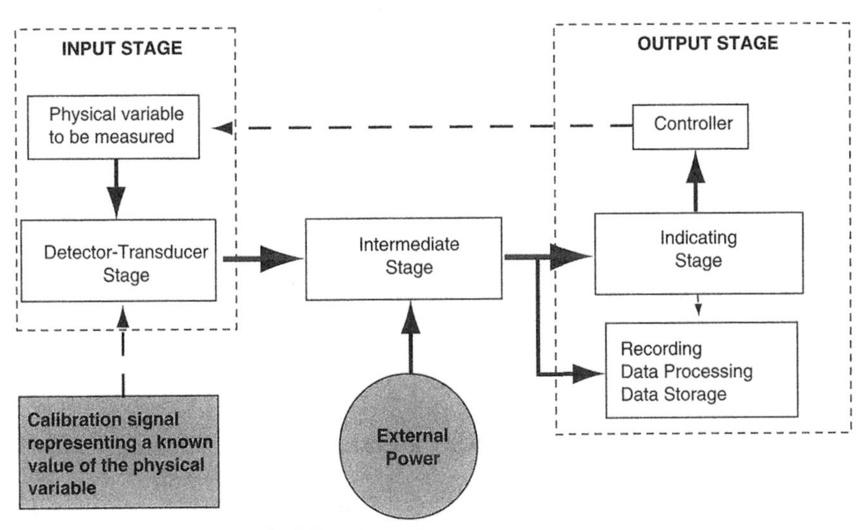

FIGURE 1.37 Generalized measurement and control.

which senses the signal; a modifying stage, which typically amplifies the relatively small signal from the sensor; and a final stage, which interprets the signal, visually displays the interpreted signal, and provides some form of data storage.

1.11.1 ELEMENT PURPOSE

The detector senses desired input to the exclusion of all other inputs and provides an analogous output. The transducer is any device that accepts one form of energy from a system and transmits that energy (typically in another form) to another system. The modifying stage is present to modify the signal received from the transducer into a form usable by the final stage. The modifying stage commonly increases amplitude and/or power. The termination stage provides indication and/or recording in a form for easy interpretation and evaluation.

The detector types presented in this text are those that are commonly associated with fatigue life testing. The detector's function is to sense input and change it to a more convenient form. Electrical transducers discussed in the previous sections included strain gages and thermocouples. Of major interest is the transducer's transfer efficiency. The ideal transducer has a transfer efficiency of 1, which means that the measured value is transferred to the modifying stage without loss. Transfer efficiency should be close to 1, but can never be higher than 1.

$$\eta_T = \frac{\text{Info delivered}}{\text{Info received}} = \frac{I_{out}}{I_{in}} \tag{1.11.1}$$

The modifying stages of interest include amplifying, filtering, attenuating, digitizing, and other signal modification techniques. The terminating stage is any component aiding the engineer in experimental data analysis, including digitized indication, recorders, digitized signals for computer data analysis, graphical displays, and possibly any controlling devices.

Output from transducers used in fatigue measurement is generally small. To facilitate proper signal handling in the A–D conversion process and post–A–D stages, the signal often needs to be amplified. The amplified signal then becomes of such magnitude that the A–D conversion can function in a useable range. Figure 1.38 presents a common arrangement for high-input impedance signal amplification.

1.11.2 SOURCE LOADING

Every attempt at measuring any variable tends, in some way, to alter the quantity being measured. An engineer or experimentalist must design instrumentation systems that minimize this system loading. Typically, minimizing the inertial effects of the transducer with respect to the quantity of interest is always the best path. Source loading is always critical when time-varying data are anticipated. This implies that if the quantity to be measured is temperature, then the mass of the thermocouple tip or the mass of the equivalent transducer should not affect the response of the measured temperature.

1.11.3 FILTERING UNWANTED SIGNALS IN THE MODIFYING STAGE

Filters discriminate between the components of an electrical signal according to their frequencies. Filters therefore alter the amount of those

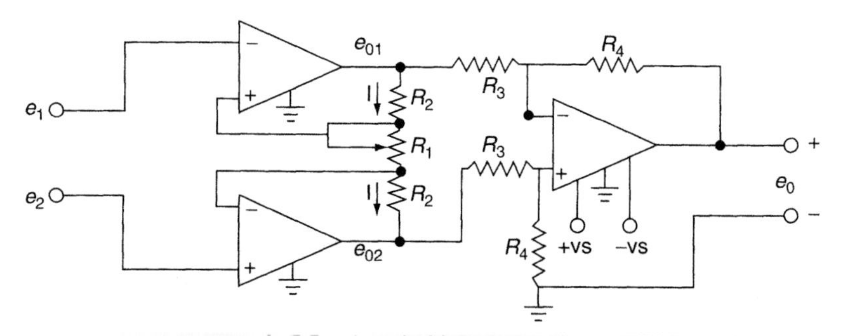

FIGURE 1.38 A typical instrumentation amplifier.

frequencies present at the output. The range of input frequencies over which the filter produces substantial output are the passbands, whereas the ranges that the filter suppresses are called the stopbands. The simplest types of filters are low-pass filters, whose passbands are at lower frequencies than the stopbands, and high-pass filters, whose passbands are at higher frequencies than the stopbands.

Low signal strengths from the transducers invite the potential for unwanted signals to alter the data. Ideally, high signal-to-noise ratios are imperative. When the transducer is in a noisy environment, where a large ambient electrical field is present near the transducer, it may be necessary to filter the unwanted noise before the signal is amplified. Filtering requires knowledge of the frequencies associated with the desired analog signal. Low-pass filters are used when the desired analog signal is below the frequency of the unwanted signal. High-pass filters are used when the anticipated analog signal frequency is above the background noise frequency. Notch or band pass filters allow only signals within a certain frequency range to pass or frequencies within a certain frequency range to be filtered out. Filter characteristics are shown in Figure 1.39.

Filters can be designed as passive filters, as shown in Figures 1.40 and 1.41. These filters consist of a resistance–capacitance (R–C) circuit or a resistance–inductance–capacitance (R–I–C) circuit. The passive implementation is less expensive; however, the input and output resistance is high, which may require the use of a buffer. Filters may also be designed as active filters, as shown in Figures 1.42 and 1.43, using operational amplifiers that allow better roll-off characteristics (sharpness) to the filter and better input and output impedance characteristics. The active implementations have low output resistance. The ideal filter cuts off unwanted frequencies, does not load the circuit, and passes the desired frequencies along undisturbed.

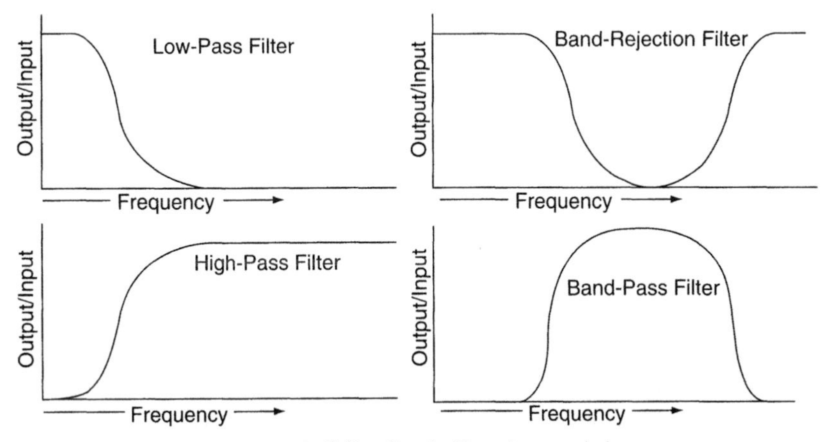

FIGURE 1.39 Simple filter characteristics.

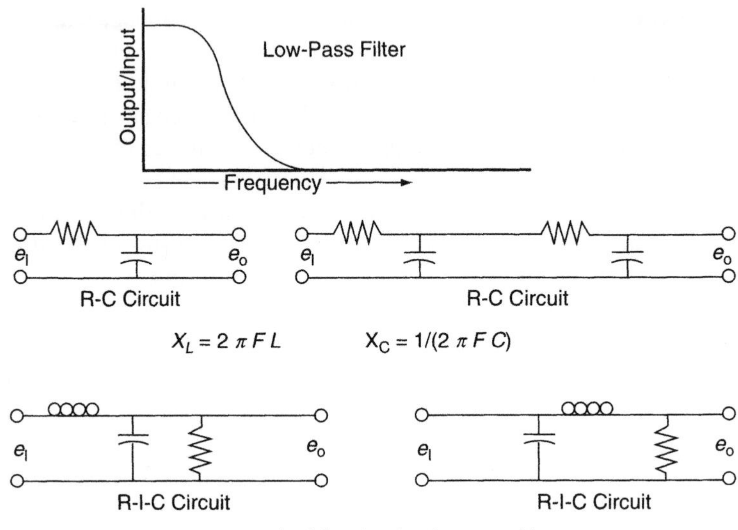

FIGURE 1.40 Passive low-pass filters.

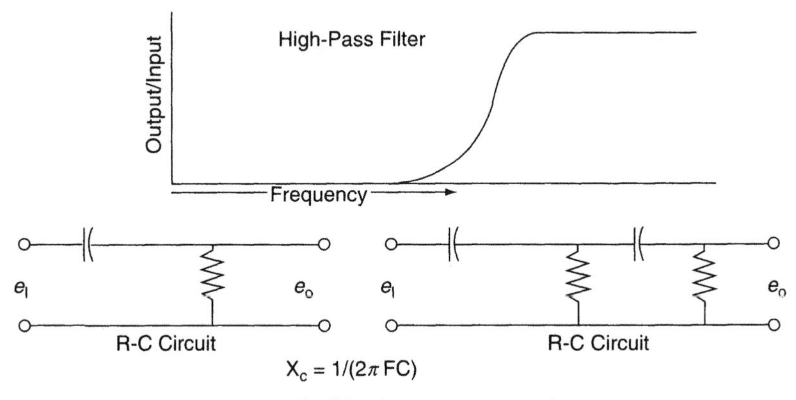

FIGURE 1.41 Passive high-pass filters.

For active low-pass (Figure 1.42) and active high-pass filters (Figure 1.43), the filter is designed with a certain cutoff frequency, as defined in Equation 1.11.2:

$$\text{cut off} = \frac{1}{2\pi RC} \tag{1.11.2}$$

In data acquisition, the most important use for filters concerns aliasing. Aliasing occurs when specific values cannot be uniquely tracked with their corresponding frequencies. Anti-aliasing filters are applied to limit the input frequency components present in the data. Otherwise, there are many possible input frequencies (the aliases), all of which can produce the same

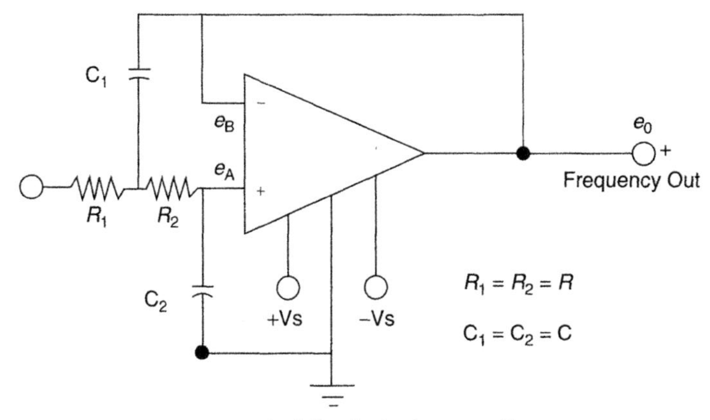

FIGURE 1.42 Active low-pass filters.

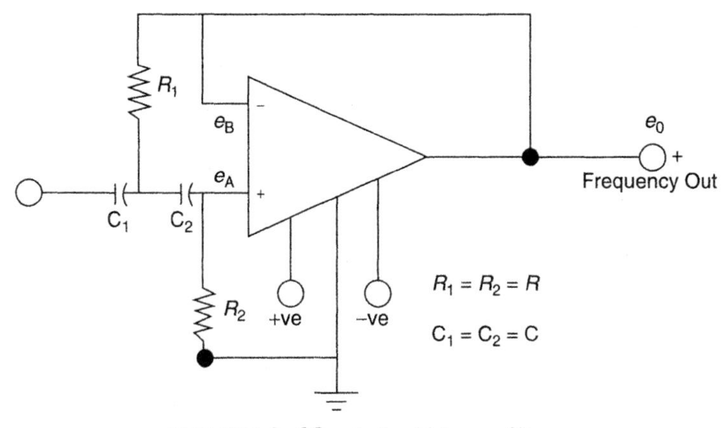

FIGURE 1.43 Active high-pass filters.

data points. Anti-aliasing filters remove all the unwanted input frequencies. Anti-aliasing filters are commonly, but not always, low-pass filters. Aliasing is covered in greater detail in Section 1.12.4.

1.12 DATA ACQUISITION

A data acquisition system allows data to be taken from an experiment and fed to a display or a system for manipulation or storage. The signal generated by the strain transducer and the load transducer must be measured to determine the strain or the load. The system takes the signals from the strain gages, load cells, and temperature sensors and converts them to a form that the instrumentation system can process. Many commercial data acquisition systems are available on the market. The price ranges and the capabilities of the systems can vary, but the basic components are almost identical. A basic data acquisition system consists of a signal conditioner, analog filters, A–D converter, digital filters, storage media, and the software to interface all the modules together. After the data are in digital form, they can be processed using common digital signal process techniques. This section gives an overview of the basic operating principle of the data acquisition system.

Analog signals are signals that vary continuously and can take on an infinite number of values in a given range. Most measuring devices (transducers) are analog devices. Digital signals vary in discrete steps and as a result can only take on a finite number of values in a given range. Common computers are digital devices that imply that both input and output are in digital form. This brings about the need for analog-to-digital (A–D) and digital-to-analog (D–A) converters.

1.12.1 SIGNAL CONDITIONING MODULE

The main function of the signal-conditioning module is to provide a means to accept various types of input signals, amplify the signals, and provide a means for filtering unwanted signals. The input signal can be the analog voltage from a strain gage Wheatstone bridge or the signal from a frequency counter. After receiving the signal, the signal conditioner amplifies the signal to the appropriate level. The module also provides an excitation voltage, balance and zero, and means for shunt calibration of the Wheatstone bridge. In the case of strain gage input, the device provides bridge excitation to the strain gage circuit. There are two ways to provide bridge excitation: constant-voltage and constant-current excitation. These have been described in Sections 1.3.2 and 1.3.3. The constant-voltage method is the most commonly used.

1.12.2 ANALOG-TO-DIGITAL CONVERSION

Digital computers and digital signal processors are used extensively for measurement and control. Their use leads to the need for A–D converters, which allow analog sensors to communicate with the digital computer (Figure 1.44). The A–D converter is responsible for converting the continuous analog signal to the digital form, allowing the digital computer to interact with the digitized representation of the original signal. There are two steps in the A–D process.

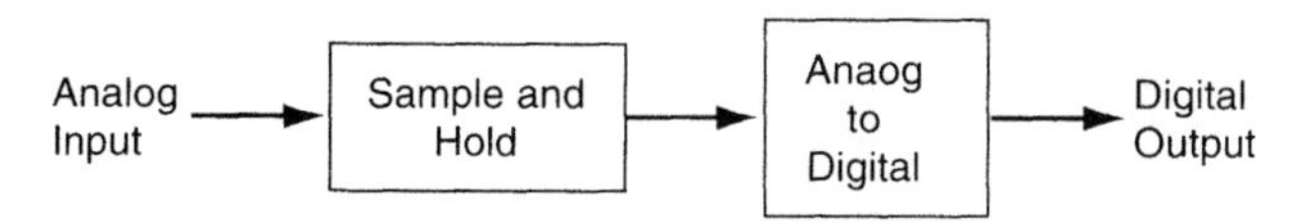

FIGURE 1.44 The analog-to-digital process.

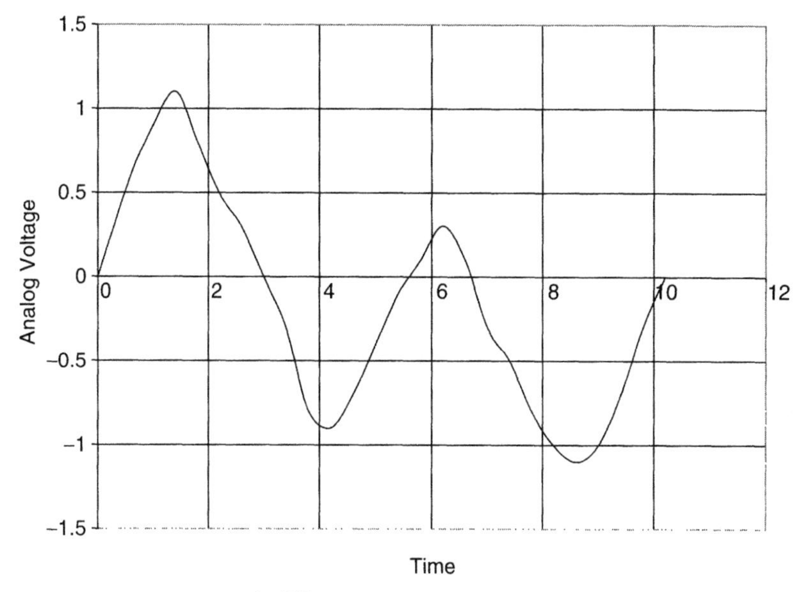

FIGURE 1.45 An analog transducer input signal.

The output of the sample-and-hold is allowed to change only at periodic intervals and maintain the instantaneous value at which the input signal was sampled. Changes in the input signal that occur between these sampling times are ignored. The sampling process is illustrated in Figure 1.45. The analog signal is multiplied by a unity magnitude sampling function, which produces the digital output signal (Figure 1.46).

Selection of the proper A–D converter aids in data resolution. An 8-bit converter has 2^8 bits of resolution, or 255 amplitude levels. A 12-bit converter has 2^{12} bits of resolution, or 4095 amplitude levels. A 16-bit converter has 2^{16} bits of resolution, or 65,535 amplitude levels. The bits of resolution are commonly used over a ± signal range resulting in the amplitude level resolution being half of the bits of resolution.

Another important selection criterion for the A–D card is the data throughput. Data throughput addresses how frequently the system reads channel data. Data throughput is a card function, not an individual channel function. This implies the sampling rate is determined by how many channels of data are being read.

Example 1.5. If a 0- to 4.095-V signal were to be digitized with a 12-bit digitizer, the corresponding digital output would vary between 0 and 4095 if the full range of the card could be used for this signal. This might lead one to believe that the original signal, when digitized, has a resolution of 4.095/4095 or 0.001 volts or 1 mV. This assumption is not quite accurate, as the A–D card is most likely to be configured for a ±5 V signal. If the input is

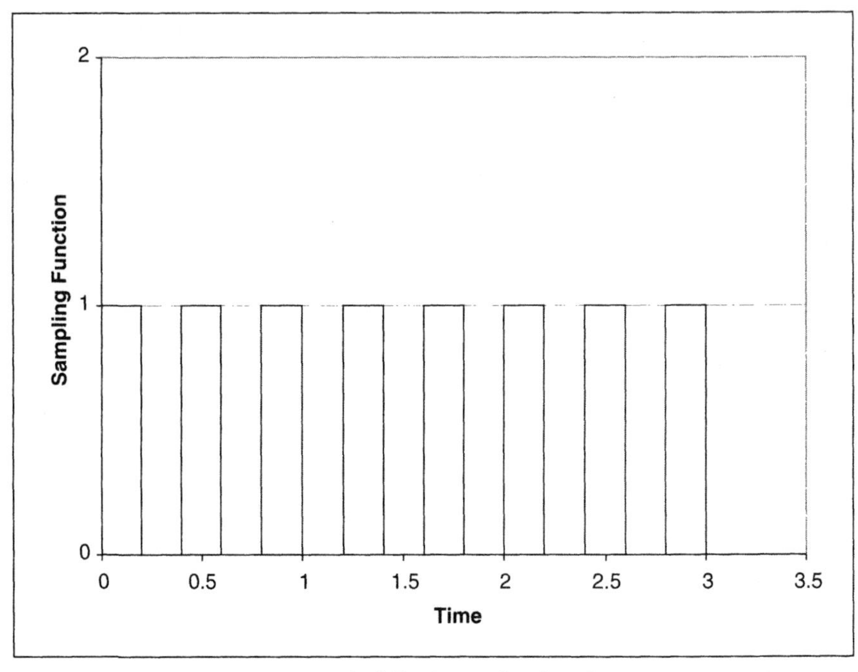

FIGURE 1.46 A sampling function.

configured in this manner, the resolution is approximately 10/4095 or 0.0024 V or 2.4 mV. If a 16-bit A–D card is used, the output range will be 0 to 65,535. For the 16-bit card with a ±5-V configuration, the resolution is 10/65,535 or 0.00015 V or 0.15 mV.

1.12.3 SAMPLING FREQUENCY

The sampling frequency (F_s), the sampling points per measurement period (N), and the frequency resolution are defined in Equations 1.12.1 and 1.12.2:

$$\text{Sampling frequency} \quad F_s = \frac{1}{\Delta t} \tag{1.12.1}$$

$$\text{Number of sample points} \quad N = \frac{T}{\Delta t} \tag{1.12.2}$$

The time interval, Δt, is the uniform sample spacing between two adjacent sampling points; N is the sampling points per measurement period; and T is the total time for the measurement.

1.12.4 ALIASING ERROR

Anti-aliasing filters are applied so that when a set of sample points is inspected, the input frequency components that are present are uniquely related to the input data. Otherwise, there are many possible input frequencies (the aliases), all of which can produce the same data points. The anti-aliasing filters theoretically should remove all but the wanted input frequencies.

Aliasing error results when the digitizing sampling rate is too low. As the input signal frequency nears the digitizing frequency, some components of the signal can be lost. The higher-frequency signal, after being digitized, can then show up as a lower-frequency signal (Figure 1.47). Because the data is no longer uniquely related to the analog signal, a unique reconstruction is impossible. Figure 1.48 shows an example of an aliasing error in the time domain. In the frequency domain, the spectrum will be shown as in Figure 1.49.

According to Shannon's sampling theorem, the sampling frequency should be a minimum of twice the signal frequency, as shown in Equation 1.12.3:

$$F_{\text{signal}} < \frac{F_{\text{sampling}}}{2} \quad (1.12.3)$$

The term on the right side of Equation 1.12.3 is exactly half of the sampling rate, which is commonly referred to as the Nyquist frequency. A continuous signal can only be properly sampled if it does not contain frequency components above the Nyquist frequency. For example, a sample rate of 1000 samples/s will be enough for an analog signal to be composed of frequencies below 500 cycles/s. If the analog signal contains frequencies

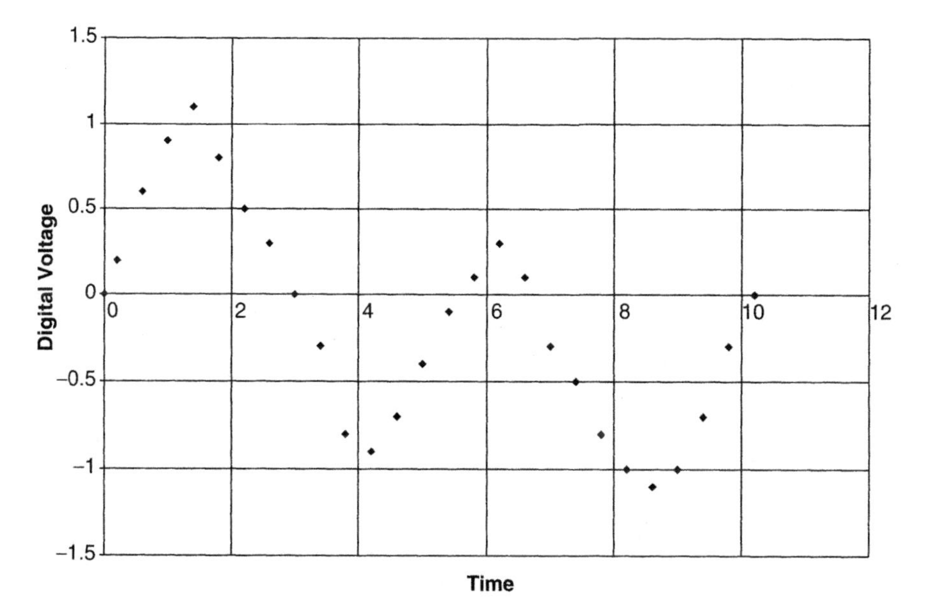

FIGURE 1.47 A resulting digitized signal.

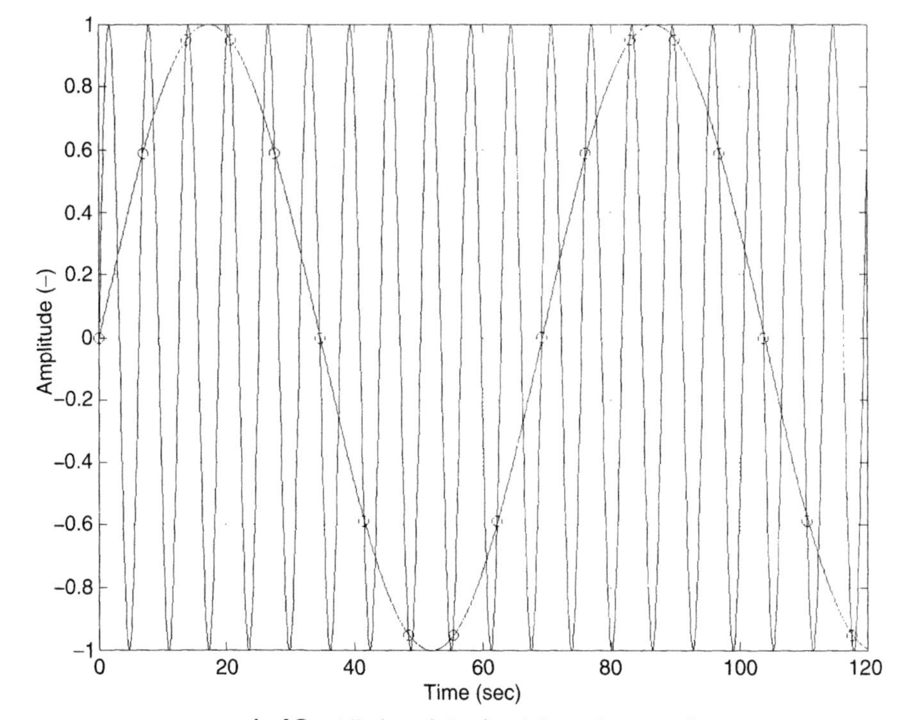

FIGURE 1.48 Aliasing of the signal due to low sampling rate.

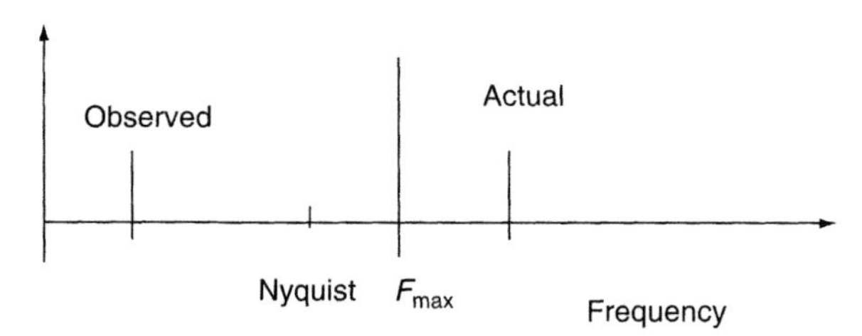

FIGURE 1.49 Frequency spectrum of an aliased signal.

above this limit, they will be aliased to frequencies between 0 and 500 cycles/s and combined, in an untraceable way, with those that were properly measured. When the signal is digitized for fatigue analysis, a sample rate of at least 10 times the maximum frequency of interest is recommended. A lower sampling rate may not capture the signal peaks.

An analog filter is required to remove the portion of the signal that can cause an aliasing error. To prevent aliasing in a complex signal, a more complex filter needs to be used. Real-world analog filters, as with digital filters, do not produce a theoretical, clean cut-off characteristic in the frequency domain. An on–off type of filter (clean cut-off) theoretically requires an infinite number of poles (implying an infinite number of amplifiers, resistors, and capacitors) and is inherently unstable in the time domain. The number of poles needed in the filter response increases with the complexity of the signal in the application. Approaching the filter circuit from the perspective of frequency will force issues such as the filter bandwidth, design, and order (number of poles) to be considered.

The first concern is with the relationship between the unwanted and desired frequencies. The challenge is the separation of wanted frequency components from nearby unwanted ones. The difficulty is related to the frequency ratio between wanted and unwanted components. As this number gets closer to unity, separation becomes more difficult and a filter is needed with more discrimination. The corner, or knee, of the cut-off frequency and the roll-off frequencies determines how cleanly this can be accomplished.

The second task is to separate desired frequency components from unwanted frequency components, which are at a much higher level. The difficulty is in determining how much the unwanted signals need to be reduced. As the need to reduce the unwanted frequency components increases, the rejection capabilities of the filter must increase.

The third challenge is to preserve the waveform accuracy of the signal, made up of only the wanted frequency components. The problem is that all filters change the signal passing through them in some way. If more accurate waveform reproduction is required, less filter vector error can be tolerated.

The fourth difficulty is in controlling the disturbance caused to the output signal by a sudden step change in the input signal. All filters take some time to recover after being shocked by a sudden change in the input signal. The faster the system can settle down, the lower the overshoot in the filter response.

A few of the more common filter designs used in data acquisition are the Butterworth, Bessel, and Chebyshev filters. These filter designs are well understood and can be realized with an operational amplifier, resistors, and capacitors.

A typical characteristic of the low-pass anti-alias filter is shown in Figure 1.50. General filter characteristics are covered in Section 1.11.3. The two main factors to be considered when selecting a filter are the cut-off frequency and the roll-off rate. The cut-off frequency, sometimes called the corner frequency, is the frequency at which the filter begins to attenuate a signal. The roll-off rate, typically defined in units of signal attenuation per frequency change (dB/octave), defines how sharply the filter attenuates the higher (low-pass filter) or lower (high-pass filter) frequencies.

Data acquisition systems also have a limited dynamic range (Section 1.12.2). This is the ratio of the largest and the smallest signals that can be recorded. Before these limits are reached, the ability of the system to discriminate between different components of a mixed input is compromised. Small important signals can be mixed in with large but irrelevant signals, resulting in erroneous data capture. A filtering process can be applied to the data once the data are in digital form, but the signal that remains will be corrupted with

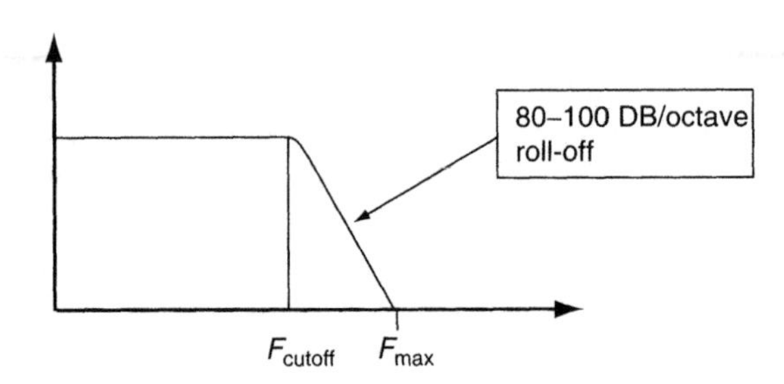

FIGURE 1.50 Typical characteristics of an anti-alias (low-pass) filter.

residues of the unwanted signal. This process, known as swamping, is caused by the failure of the acquisition process.

1.12.5 SAMPLING RATE CONSIDERATION

High sampling rates allow more accurate capture of the peak values of the analog signal. This leads to a more precise analog-to-digital conversion. This provides not only a better time base resolution but also good resolution in the frequency domain. The negative effect of the high sampling rate is that the number of the data points is larger for a given data acquisition interval. More data occupies more storage space and takes longer for the subsequent processing. An optimal sampling rate should be chosen based on the maximum frequency of the expected input signal. When the data is to be used for fatigue analysis, it is good practice to start with a high sampling rate. The common practice is to use a sampling rate at least 10 times the frequency of interest.

1.12.6 FREQUENCY DOMAIN ANALYSIS

The frequency composition of a time series can be described in terms of the power spectrum or the PSD. Analyzing the PSD of the digital data is also useful in confirming the sampling rate used. If a lower frequency is desired after the data is digitized, a digital filter can be used to remove frequencies out of the range of interest. However, the digital filter cannot remove the alias error components because it cannot identify which frequency is the true frequency and which is the alias frequency. This lack of discrimination is because the aliased components are dropped into the lower frequency data. The frequency content of a time series provides information about its characteristics. It is always possible to compare two time histories and determine the correlation based on the PSD.

1.13 DATA VERIFICATION

Data verification is the process used to determine whether the data are valid. To perform data verification, a statistical examination of the data is frequently required. This section overviews discrete data statistics and then examines continuous data techniques.

1.13.1 STATISTICAL DECISION MAKING

The measure of central tendency of data is accomplished through the mean and the median. The mean for a population is the sum of the values in a data set divided by the number of elements in the set, as shown in Equation 1.13.1. The median is the central value in a set of data arranged in order of magnitude:

$$\mu = \frac{\sum_{i=1}^{n} (X_i)}{n} \tag{1.13.1}$$

The mode can be defined as the observation that occurs most frequently in a sample of data. If the data are symmetrically distributed, the mean, the median, and the mode are the same. The measure of variability of the data is the range or dispersion. Range (R) is the absolute difference between the lowest and highest values in a data set.

$$R = |X_{\text{high}} - X_{\text{low}}| \tag{1.13.2}$$

The mean deviation and the standard deviation of the data are presented in Equations 1.13.3 and 1.13.4 respectively:

$$\text{Mean Dev} = \frac{\sum_{i=1}^{n} |X_i - \overline{X}|}{n} \tag{1.13.3}$$

$$\text{Standard Dev} = \sigma = \sqrt{\frac{\sum_{i=1}^{n} (X_i - \overline{X})^2}{n-1}} = \sqrt{\frac{\sum_{i=1}^{n} (X_i)^2 - n\overline{X}^2}{n-1}} \tag{1.13.4}$$

where n is the number of observations and X_i the values of the random sample variables. Variance (S^2) indicates the spread of the distribution and is defined by the square of the standard deviation:

$$S^2 = \frac{\sum_{i=1}^{n} (X_i - \overline{X})^2}{n-1} \tag{1.13.5}$$

The distribution of the data about the mean is its measure of skewness, as shown in Equation 1.13.6:

$$\text{Skew} = \frac{3(\mu - \text{median})}{\sigma} \tag{1.13.6}$$

1.13.2 EXPERIMENTAL DATA VERIFICATION

Most transducers and data acquisition systems are susceptible to external interference such as electrical noise or even bad contacts on a transducer. Removing unwanted signals that are larger than the wanted ones allows the desired signal to be amplified and captured with greater security, independent of the type of data acquisition or recording being used. The data shows up as a dropout, spike, or drift. A typical data set with a data spike is shown in Figure 1.51. The spike in the data generated can cause numerous problems in the subsequent analysis in both the time domain and the frequency domain. For example, the fatigue life prediction will be invalid with the presence of the bogus data.

After a set of data is collected, it is necessary to verify the integrity of the data by calculating the statistics of the data and examining any questionable data. The following parameters are commonly used to determine the statistics.

The maximum and minimum values are the maximum and minimum of the entire data set. The difference between the maximum and minimum was previously defined as the range or dispersion.

The mean of the sample is defined as the area under the data curve divided by the total time interval, as shown in Equation 1.13.7. Functionally, it is the summation of each discrete data value at each time interval divided by the number of time intervals in the total sample period, as shown in Equation 1.13.1.

$$\bar{x} = \frac{1}{T}\int_0^T x(t)dt$$
$$\bar{x}^2 = \frac{1}{T}\int_0^T x^2(t)dt \tag{1.13.7}$$

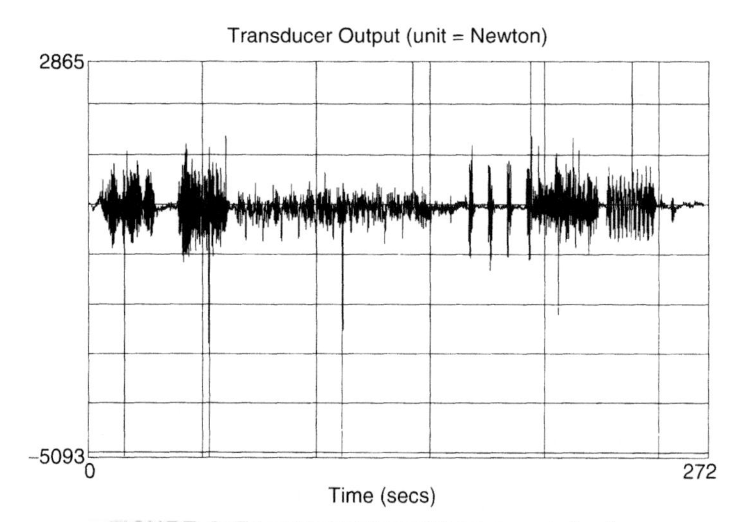

FIGURE 1.51 Typical data with an unwanted spike.

The square of the standard deviation is referred to as the variance. The value of the variance of a set of data is given in Equation 1.13.8:

$$S^2(x) = \frac{1}{T}\int_0^T [x(t) - \bar{x}]dt \tag{1.13.8}$$

The root mean square (RMS) is defined as the absolute value of the square root of the mean squares. If the mean is equal to zero, the RMS value is equal to the standard deviation (S). The crest or crest factor is the ratio of the maximum or minimum value to the standard deviation of the data set, as defined in Equation 1.13.9:

$$\text{Crest} = \frac{\max(|\text{maximum}|, |\text{minimum}|)}{S} \tag{1.13.9}$$

A spike or dropout is when there are data with significant amplitude above (spike) or below (dropout) the mean of the data sample. A spike or dropout can be detected by observing the data maximums and minimums, by statistically analyzing the data, and by examining the amplitude change rate by differentiation. The maximum and minimum method is simply that when a spike or dropout occurs in the data, its magnitude will always tend toward full scale. The maximum and minimum magnitude of the data set is usually a good way to check for spikes and dropouts.

The statistical method of identifying erroneous data uses the standard deviation (S) of the data. The crest factor is used to detect the spike or dropout. If the crest calculated from the data is higher than a predetermined level, it may be indicative of a spike or dropout in the data.

The differentiation method examines the data from the perspective of the rate of change between adjacent data or data blocks in a stream of data. If the amplitude of the data changes too rapidly over a small time interval, it may be indicative of a spike or dropout.

The methods described here demonstrate the basic concepts used in detecting a spike in the data. In actual applications, computer programs are written to implement spike detection. It is a very tedious job to detect a spike manually.

Another useful way to verify the data is in the frequency domain. The frequency composition of a time series can be described in terms of the PSD, which describes how the signal's power is distributed among the different frequencies (Figure 1.52).

The Fourier transform is the background for modern time series analysis. In most modern instrumentation and software analysis for spectral analysis, the calculation involves determining the amplitude and phase of a given data set. There exist two important definitions of the Fourier transform, the integral Fourier transform and the discrete Fourier transform. The integral transform, with its limits of integration ($-\infty$, $+\infty$), forms the

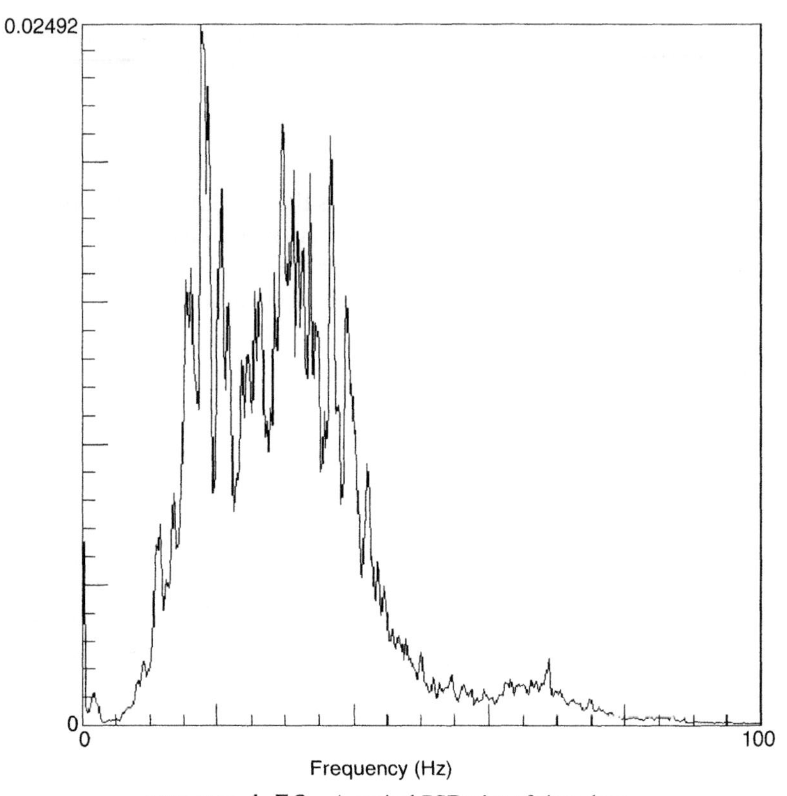

FIGURE 1.52 A typical PSD plot of time data.

basis for all Fourier analysis. The integral Fourier transform is shown in Equation 1.13.10:

$$X(f) = \int_{-\infty}^{\infty} x(t)e^{-j2\pi ft}\,dt \tag{1.13.10}$$

As shown in this expression, to evaluate a function, the function has to be continuous and exist from $-\infty$ to $+\infty$. In reality, data are available in a discrete form and over a limited time period.

The discrete Fourier transform is based on the assumption that the signal is transient with respect to the period observed, or, if it is not, the signal must then be composed of the harmonics of the time period. The discrete Fourier transform is shown in Equation 1.13.11:

$$X(m\Delta f) = \frac{2}{T}\Delta t \sum_{n=0}^{N-1} x(n\Delta t)e^{-j2\pi m\frac{n}{N}} \tag{1.13.11}$$

The PSD provides information for comparing the time series. If the frequency measured by the transducer is near the natural frequency of the structure or if the structure is excited by a dominant frequency, the PSD will show a large amplitude at the frequency line corresponding to the excitation frequency. If the time series are similar, the PSD will be almost similar. It is therefore possible to compare two time histories and determine the correlation based on the PSD.

1.14 DATA TRANSMISSION FROM ROTATING SHAFTS

Data from rotating assemblies must have an interface between the rotating and stationary components of the data acquisition system. The most common forms of interfacing are slip rings and telemetry systems (Dally and Riley, 1991).

1.14.1 SLIP RING ASSEMBLIES

Strain gage transducers are often used on rotating components in which it is impossible to use ordinary wire to connect them to the recording system. To enable the data from the rotating assembly to be transmitted to the stationary data acquisition system, a coupling system called a slip ring assembly is used (Figure 1.53). Slip ring assemblies are used to connect the transducers on the rotating structure to the power supply and recording system. The slip rings of the assembly are mounted on the rotating part. The shell of the slip ring assembly is stationary and consists of a series of isolated brushes. Bearings are used to keep the brushes and the rings aligned and concentric.

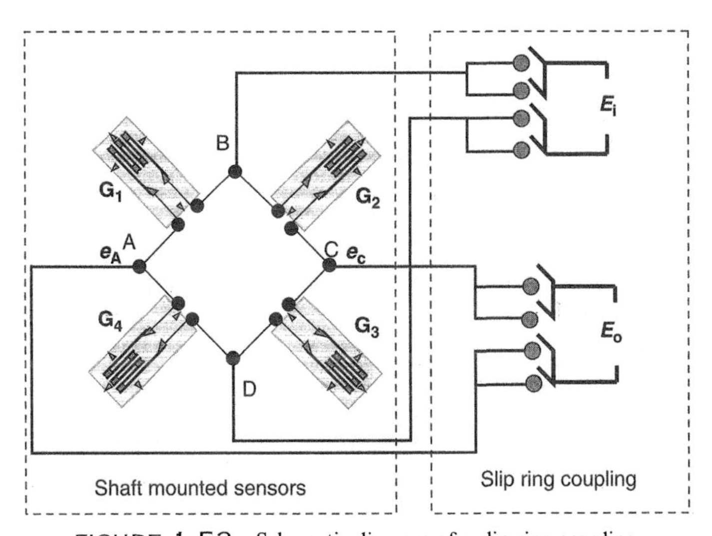

FIGURE 1.53 Schematic diagram of a slip ring coupling.

Slip ring couplings must be mounted at the end of the rotating shaft or assembly from which the data is being collected. Slip ring assemblies also tend to introduce noise into the measurement system. The inherent tendency toward bouncing between the ring and the brush can be minimized if all parts are aligned. Dirt and excessive wear on the slip ring and brush tend to generate electrical noise. The use of multiple rings and corresponding brushes for each connection helps minimize the noise.

1.14.2 TELEMETRY SYSTEMS

Telemetry systems are an alternative method of transmitting data from the rotating assembly to the stationary data acquisition system. Basic telemetry systems consist of a modulator, a voltage-controlled oscillator (VCO), and a power supply for the strain gage bridge. The signal from the strain gage bridge is used to pulse modulate a constant-amplitude square wave. The output pulse width is proportional to the voltage from the bridge. This square wave serves to vary the frequency of the voltage-controlled oscillator, which has a center frequency (f_c). The VCO signal is transmitted by an antenna mounted on the rotating shaft and is received by a stationary loop antenna, which encircles the shaft. After the signal is received, it is demodulated, filtered, and amplified before recording. Most of the transmitting unit of a telemetry system is completely self-contained. The power supply to the components on the rotating shaft is obtained by inductively coupling the power supply through the stationary loop antenna. Figure 1.54 is a schematic diagram of the operating principle of a typical single channel telemetry system.

When more than one channel is to be transmitted, two different methods can be used: frequency-division multiplexing and time-division multiplexing. In addition to more VCOs, the frequency-division multiplexing uses multiplexing equipment to attach specific channel data to specific frequencies before transmitting. The VCO is designed so that the VCO output frequency range of each channel is not overlapped. At the receiving end, demultiplexing equipment is required to separate the frequencies so that the data in each channel are restored.

In time-division multiplexing, all the channels use the same frequency spectrum, but at different times. Each channel is sampled in a repeated

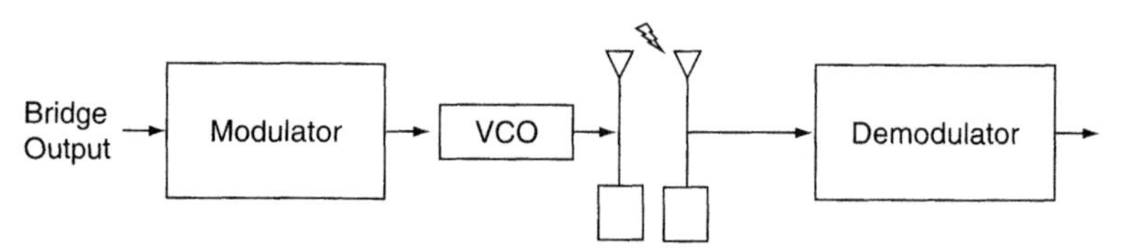

FIGURE 1.54 Schematic diagram of a simplified telemetry system.

sequence by a commuter building a composite signal containing the time-spaced signals from each channel. A decommuter, operating at the same frequency as that of the commuter, separates the channels at the receiving end. Because the channels are not monitored continuously, the sample rate must be sufficient to ensure that the signal amplitude does not change appreciably during the time between samples. Most telemetry systems use sample rates at least five times higher than the highest expected frequency.

1.15 VIRTUAL INSTRUMENT SYSTEMS AND THEIR HIERARCHICAL STRUCTURE

Virtual instrumentation systems are software systems that are assembled on the computer to perform all the functions of a stand-alone system, with the additional advantage of taking all the different measured data through the same system. Virtual systems can usually perform both measurement and control functions. The instruments are called virtual because they are created as software only. The virtual instruments have an appearance and operation that can imitate conventional instruments. The software then controls cards that take transducer signals into the computer, making the virtual instruments perform as functional instruments. Many virtual instrumentation packages use a graphical interface and a graphical programming language and adhere to the concept of modular programming to simplify set-up.

The virtual instrument (VI) has a front panel that is an interactive user interface that allows the operator to control the instrument. The operator can view the measurements as an analog, digital, or graphical representation on the computer monitor.

The VI receives its instructions from a block diagram, which the user creates in graphical format. The block diagram presents a pictorial representation to the problem solution and supplies the source code for the interpretation.

Sub VIs are virtual instruments within other VIs. This allows modular system setup and provides a hierarchical structure to the program. Debugging is also simplified, as each VI can be verified individually. Data can be passed from one VI to the next in the system. Icon or connector planes are used to turn the VI into an object to be used in the block diagrams of other VIs as subroutines or functions.

REFERENCES

Brophy, J. J., *Basic Electronics for Scientists*, 3rd ed., McGraw-Hill, New York, 1997.

Dally, J. W., and Riley, W. F., *Experimental Stress Analysis*, 3rd ed., McGraw-Hill, New York, 1991.

Dally, J. W., Riley, W. F., and McConnell, K. G., *Instrumentation for Engineering Measurements*, 2nd ed., Wiley, New York, 1993.

Hung, Y. Y., Shearography: A new optical method for strain measurement and nondestructive testing, *Optical Engineering*, May/June, 1982, pp. 391–395.

Lee, Y. L., Lu, M. W., and Breiner, R. W., Load acquisition and damage assessment of a vehicle bracket component, *International Journal of Materials and Product Technology*, Vol.12, Nos. 4–6, 1997, pp. 447–460.

Kobayashi, A. S., *Handbook on Experimental Mechanics*, 2nd rev. ed., VCH Publishers, New York, 1993.

Perry, C. C., and Lissner, H. R., *The Strain Gage Primer*, 2nd ed., McGraw-Hill, New York, 1962.

2

FATIGUE DAMAGE THEORIES

YUNG-LI LEE
DAIMLERCHRYSLER

2.1 INTRODUCTION

Predicting fatigue damage for structural components subjected to variable loading conditions is a complex issue. The first, simplest, and most widely used damage model is the linear damage. This rule is often referred to as Miner's rule (1945). However, in many cases the linear rule often leads to nonconservative life predictions. The results from this approach do not take into account the effect of load sequence on the accumulation of damage due to cyclic fatigue loading. Since the introduction of the linear damage rule many different fatigue damage theories have been proposed to improve the accuracy of fatigue life prediction. A comprehensive review of many fatigue damage approaches can be found elsewhere (Fatemi and Yang, 1998). This chapter addresses (1) underlying fatigue damage mechanisms, (2) fatigue damage models commonly used in the automotive industry, and (3) postulations and practical implementations of these damage rules.

2.2 FATIGUE DAMAGE MECHANISM

Fatigue is a localized damage process of a component produced by cyclic loading. It is the result of the cumulative process consisting of crack initiation, propagation, and final fracture of a component. During cyclic loading, localized plastic deformation may occur at the highest stress site. This plastic deformation induces permanent damage to the component and a crack develops. As the component experiences an increasing number of loading cycles, the length of the crack (damage) increases. After a certain number of cycles, the crack will cause the component to fail (separate).

In general, it has been observed that the fatigue process involves the following stages: (1) crack nucleation, (2) short crack growth, (3) long crack growth, and (4) final fracture. Cracks start on the localized shear plane at or near high stress concentrations, such as persistent slip bands, inclusions, porosity, or discontinuities. The localized shear plane usually occurs at the surface or within grain boundaries. This step, crack nucleation, is the first step in the fatigue process. Once nucleation occurs and cyclic loading continues, the crack tends to grow along the plane of maximum shear stress and through the grain boundary.

A graphical representation of the fatigue damage process shows where crack nucleation starts at the highest stress concentration site(s) in the persistent slip bands (Figure 2.1). The next step in the fatigue process is the crack growth stage. This stage is divided between the growth of Stage I and Stage II cracks. Stage I crack nucleation and growth are usually considered to be the initial short crack propagation across a finite length of the order of a couple of grains on the local maximum shear stress plane. In this stage, the crack tip plasticity is greatly affected by the slip characteristics, grain size, orientation, and stress level, because the crack size is comparable to the material microstructure. Stage II crack growth refers to long crack propagation normal to the principal tensile stress plane globally and in the maximum shear stress direction locally. In this stage, the characteristics of the long crack are less affected by the properties of the microstructure than the Stage I crack. This is because the crack tip plastic zone for Stage II crack is much larger than the material microstructure.

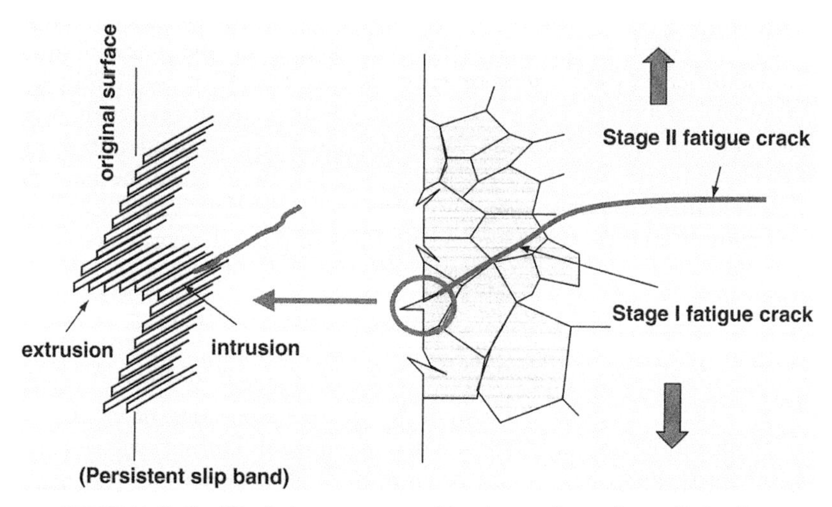

FIGURE 2.1 The fatigue process: a thin plate under cyclic tensile loading.

In engineering applications, the amount of component life spent on crack nucleation and short crack growth is usually called the *crack initiation period*, whereas the component life spent during long crack growth is called the *crack propagation period.* An exact definition of the transition period from initiation to propagation is usually not possible. However, for steels the size of a crack at the end of the initiation stage, a_0, is of the order of a couple of grains of the material. This crack size typically ranges from about 0.1 to 1.0 mm. The crack initiation size can be estimated using the linear elastic fracture mechanics approach for smooth specimens by Dowling (1998):

$$a_0 = \frac{1}{\pi}\left(\frac{\Delta K_{th}}{\Delta S_e}\right)^2 \tag{2.2.1}$$

or by 0.1 to 0.2 times the notch-tip radius for notched specimens (Dowling, 1998), or by twice the Peterson empirical material constant for steels (Peterson, 1959):

$$a_0(\text{mm}) = 2 \times 0.0254 \times \left(\frac{2079}{S_u(\text{MPa})}\right)^{1.8} \tag{2.2.2}$$

where S_u is the ultimate tensile strength of a material, ΔS_e is the stress range at the fatigue limit, and ΔK_{th} is the range of the threshold intensity factor for $R = -1$.

Typically, the crack initiation period accounts for most of the fatigue life of a component made of steels, particularly in the high-cycle fatigue regime (approximately >10,000 cycles). In the low-cycle fatigue regime (approximately <10,000 cycles), most of the fatigue life is spent on crack propagation.

Once a crack has formed or complete failure has occurred, the surface of a fatigue failure can be inspected. A bending or axial fatigue failure generally leaves behind clamshell or beach markings. The name for these markings comes from the appearance of the surface. An illustration of these markings is shown in Figure 2.2. The crack nucleation site is the center of the shell, and the crack appears to propagate away from the nucleation site, usually in a radial manner. A semielliptical pattern is left behind. In some cases, inspection of the size and location of the beach marks left behind may indicate where a different period of crack growth began or ended.

Within the beach lines are striations. The striations shown in Figure 2.2 appear similar to the rings on the cross-section of a tree. These striations represent the extension of the crack during one loading cycle. Instead of a ring for each year of growth, there is a ring for each loading cycle. In the event of a failure, there is a final shear lip, which is the last bit of material supporting the load before failure. The size of this lip depends on the type of loading, material, and other conditions.

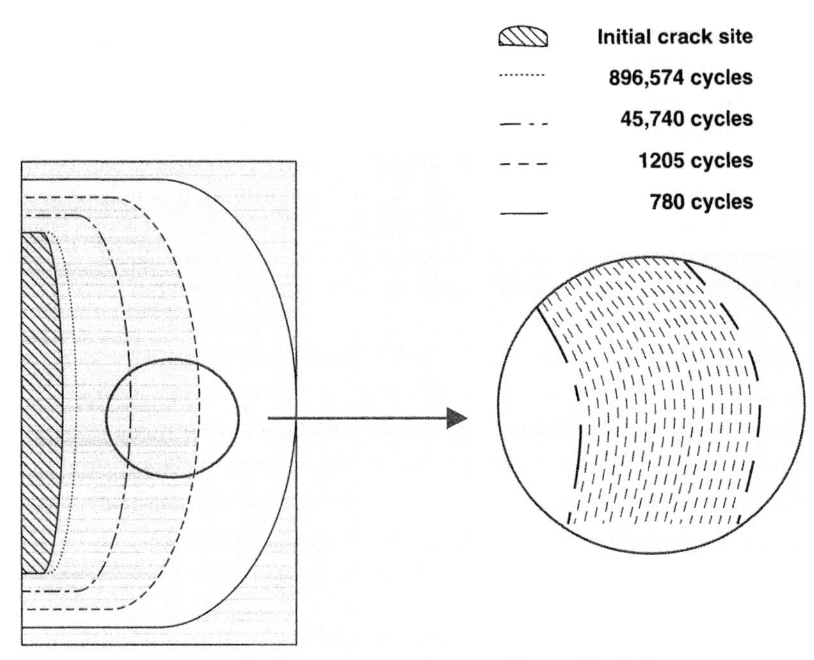

FIGURE 2.2 Fracture surface markings and striations.

2.3 CUMULATIVE DAMAGE MODELS—THE DAMAGE CURVE APPROACH

The component's damage can be expressed in terms of an accumulation of the crack length toward a maximum acceptable crack length. For example, a smooth specimen with a crack length at fracture of a_f is subjected to cyclic loading that results in a crack length of a. The amount of damage, D, at a given stress level S_1, would be the ratio of a to a_f. To illustrate the cumulative damage concept, a crack growth equation developed by Manson and Halford (1981) is adopted:

$$a = a_0 + (a_f - a_0)\left(\frac{n}{N_f}\right)^{\alpha_f} \tag{2.3.1}$$

Equation 2.3.1 was derived based on early crack growth fracture mechanics and fitted with a large amount of test data for loading with two-step life/stress levels, where n is the number of loading cycles applied to achieve a crack length of a, and a_0 is the initial crack length. The value N_f represents the number of cycles applied to achieve the crack length a_f at final fracture. The exponent α_f is empirically determined and has the following form:

$$\alpha_f = \frac{2}{3}N_f^{0.4} \tag{2.3.2}$$

Cumulative damage (D) is the ratio of instantaneous to final crack length and can be expressed as follows:

$$D = \frac{a}{a_f} = \frac{1}{a_f}\left[a_0 + (a_f - a_0)\left(\frac{n}{N_f}\right)^{\alpha_f}\right] \tag{2.3.3}$$

This damage equation implies that fatigue failure occurs when D is equal to unity (i.e., $a = a_f$).

Consider a two-step high–low sequence loading in Figure 2.3, where n_1 denotes the initial applied load cycles with a higher stress or load level and $n_{2,f}$ the remaining cycles to eventual fatigue failure with a lower stress or load level. Note that the subscripts 1 and 2 refer to sequence of the applied loading: 1 is the first and 2 is the second load level. The S–N curve is used to obtain the fatigue lives $N_{1,f}$ and $N_{2,f}$ for each load level. Nonlinear damage curves for two different loads are shown schematically in Figure 2.4. Each of these curves represents a different loading condition that leads to a different time to failure (or life level). At each load level or life level, the relation between the damage value and the applied cycles or the cycle ratio follows the power law equation in Equation 2.3.3. If a cycle ratio $n_1/N_{1,f}$ is first applied along the curves representing the life level $N_{1,f}$ to point OA, the damage accumulation process will be represented by the life level curve $N_{1,f}$ from zero to point A. If at this point a new loading level with a life of $N_{2,f}$ is introduced and this loading is applied, the damage process will proceed from point A to point A′ from the same damage value.

If the load corresponding to the cycle ratio $n_{2,f}/N_{2,f}$ is applied from A′ to B′ at the life level N_{2f}, failure takes place if $D = 1.0$ is reached at point B′. From this figure, it is clear that if a higher load level with a lower life along OA is first applied and followed by the lower load magnitude with a higher

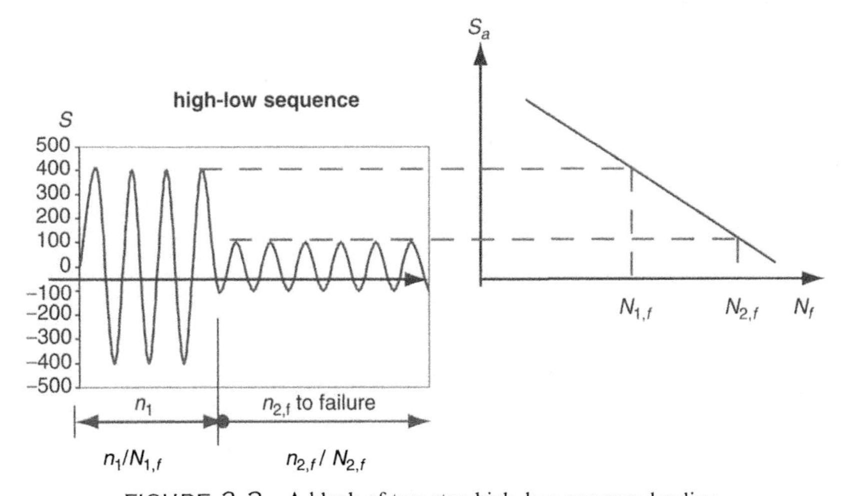

FIGURE 2.3 A block of two-step high–low sequence loading.

life along A′B′, the sum of cycle ratios will be smaller than unity. Thus, the estimated fatigue life depends on the sequence of loading. However, if a lower load level is applied first along OA′ and is followed by the higher load level along AB, the summation of the cycle ratios is greater than unity because the cycle ratio AA′ is accounted for twice.

Based on the equal damage at A and A′ in Figure 2.4 for the two load levels, the following equations hold true for the relation between the cycle ratio $n_1/N_{1,f}$ and the equivalent damage cycle ratio $n_2/N_{2,f}$:

$$\left(\frac{n_{2,f}}{N_{2,f}}\right) = 1 - \left(\frac{n_1}{N_{1,f}}\right)^{\left(N_{1,f}/N_{2,f}\right)^{0.4}} \tag{2.3.4}$$

and

$$\frac{n_1}{N_{1,f}} = \left[\frac{n_2}{N_{2,f}}\right]^{\left(N_{2,f}/N_{1,f}\right)^{0.4}} \tag{2.3.5}$$

where n_2 is the number of cycles at the life level $N_{2,f}$, equivalent damage to the initial cycle ratio $n_1/N_{1,f}$.

It is clear that Equation 2.3.5 is independent of material and geometric parameters (e.g., a_0, a_f, and α_f) that were introduced in the damage accumulation equation (Equation 2.3.3). Thereby, a nonlinear damage curve for a reference life level ($N_{1,f}$) can be linearized by replacing a_0 by zero and 2/3 by $\left(1/N_{1,f}\right)^{0.4}$. Thus, the damage function for the reference life level can be simplified as a linear line connecting (0,0) with (1,1), i.e.,

$$D_1 = \frac{n_1}{N_{1,f}} \tag{2.3.6}$$

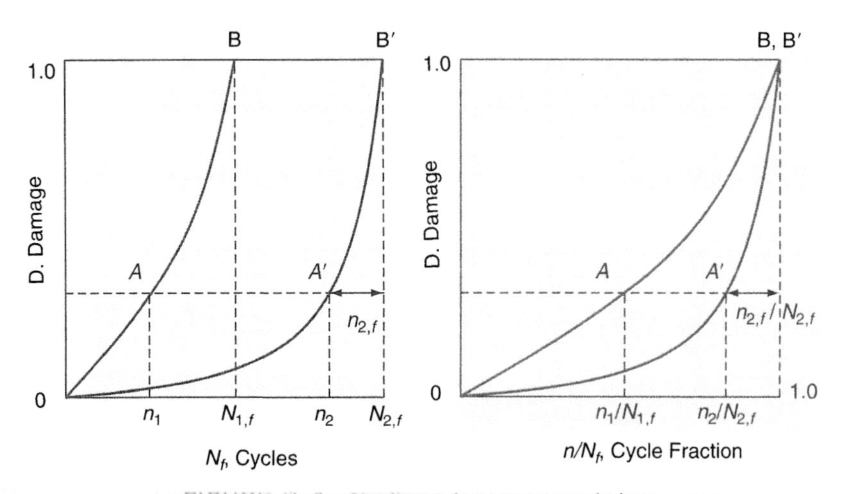

FIGURE 2.4 Nonlinear damage accumulation.

Therefore, the damage curve for another life level ($N_{2,f}$) is then given by the power law damage equation defined as:

$$D_2 = \left[\frac{n_2}{N_{2,f}}\right]^{\left(N_{2,f}/N_{1,f}\right)^{0.4}} \quad (2.3.7)$$

For multiple life levels (e.g., $N_{1,f} < N_{2,f} < \cdots < N_{n,f}$), the damage curves can be constructed expeditiously by letting the damage curve for the lowest life level be the reference life.

Two other methods have been developed to determine the power law damage equation. Based on the experimental observations that the equivalent damage S–N lines converge to the fatigue limit, Subramanyan (1976) calculated damage by referring to the reference stress amplitude S_{ref} and the fatigue limit S_e as follows:

$$D_N = \left[\frac{n_n}{N_{n,f}}\right]^{(S_n - S_e)/(S_{ref} - S_e)} \quad (2.3.8)$$

On the other hand, Hashin (1980) expressed this differently by using the fatigue life N_e at the fatigue limit S_e:

$$D_N = \left[\frac{n_n}{N_{n,f}}\right]^{\log(N_{n,f}/N_e)/\log(N_{ref}/N_e)} \quad (2.3.9)$$

The power law accumulation theories using the three different methods are compared with the experimental data by Manson et al. (1967). The high-to-low (H-L) step-stress series was applied to smooth components made of SAE 4130 steel with a soft heat treatment. The two stress levels are 881 and 594 MPa, which correspond to fatigue lives of 1700 and 81,250 cycles, respectively. A fatigue limit of 469 MPa was also determined at 800,000 cycles. By applying Equations 2.3.7–2.3.9, the three power law exponent values are obtained (i.e., 0.213 on Manson and Halford, 0.303 on Subramanyan, and 0.372 on Hashin). Figure 2.5 compares the predicted fatigue behavior of the three power law damage rules with experimental data for SAE 4130. The curve derived from the Mason rule is close to the experimental data and the Subramanyan and the Hashin curves are slightly nonconservative.

Example 2.1. An unnotched component is subjected to a four-level step-stress fatigue test, which starts with the highest load level, ±800 MPa, to the lowest level, ±200 MPa. At each load level, a cycle fraction of 0.01 is added before proceeding to the next level. The sequence 1, 2, 3, 4 is repeated until failure occurs when $D = 1.0$. The specific four-level step stresses (S_i), the applied cycles (n_i), and the associated fatigue lives ($N_{i,f}$) are listed in Table 2.1. Estimate the fatigue life of the specimen based on the damage curve approach by Manson and Halford.

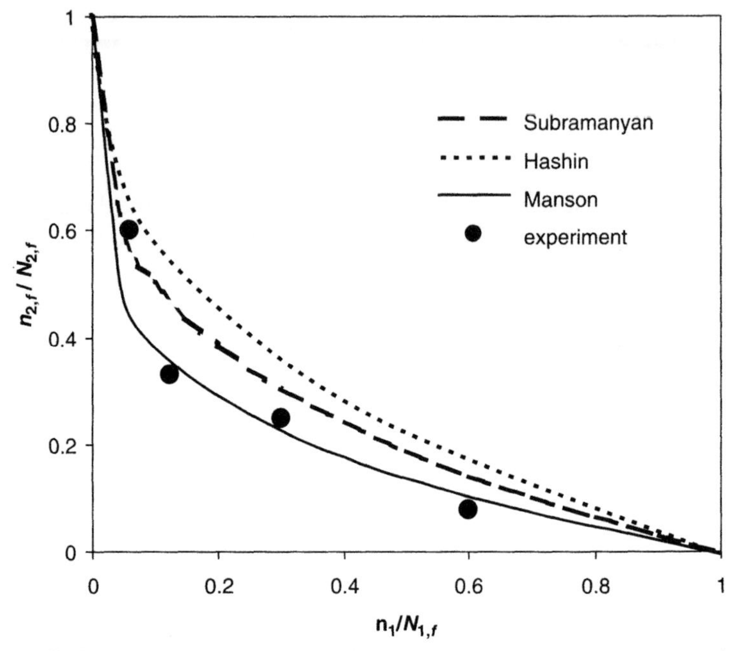

FIGURE 2.5 Comparison of predicted fatigue behavior with experimental data for SAE 4130 steel for two-step loading.

TABLE 2.1 Four-Level Step-Stress Fatigue Test Data

Four-Level Step-Stress S_i (Mpa)	n_i (cycles)	$N_{i,f}$ (cycles)
±800	10	1000
±600	100	10,000
±400	1000	100,000
±200	10,000	1,000,000

Solution. Choose $N_{ref} = 10^3$ cycles for the reference life because it has the shortest life of the four life levels and so the linear damage rule applies for this life level. For the first block of loading, proceed along the straight line curve until $D_1 = 10/10^3 = 0.01$. Then stop and traverse horizontally until on the damage curve for $N_{2,f} = 10^4$ cycles at $D_1 = 0.01$. The equivalent damage cycles for $N_{2,f} = 10^4$ cycles are $n_{2,eq} = N_{2,f} \times [D_1]^{(N_{ref}/N_{2,f})^{0.4}} = 10^4 \times (0.01)^{(10^3/10^4)^{0.4}} = 1599$ cycles.

With the additional $n_2 = 10^2$ cycles applied in the second block of loading, the accumulated damage value becomes

$$D_{1+2} = \left[\frac{n_{2,eq} + n_2}{N_{2,f}}\right]^{[N_{2,f}/N_{ref}]^{0.4}} = \left[\frac{1599 + 100}{10^4}\right]^{[10^4/10^3]^{0.4}} = 0.01165$$

Proceed horizontally to the next damage curve for $N_{3,f} = 10^5$ cycles at $D_{1+2} = 0.01165$. The corresponding number of cycles are

$$n_{3,eq} = N_{3,f} \times [D_{1+2}]^{(N_{ref}/N_{3,f})^{0.4}} = 10^5 \times (0.01165)^{(10^3/10^5)^{0.4}} = 49378 \text{ cycles}$$

By adding $n_3 = 10^3$ cycles in the third block of loading, the damage accumulation so far is

$$D_{1+2+3} = \left[\frac{n_{3,eq} + n_3}{N_{3,f}}\right]^{[N_{3,f}/N_{ref}]^{0.4}} = \left[\frac{49378 + 1000}{10^5}\right]^{[10^5/10^3]^{0.4}} = 0.01322$$

Finally, move horizontally to the last damage curve for $N_{4,f} = 10^6$ cycles at $D_{1+2+3} = 0.01322$. The equivalent damage cycles are

$$n_{4,eq} = N_{4,f} \times [D_{1+2+3}]^{(N_{ref}/N_{4,f})^{0.4}} = 10^6 \times (0.01322)^{(10^3/10^6)^{0.4}} = 761{,}128$$

The accumulated damage for additional $n_4 = 10^4$ cycles is calculated:

$$D_{1+2+3+4} = \left[\frac{n_{4,eq} + n_4}{N_{4,f}}\right]^{[N_{4,f}/N_{ref}]^{0.4}} = \left[\frac{761128 + 10000}{10^6}\right]^{[10^6/10^3]^{0.4}} = 0.01626$$

This completes the calculation of the total damage accumulation for the four-level step loading. Then traverse horizontally back to the first curve where you move up along the linear line from the damage value of 0.01626 to 0.02626 before going back horizontally to the second curve, etc. Advance alternately up these curves until a total damage value of $D = 1$ is eventually reached. The complete list of iterations is illustrated in Figure 2.6 and Table 2.2. In this example, it takes 11 blocks to reach the failure point.

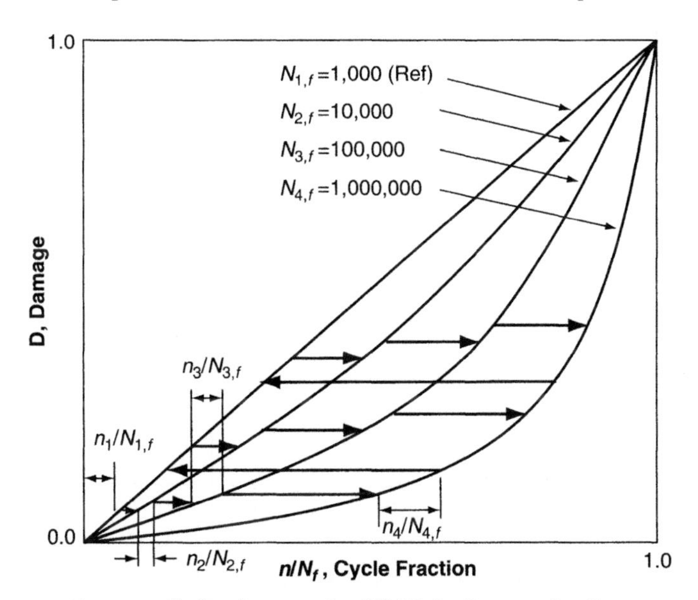

FIGURE 2.6 An example of DCA for four-step loading.

TABLE 2.2 Procedures for Fatigue Life Estimations Based on the Manson and Halford Damage Curve Approach

No. of Blocks	$n_{eq}+n_1$	D_1	$n_{eq}+n_2$	D_{1+2}	$n_{eq}+n_3$	D_{1+2+3}	$n_{eq}+n_4$	$D_{1+2+3+4}$
1	10	0.01000	1699	0.01165	50378	0.01322	771128	0.01625
2	26	0.02625	2448	0.02915	58105	0.03253	815621	0.03955
3	50	0.04955	3123	0.05377	63923	0.05940	846818	0.07170
4	82	0.08170	3789	0.08738	68956	0.09582	872450	0.11503
5	125	0.12503	4470	0.13234	73577	0.14427	895010	0.17239
6	182	0.18239	5179	0.19155	77957	0.20780	915622	0.24731
7	257	0.25731	5925	0.26855	82191	0.29012	934891	0.34403
8	354	0.35403	6714	0.36763	86334	0.39567	953177	0.46765
9	478	0.47765	7552	0.49392	90423	0.52982	970713	0.62432
10	634	0.63432	8443	0.65354	94482	0.69898	987656	0.82131
11	831	0.83131	9391	0.85397	98529	0.91074	1004118	1.06730

2.4 LINEAR DAMAGE MODELS

If the damage curves in the cycle $N_{i,f}$ coordinate are linearized, the linear damage rule is developed by reducing the damage curves to a single line in the cycle ratio $n_i/N_{i,f}$ domain, as shown in Figure 2.7. In this case, the fatigue damage has a unique, linear relation with the cycle ratio regardless of the stress levels. Hence, at a given level of damage, the cycle ratio for two different damage curves will be the same. This is illustrated by Figure 2.7, in which two linear damage curves plotted on a graph of damage versus cycles to failure are equal to each other when plotted on a graph of damage versus cycle ratio. Failure will occur when the sum of the ratios at each stress

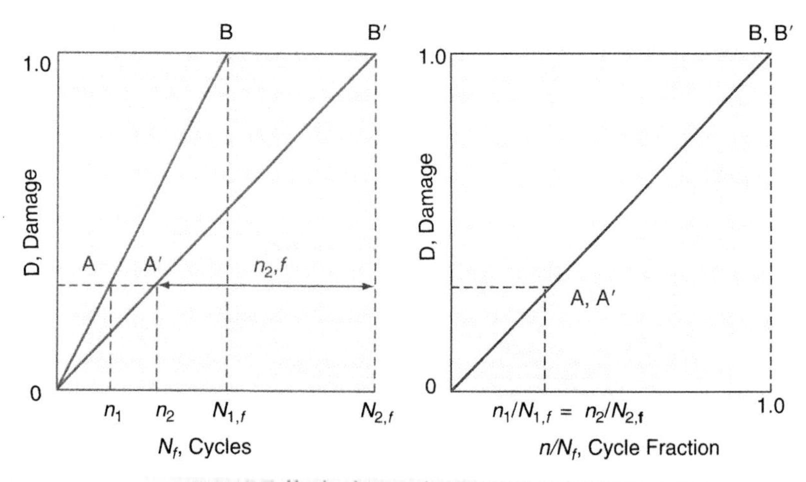

FIGURE 2.7 Linear damage accumulation.

level reaches a value of 1.0. In terms of mathematics, the linear damage rule can be expressed as follows:

$$D_i = \frac{n_i}{N_{i,f}} \tag{2.4.1}$$

Failure is predicted when

$$\Sigma D_i = \sum \frac{n_i}{N_{i,f}} \geq 1.0 \tag{2.4.2}$$

The universally used linear damage assessment model was first proposed by Palmgren (1924) for application to the Swedish ball bearing industry. Langer (1937), working for Westinghouse in the electric power generation area, independently proposed a similar linear rule for pressure vessel and piping components made of steel. Miner (1945) of Douglas Aircraft built on Langer's work and applied the linear damage rule to tension–tension axial fatigue data for aircraft skin material (aluminum alloy 24S-T ALCLAD). Miner demonstrated excellent agreement between the predictions from the linear damage rule and his experimental results. This success led to the strong association between Miner and the linear damage rule, and the linear damage rule is commonly referred to as Miner's linear damage rule.

Since Miner's work was conducted, the linear damage rule has been demonstrated to be unreliable. Studies by Wirshing et al. (1995) and Lee et al. (1999) are listed in Table 2.3 and show that the median damage values to test specimens under certain loading conditions range from 0.15 to 1.06. This is attributable to the fact that the relationship between physical damage (i.e., crack size or crack density) and cycle ratio is not unique and varies from one stress level to another.

TABLE 2.3 Statistics on Damage Values

	Median	Coefficient of Variation	Statistical Distribution
Miner: Original work	0.95	0.26	Lognormal
Schutz: Crack initiation			
29 Random sequence test series	1.05	0.55	Lognormal
Test with large mean load change	0.60	0.60	Lognormal
Significant notch plastic strain	0.37	0.78	Lognormal
Automotive axle spindle	0.15	0.60	Lognormal
Shin and Lukens: Extensive random test data	0.90	0.67	Lognormal
Gurney: Test data on welded joints	0.85	0.28	Lognormal
Lee: Mean stress effect—SAE cumulative fatigue test data	1.06	0.47	Normal

Example 2.2. Repeat Example 2.1 to determine the damage value for the four-level step loading, using the linear damage rule.

Solution. The damage value due to the four-level step stresses is

$$\sum D_i = \frac{n_1}{N_{1,f}} + \frac{n_2}{N_{2,f}} + \frac{n_3}{N_{3,f}} + \frac{n_4}{N_{4,f}} = \frac{10}{10^3} + \frac{10^2}{10^4} + \frac{10^3}{10^5} + \frac{10^4}{10^6} = 0.04$$

The total number of repetitions required for the four-step loading to reach the fatigue failure point can be determined as follows:

$$\text{Repetitions} = \frac{1.0}{\sum D_i} = \frac{1.0}{0.04} = 25$$

The estimated 25 repetitions based on the linear damage rule are nonconservative compared with the estimated 11 repetitions from the nonlinear damage theory.

2.5 DOUBLE LINEAR DAMAGE RULE BY MANSON AND HALFORD

The tedious iteration process using the nonlinear damage theory and the deficiency in damage assessment using the simple linear damage rule have motivated researchers to look for a better way to overcome the disadvantages of each method. Based on the observation that fatigue is at least a two-phase process—crack initiation and crack propagation—the models for the damage curves can be assumed to be bilinear.

Examples of such curves are shown in Figure 2.8. The bilinear model represents an equivalent damage model to the damage curve accumulation rule.

Manson and Halford (1981) derived the required criteria to determine the coordinates of the knee point, i.e., the intersection between the two straight lines of the bilinear curves. It is suggested that the straight line connecting (0, 0) and (1, 1) be the reference damage line for the lowest life level. Because of the nonlinear nature of damage and the accumulation of damage being modeled as a bilinear process, the two regions of damage are identified. The region of damage from the origin to the level of AA′ is designated as Phase I (D_I), and the region of damage from AA′ to BB′ is defined as Phase II (D_{II}). Using an approach similar to the method presented to normalize the cycle ratio for linear damage accumulation, the cycle ratios for Phase I and Phase II damage accumulations are constructed in a linear fashion, as shown in Figure 2.9.

Figure 2.9 illustrates that the total damage can be decomposed into Phase I damage (D_I) and Phase II damage (D_{II}). The Phase I linear damage

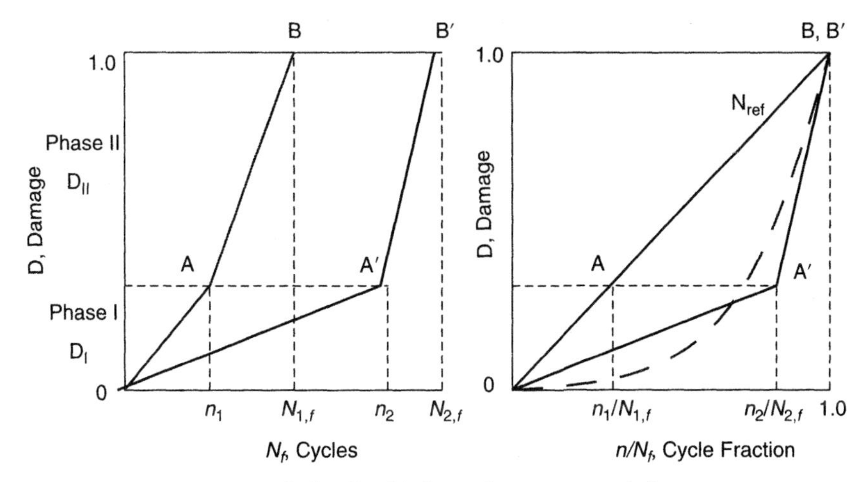

FIGURE 2.8 Double linear damage accumulation.

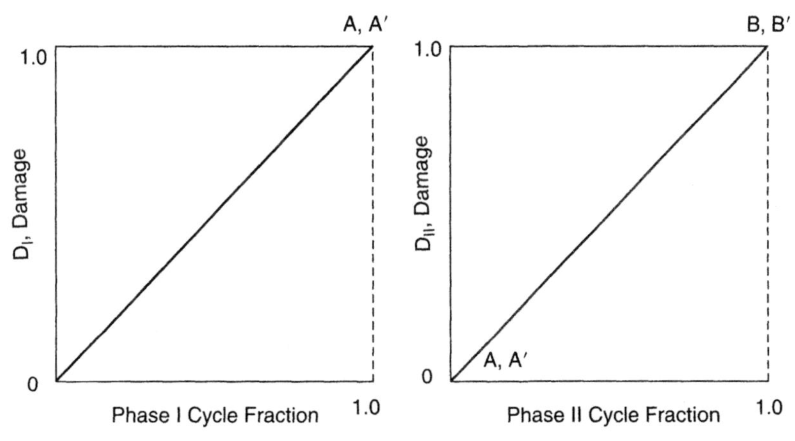

FIGURE 2.9 Phase I and Phase II linear damage rules.

accumulation rule states that prior to reaching the damage D_I, the cycle ratios can be summed linearly and are independent of the loading sequence (i.e., from OA to OA′ or from OA′ to OA). When the sum of the cycle ratios reaches unity, Phase I damage is completed. After the total damage beyond D_I, the Phase II linear damage accumulation rule applies. Regardless of the loading sequence, the damage accumulation depends only on the total sum of cycle fractions at each level. Based on the considerable amount of test data developed for two-step loading on many materials, Manson and Halford discovered that the knee point between Phase I and Phase II damage depends on the ratio of $N_{1,f}/N_{2,f}$ instead of the physical significance of crack initiation and crack propagation. Figure 2.10 shows the linear damage rule for the H-L step stress loading with the initial applied cycle fraction $n_1/N_{1,f}$ and the remaining cycle fraction $n_{2,f}/N_{2,f}$. Figure 2.10 illustrates the double linear damage rule and damage curve accumulations for the two-step loading and the relationship between $n_1/N_{1,f}$ and $n_{2,f}/N_{2,f}$. Equation 2.3.4 is the mathematical model for the description of the relationship between $n_1/N_{1,f}$ and $n_{2,f}/N_{2,f}$. To meet the condition that the bilinear model is equivalent to the damage curve accumulation rule, the knee point coordinates were derived and are found to depend on the ratio of $N_{1,f}/N_{2,f}$. Figure 2.11 shows that the test data on the knee point coordinates correlates well with $N_{1,f}/N_{2,f}$. The coordinates of the knee point are empirically determined as follows:

$$\left[\frac{n_1}{N_{1,f}}\right]_{\text{knee}} = 0.35 \times \left(\frac{N_{1,f}}{N_{2,f}}\right)^{0.25} \tag{2.5.1}$$

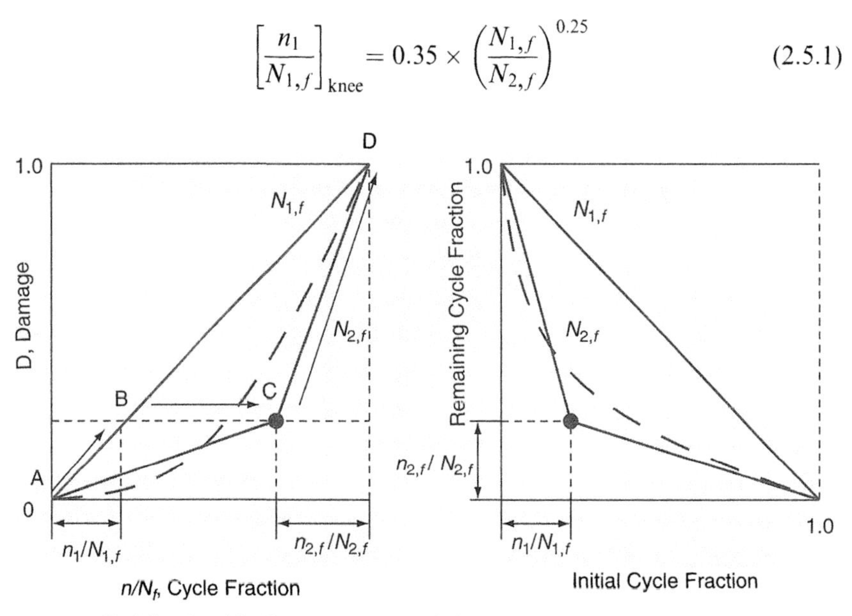

FIGURE 2.10 Double linear damage and damage curve accumulations for two-step loading.

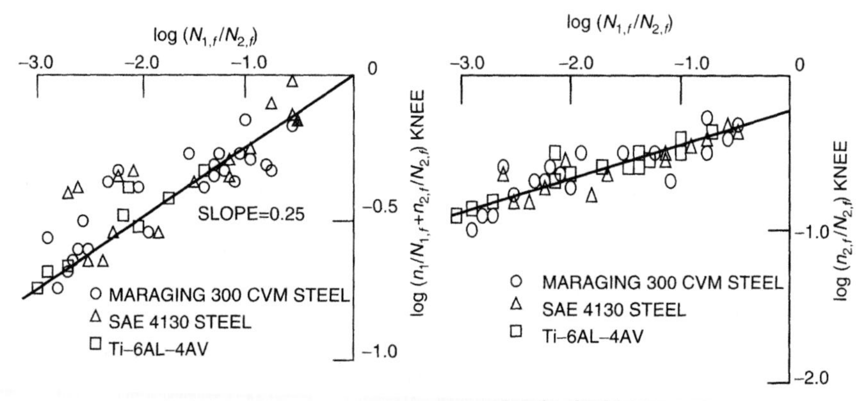

FIGURE 2.11 The knee point coordinates for the double linear damage rule.

$$\left[\frac{n_{2,f}}{N_{2,f}}\right]_{\text{knee}} = 0.65 \times \left(\frac{N_{1,f}}{N_{2,f}}\right)^{0.25} \tag{2.5.2}$$

It is important to note that the knee coordinates are independent of the specific material. Hence, the knee points would be the same for all materials. Their location is dependent only on maximum and minimum lives.

Example 2.3. Block loads are successively applied to a component until failure occurs. Each block has a two-step loading in which 10 cycles of the first loading are applied and followed by 10^3 cycles of the second loading. The first and second loading alone will fail the component to 10^3 cycles and 10^5 cycles, respectively. Determine how many blocks the component can sustain.

Solution. The first step is to construct the double linear damage rule. As shown in the double linear damage accumulation plot (Figure 2.12), a straight line connecting (0,0) with (1,1) is chosen as the reference damage curve for the lower life level ($N_{1,f} = 10^3$ cycles). Along this line, the damage and the coordinate at the breakpoint are defined by Equation 2.5.1, i.e.,

$$D_{\text{knee}} = \left[\frac{n_1}{N_{1,f}}\right]_{\text{knee}} = 0.35 \times \left(\frac{N_{1,f}}{N_{2,f}}\right)^{0.25} = 0.35 \times \left(\frac{10^3}{10^5}\right)^{0.25} = 0.111$$

Therefore, the number of cycles to Phase I damage for the 10^3-cycle-life loading is 111 cycles (i.e., $N_{I1,f} = 0.111 \times 10^3$), and the number of cycles to Phase II damage is 889 cycles (i.e., $N_{II1,f} = 1000 - 111$). The remaining cycle

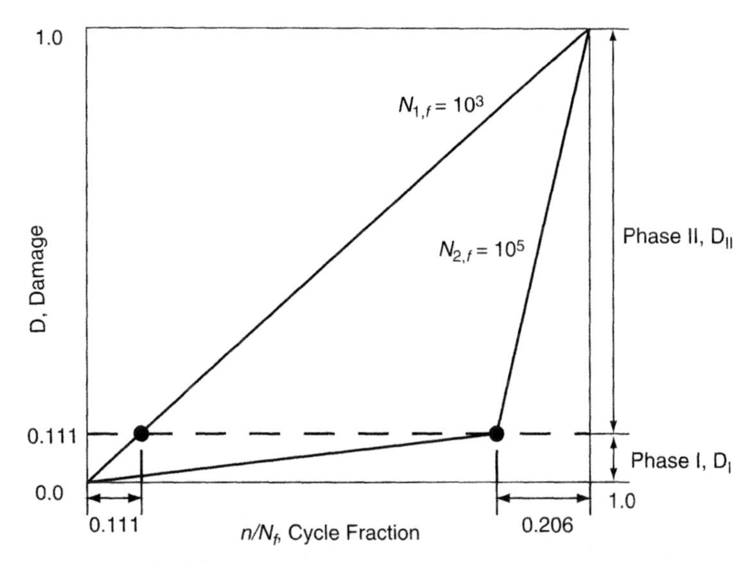

FIGURE 2.12 Double linear damage rule for two-step load levels.

fraction of the break point at the equivalent damage of 0.111 for 10^3-cycle-life loading can be defined by Equation 2.5.2:

$$\left[\frac{n_{2,f}}{N_{2,f}}\right]_{\text{knee}} = 0.65 \times \left(\frac{N_{1,f}}{N_{2,f}}\right)^{0.25} = 0.65 \times \left(\frac{10^3}{10^5}\right)^{0.25} = 0.206$$

The number of cycles to Phase II damage for the 10^5-cycle-life loading is 20,600 cycles (i.e., $N_{II2,f} = 0.206 \times 10^5$), and the number of cycles to Phase I damage is 79,400 cycles (i.e., $N_{I2,f} = 100{,}000 - 20{,}600$).

The next step is to track the damage accumulation as successive loading blocks are applied. To reach Phase I damage the number of cycles for the 10^3-cycle-life loading alone is 111 cycles and for the 10^5-cycle-life loading alone is 79,400 cycles. Therefore, the number of blocks (B_I) to complete Phase I damage can be calculated as follows:

$$B_I \times \left[\frac{10}{111} + \frac{1000}{79400}\right] = 1.0; \quad B_I = 9.7 \text{ blocks}$$

In the similar fashion, the number of blocks (B_{II}) to complete Phase II damage is estimated in the following:

$$B_{II} \times \left[\frac{10}{899} + \frac{1000}{20600}\right] = 1.0; \quad B_{II} = 17.0 \text{ blocks}$$

Thus, based on the double linear damage rule the total blocks to failure ($B_I + B_{II}$) are 26.7 blocks ($= 9.7 + 17.0$). For reference purpose, the Miner linear damage rule would have unconservatively predicted 50 blocks for this sequence of loading.

For block loading involving more than two load levels, the following procedures (Manson and Halford, 1981) for the double linear damage rule were developed. It is conservative to assume that if the individual loading within the block has lives from $N_{\text{low}} = N_{1,f}$ to $N_{\text{high}} = N_{2,f}$, the bilinear damage curve for other loading can be interpolated from the double linear damage rule established by Equations 2.5.1 and 2.5.2. It is stated that the total fatigue life (N_f) can be decomposed into Phase I fatigue life (N_I) and Phase II fatigue life (N_{II}), i.e.,

$$N_f = N_I + N_{II} \tag{2.5.3}$$

One chooses the following form for the relationship between Phase I fatigue life and the total fatigue life:

$$N_I = N_f \exp(Z N_f^{\phi}) \tag{2.5.4}$$

where Z and ϕ are constants. The two parameters can be determined from the two knee points on the N_I curve for the double linear damage rule. Applying Equations 2.5.1 and 2.5.2 for the knee points leads to the following equations for the number of cycles to Phase I damage for the two loads with the N_{1f} and N_{2f} life levels:

$$N_{I,N_{1,f}} = N_{1,f}\left(\frac{n_1}{N_{1,f}}\right)_{\text{knee}} = 0.35N_{1,f}\left(\frac{N_{1,f}}{N_{2,f}}\right)^{0.25} \quad (2.5.5)$$

$$N_{I,N_{2,f}} = N_{2,f}\left(1-\left(\frac{n_{2,f}}{N_{2,f}}\right)_{\text{knee}}\right) = N_{2,f}\left(1-0.65\left(\frac{N_{1,f}}{N_{2,f}}\right)^{0.25}\right) \quad (2.5.6)$$

Substituting into Equation 2.5.4 allows the solution for Z and ϕ as follows:

$$\phi = \frac{1}{Ln(N_{1,f}/N_{2,f})}Ln\left[\frac{Ln(0.35(N_{1,f}/N_{2,f})^{0.25})}{Ln(1-0.65(N_{1,f}/N_{2,f})^{0.25})}\right] \quad (2.5.7)$$

$$Z = \frac{Ln(0.35(N_{1,f}/N_{2,f})^{0.25})}{N_{1,f}^{\phi}} \quad (2.5.8)$$

The equation for N_{II} becomes

$$N_{II} = N_f - N_I = N_f(1-\exp(ZN_f^{\phi})) \quad (2.5.9)$$

Example 2.4. An additional loading with 100 cycles is inserted into the previous two-step loading in Example 2.3. This loading will produce the intermediate fatigue life of 10^4 cycles. Determine how many blocks the component can sustain.

Solution. Because the double linear damage rule based on $N_{\text{low}} = N_{1,f} = 10^3$ cycles and $N_{\text{high}} = N_{2,f} = 10^5$ cycles was constructed in Example 2.3, the interpolation formulas will be used to generate the bilinear damage model for the 10^4-cycle-life loading (as shown in Figure 2.13). From Equations 2.5.7 and 2.5.8, the two parameters Z and ϕ are determined as follows:

$$\phi = \frac{1}{Ln(10^3/10^5)}Ln\left[\frac{Ln\left(0.35(10^3/10^5)^{0.25}\right)}{Ln\left(1-0.65(10^3/10^5)^{0.25}\right)}\right] = -0.4909$$

$$Z = \frac{Ln\left(0.35(10^3/10^5)^{0.25}\right)}{(10^3)^{-2.077}} = -65.1214$$

Based on Equation 2.5.4, the number of cycles to Phase I damage for the 10^4-cycle-life loading is

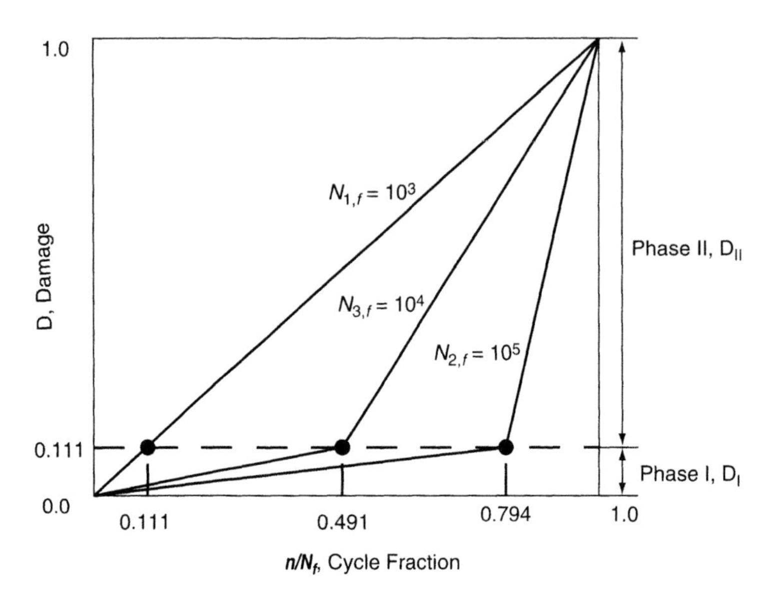

FIGURE 2.13 Double linear damage rule for three-step load levels.

$$N_{I,3,f} = N_f \exp(ZN_f^{\phi}) = 10^4 \exp(-65.1214 \times (10000)^{-0.4904}) = 4910 \text{ cycles}$$

The number of cycles to Phase II damage is then calculated:

$$N_{II,3,f} = N_f - N_I = 10{,}000 - 4909 = 5090 \text{ cycles}$$

Therefore, the number of blocks (B_I) to complete Phase I damage can be calculated as follows:

$$B_I \times \left[\frac{10}{111} + \frac{100}{4910} + \frac{1000}{79400}\right] = 1.0; \quad B_I = 8.1 \text{ blocks}$$

The number of blocks (B_{II}) to complete Phase II damage is estimated in the following:

$$B_{II} \times \left[\frac{10}{899} + \frac{100}{5910} + \frac{1000}{20600}\right] = 1.0; \quad B_{II} = 13.1 \text{ blocks}$$

The total number of blocks to failure is estimated to be 21.2 blocks (= 8.1 + 13.1). The Miner linear damage rule would have unconservatively predicted 33 blocks.

Example 2.5. Repeat Example 2.1 by using the double linear damage rule for fatigue life estimation.

Solution. To construct the double linear damage rule, $N_{\text{low}} = N_{1,f} = 10^3$ cycles and $N_{\text{high}} = N_{2,f} = 10^6$ cycles were chosen. The bilinear damage models for the 10^4-cycle-life and the 10^5-cycle-life loads are interpolated

TABLE 2.4 Four-Level Step-Stress Fatigue Test Data

S_i (MPa)	n_i (cycles)	$N_{i,f}$ (cycles)	$N_{I,i,f}$ (cycles)	$N_{II,i,f}$ (cycles)	$D_{I,i}$	$D_{II,i}$
±800	10	1000	62	938	0.1613	0.0107
±600	100	10000	3750	6250	0.0267	0.0160
±400	1000	100000	70700	29300	0.0141	0.0341
±200	10000	1000000	883000	117000	0.0113	0.0855
Total					0.2134	0.1463

Note: Blocks to complete Phase I damage = 1/0.2134 = 4.7 blocks; blocks to complete Phase II damage = 1/0.1463 = 6.8 blocks; total blocks to failure = 4.7 + 6.8 = 11.5 blocks.

later with the two calculated parameters ($\phi = -0.4514$ and $Z = -62.7717$). The coordinates of the knee points and the corresponding Phase I and Phase II damage values for four-step step-stress loads are presented in Table 2.4. The blocks required to complete Phase I and Phase II are calculated as 4.7 blocks and 6.8 blocks, respectively, leading to a total of 11.5 blocks to produce failure. When compared with the estimated 11 blocks to failure using the damage curve approach as illustrated in Example 1.1, this shows a very good correlation. The predicted 11.5 blocks to failure by using the double linear damage rule is very close to the prediction (11 blocks) by the damage curve accumulation rule.

2.6 CONCLUSIONS

The double linear damage rule by Manson and Halford is recommended for use in engineering design for durability because of the tedious iteration process by the nonlinear damage theory and the deficiency in linear damage assessment. Note that the knee coordinates in the double linear damage rule are independent of the specific material. Hence, the knee points would be the same for all materials. Their location depends only on the maximum and minimum lives.

REFERENCES

Dowling, N. E., *Mechanical Behavior of Materials: Engineering Methods for Deformation, Fracture, and Fatigue*, 2nd ed., Prentice Hall, New York, 1998.

Fatemi, A., and Yang, L., Cumulative fatigue damage and life prediction theories: a survey of the state of the art for homogeneous materials, *International Journal of Fatigue*, Vol. 20, No. 1, 1998, pp. 9–34.

Hashin, Z. A., A reinterpretation of the Palmgren-Miner rule for fatigue life prediction, *Journal of Applied Mechanics*, Vol. 47, 1980, pp. 324–328.

Langer, B. F., Fatigue failure from stress cycles of varying amplitude, *Journal of Applied Mechanics*, Vol. 59, 1937, pp. A160–A162.

Lee, Y., Lu, M., Segar, R. C., Welch, C. D., and Rudy, R. J., Reliability-based cumulative fatigue damage assessment in crack initiation, *International Journal of Materials and Product Technology*, Vol. 14, No. 1, 1999, pp. 1–16.

Manson, S. S., Freche, J. C., and Ensign, C. R., Application of a double linear damage rule to cumulative fatigue, *Fatigue Crack Propagation*, ASTM STP 415, 1967, pp. 384–412.

Manson, S. S., and Halford, G. R., Practical implementation of the double linear damage rule and damage curve approach for testing cumulative fatigue damage, *International Journal of Fracture*, Vol. 17, No. 2, 1981, pp. 169–192.

Miner, M. A., Cumulative damage in fatigue, *Journal of Applied Mechanics*, Vol. 67, 1945, pp. A159–A164.

Palmgren, A., Die Lebensdauer von Kugellagern, Zeitschrift des Vereinesdeutscher Ingenierure, Vol. 68, No. 14, 1924, pp. 339–341.

Peterson, R. E., Analytical approach to stress concentration effects in aircraft materials, Technical Report 59-507, U. S. Air Force—WADC Symposium on Fatigue Metals, Dayton, Ohio, 1959.

Subramanyan, S., A cumulative damage rule based on the knee point of the S-N curve, *Journal of Engineering Materials and Tech*nology, Vol. 98, 1976, pp. 316–321.

Wirshing, P. H., Paez, T. L., and Ortiz, H., *Random Vibration: Theory and Practice*, Wiley, New York, 1995.

3

CYCLE COUNTING TECHNIQUES

YUNG-LI LEE
DAIMLERCHRYSLER
DARRYL TAYLOR
DAIMLERCHRYSLER

3.1 INTRODUCTION

The fatigue damage theories discussed in Chapter 2 indicate that fatigue damage is strongly associated with the cycle ratio, $n_i/N_{i,f}$, where n_i and $N_{i,f}$ are, respectively, the number of applied stress and/or strain cycles and the fatigue life at a combination of stress and/or strain amplitude and mean stress levels. The fatigue life, $N_{i,f}$, can be obtained from baseline fatigue data generated from constant-amplitude loading tests. There are three commonly used methods—the stress-life (S-N) method, the strain-life (ε-N) method, and the linear elastic fracture mechanics (LEFM)—to characterize the baseline fatigue data. These methods are addressed in Chapters 4–6. Each of these methods relates fatigue life to cycles of loading. The engineering methods of determining n_i are presented in this chapter.

In the service lives of typical structural components, the components are subjected to cyclic loads. These loads may have a constant amplitude or an amplitude that varies with time. For cases that exhibit constant amplitude loading with or without mean offset loading, the determination of the amplitude of a cycle and the number of cycles experienced by a component is a straightforward exercise. However, if the amplitude of the loading changes with time, it is more difficult to determine what constitutes a cycle and the corresponding amplitude of that cycle.

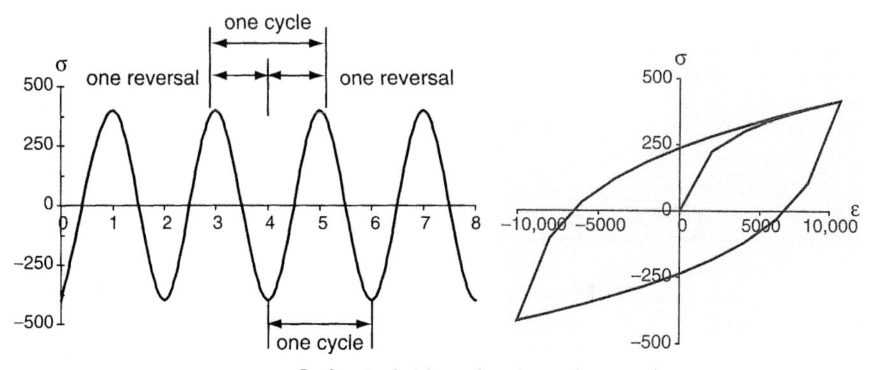

FIGURE 3.1 Definition of cycles and reversals.

Cycles can be counted using time histories of the loading parameter of interest, such as force, torque, stress, strain, acceleration, or deflection. In Figure 3.1, one complete stress cycle in a time domain is related to a closed hysteresis loop in the local stress–strain coordinate and consists of two reversals. The reversal can be described as the event of unloading or loading.

The objective of this chapter is to introduce several cycle counting techniques that can be used to reduce a complicated variable amplitude loading history into a number of discrete simple constant amplitude loading events, which are associated with fatigue damage. The one-parameter cycle counting and the two-parameter cycle counting techniques are discussed in detail. A technique to reconstruct load-time histories based on cycle counting results is then introduced.

3.2 ONE-PARAMETER CYCLE COUNTING METHODS

Over the years, one-parameter cycle counting methods such as level crossing, peak-valley, and range counting have been commonly used for extracting the number of cycles in a complex loading history. These methods are unsatisfactory for the purpose of describing a loading cycle and fail to link the loading cycles to the local stress–strain hysteresis behavior that is known to have a strong influence on fatigue failure. Thus, these methods are considered inadequate for fatigue damage analysis. However, the following sections present descriptions of the one-parameter cycle counting methods as an illustration of the techniques used in cycle counting and for comparison to the more effective two-parameter cycle counting methods.

3.2.1 LEVEL CROSSING CYCLE COUNTING

In this counting method, the magnitude of the loads in the load-time history has to be divided into a number of levels. This process is shown in

Figure 3.2(a). One count at a specific level is defined when a portion of the load-time history with a positive slope crosses through this level above a reference load or when a portion of the load-time history with a negative slope passes through this level below a reference load. The reference load level is usually determined by the mean of the complete load-time history. A variation of this method is to count all of the levels crossing with the positive-sloped portion of the load-time history. Table 3.1 and Figure 3.2(b) show the tabulated and plotted results using the level-crossing count from the load-time history in Figure 3.2(a).

Once all the counts are determined, they are used to form cycles. Based on the *SAE Fatigue Design Handbook* (Rice et al., 1997), the cycle extraction rule follows that the most damaging fatigue cycles can be derived by first constructing the largest possible cycle, followed by the second largest possible cycle, and so on. This process is repeated until all available counts are used up. Figure 3.2(c) illustrates this process in which the cycles are formed. Table 3.2 summarizes the cycle counting results.

3.2.2 PEAK-VALLEY CYCLE COUNTING

This counting technique first identifies the counts of peaks and valleys in a load-time history and subsequently constructs the possible cycles from the most to the least damaging events according to the extracted peak-valley counts. The peak is the transition point where a positive-sloped segment turns into a negative-sloped segment, and the valley is the point where a negative-sloped segment changes to a positive-sloped one. Peaks above and

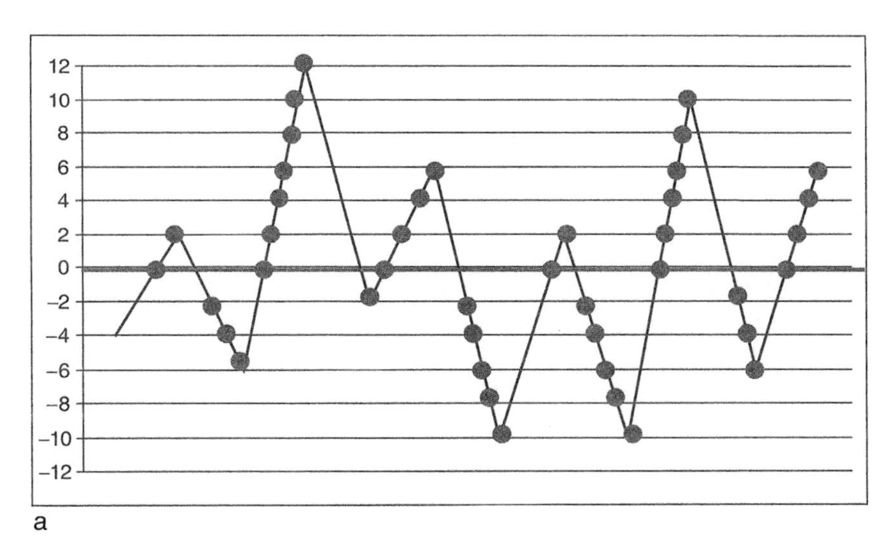

a

FIGURE 3.2 (a) Level crossing counting of a service load-time history.

Continued

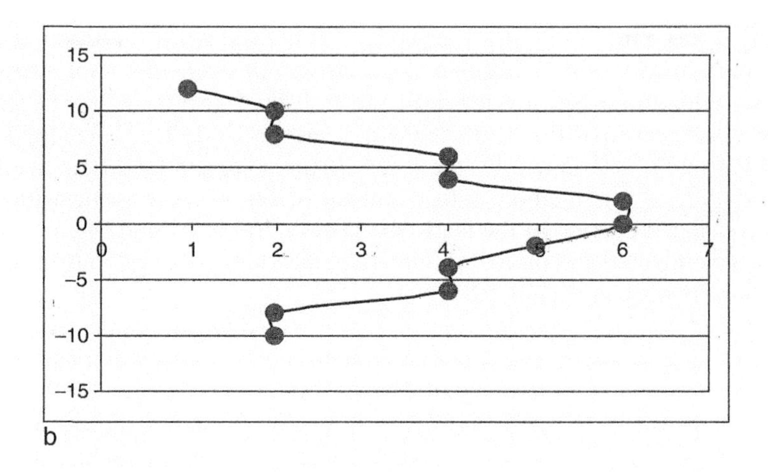

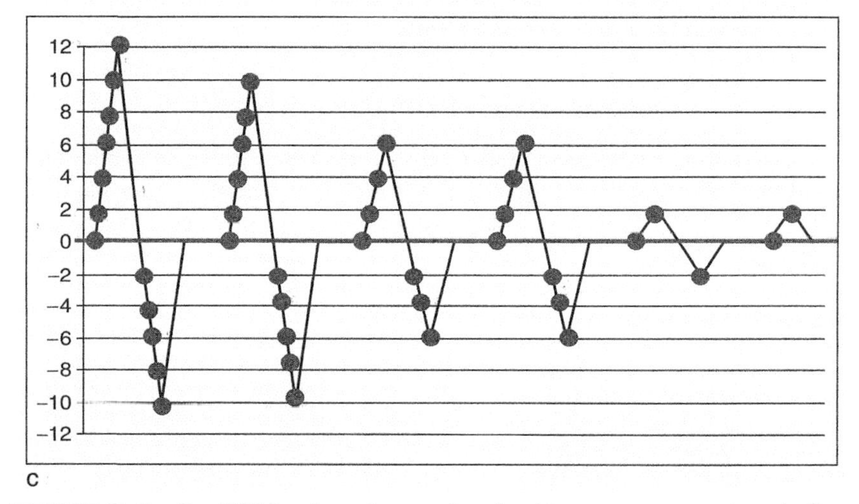

FIGURE 3.2 *Cont'd* (b) Level crossing counting plot; (c) a process to generate cycles from level crossing counts.

valleys below a reference load level are counted. Table 3.3 shows the tabulated results from the peak-valley counting in the load-time history in Figure 3.3(a). The process to generate the cycles from the peak-valley counts is illustrated in Figure 3.3(b). Table 3.4 summarizes the final cycle counting results.

3.2.3 RANGE COUNTING

This counting technique defines one count as a range, the height between a successive peak and valley. According to the *SAE Fatigue Design Handbook* (Rice et al., 1997), a sign convention is assigned to a range. Positive ranges

TABLE 3.1 Tabulated Results from the Level Crossing Counting Method

Level	Counts
12	1
10	2
8	2
6	4
4	4
2	6
0	6
−2	5
−4	4
−6	4
−8	2
−10	2

TABLE 3.2 Tabulated Cycles Extracted from the Level Crossing Counts

Range	Cycles
22	1
20	1
12	2
4	1

TABLE 3.3 Tabulated Results from the Peak-Valley Counting Method

Peaks/Valleys	Counts
12	1
10	1
6	2
2	2
−2	1
−4	1
−6	2
−10	2

and negative ranges are defined on positively sloped reversals and negatively sloped reversals, respectively. Each range represents on one-half cycle (reversal). Figure 3.4 illustrates the counts of positive and negative ranges. Table 3.5 lists the summary of the range counts and Table 3.6 shows the final cycles extracted from the range counts.

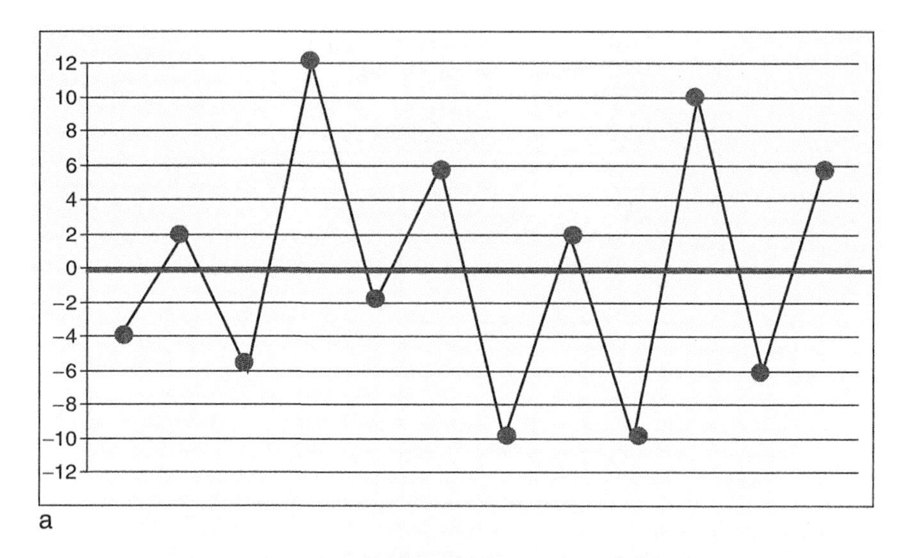

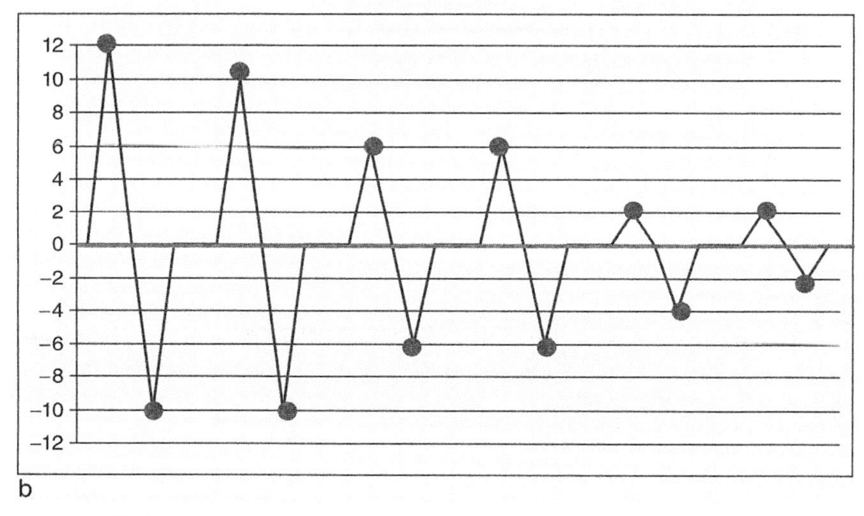

FIGURE 3.3 (a) Peak-valley counting of a service load-time history; (b) a process to derive cycles from peak-valley counting.

3.3 TWO-PARAMETER CYCLE COUNTING METHODS

Two-parameter cycle counting methods, such as the rainflow cycle counting method, can faithfully represent variable-amplitude cyclic loading. Dowling (1979) states that the rainflow counting method is generally regarded as the method leading to better predictions of fatigue life. It can identify events in a complex loading sequence that are compatible with constant-amplitude fatigue data.

TABLE 3.4 Tabulated Cycles Extracted from Peak-Valley Counts

Range	Cycles
22	1
20	1
12	2
6	1
4	1

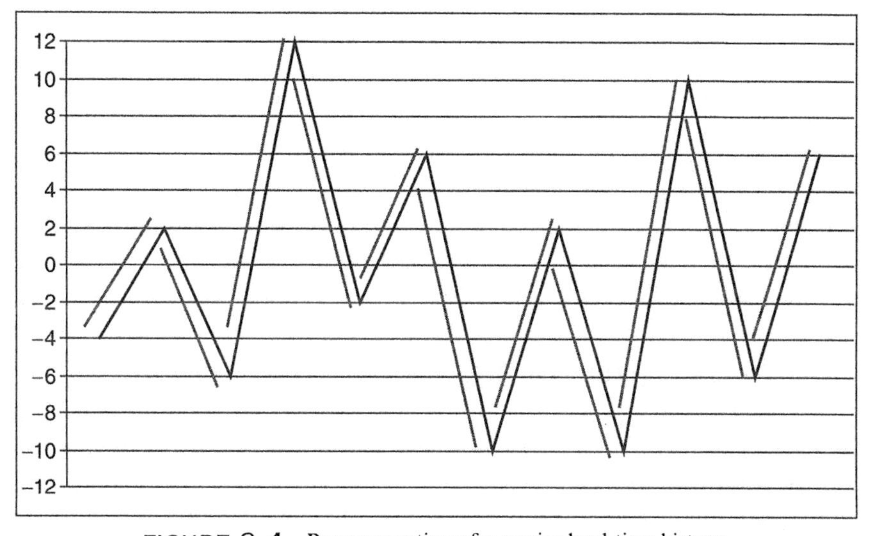

FIGURE 3.4 Range counting of a service load-time history.

TABLE 3.5 Tabulated Range Counts Results from the Range Counting Method

Range	Counts
+20	1
+18	1
+12	2
+8	1
+6	1
−8	1
−12	1
−14	1
−16	2

TABLE 3.6 Tabulated Cycle Extracted from the Range Counts

Range	Cycles
20	0.5
18	0.5
16	1
14	0.5
12	1.5
8	1
6	0.5

Matsuishi and Endo (1968) originally developed the rainflow cycle counting method based on the analogy of raindrops falling on a pagoda roof and running down the edges of the roof. A number of variations of this original scheme have been published for various applications. Among these variants are the three-point cycle counting techniques (including the SAE standard [Rice et al., 1997], the ASTM standard [ASTM E1049, 1985], and the range-pair method [Rice et al., 1997]) and the four-point cycle and counting rule (Amzallag et al., 1994). The following sections describe the two-parameter cycle counting methods.

3.3.1 THREE-POINT CYCLE COUNTING METHOD

As per the SAE and the ASTM standards, the three-point cycle counting rule uses three consecutive points in a load-time history to determine whether a cycle is formed. Figure 3.5 shows the rules that identify the two possible closed cycles in a time history where stress is the load parameter. The three consecutive stress points (S_1, S_2, S_3) define the two consecutive ranges as $\Delta S_1 = |S_1 - S_2|$ and $\Delta S_2 = |S_2 - S_3|$. If $\Delta S_1 \leq \Delta S_2$, one cycle from S_1 to S_2 is

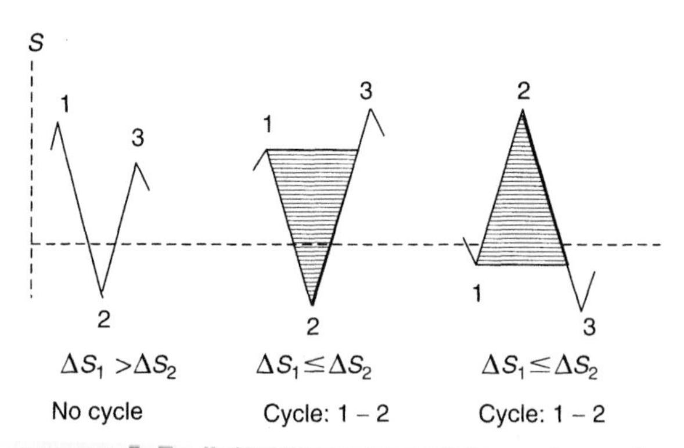

FIGURE 3.5 Rules of three-point rainflow cycle counting.

extracted, and if $\Delta S_1 > \Delta S_2$, no cycle is counted. The three-point cycle counting method requires that the stress time history be rearranged so that it contains only the peaks and valleys and it starts with either the highest peak or the lowest valley, whichever is greater in absolute magnitude. Then, the cycle identification rule is applied to check every three consecutive points from the beginning until a closed loop is defined. The two points forming the cycle are discarded and the remaining points are connected to each other. This procedure is repeated from the beginning until the remaining data are exhausted.

A variation of the three-point counting is the range-pair counting method, which applies the same rule for a cycle extraction to a load-time history and does not require the load sequence to start with either the maximum peak or the minimum valley.

Example 3.1. Use the three-point rainflow cycle counting method to determine the number of cycles in the load-time history shown in Figure 3.6.

Solution. First, the load-time history needs to be rearranged so that the cycle starts with the maximum peak or the minimum valley, whichever is greater in absolute magnitude. In this example, the highest peak with a magnitude of 12 occurs first as opposite to the lowest valley and is therefore chosen as the beginning point for the rearranged load-time history. The new time history shown in Figure 3.7(a) is generated by cutting all the points prior to and including the highest peak and by appending these data to the end of the original history. An additional highest peak is included in the new time history to close the largest loop for conservatism.

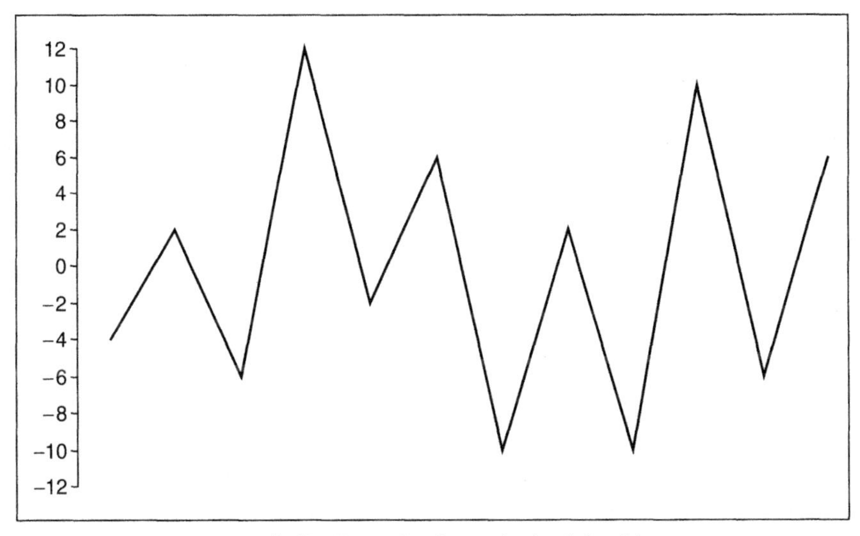

FIGURE 3.6 Example of a service load-time history.

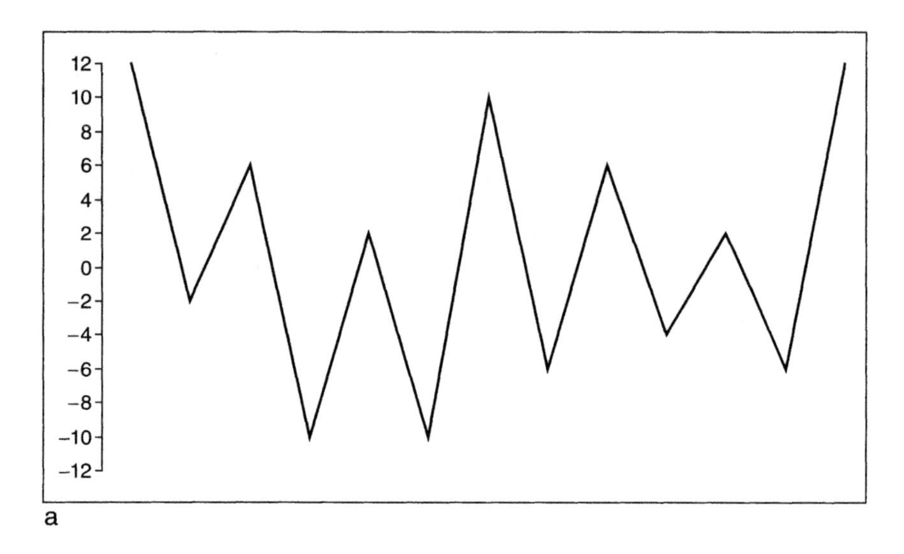

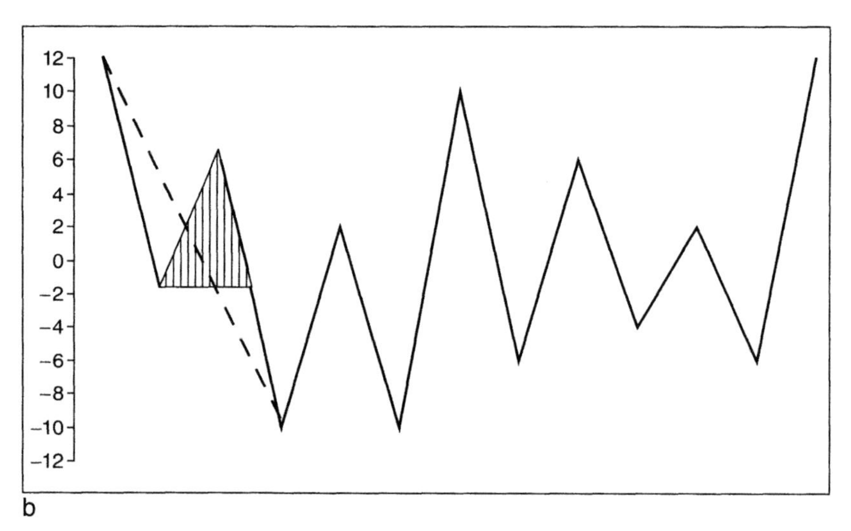

FIGURE 3.7 (*Cont'd*) Rearrangement of the service load-time history; (b) extracted cycle: from −2 to 6.

The three-point rainflow cycle counting rule is then applied to every three consecutive load points in the rearranged load-time history. The first cycle formed by two data points from −2 to 6 is extracted. A new load-time history is generated by connecting the point before −2 and the point after 6 to each other. This is illustrated in Figure 3.7(b). The same process is repeated until all the cycles are identified. This repetition is illustrated in Figures 3.7(c)–(g).

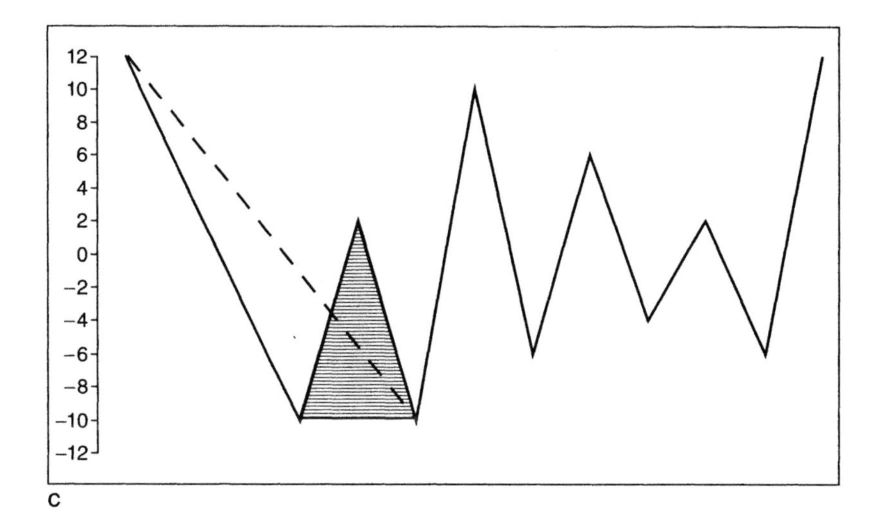

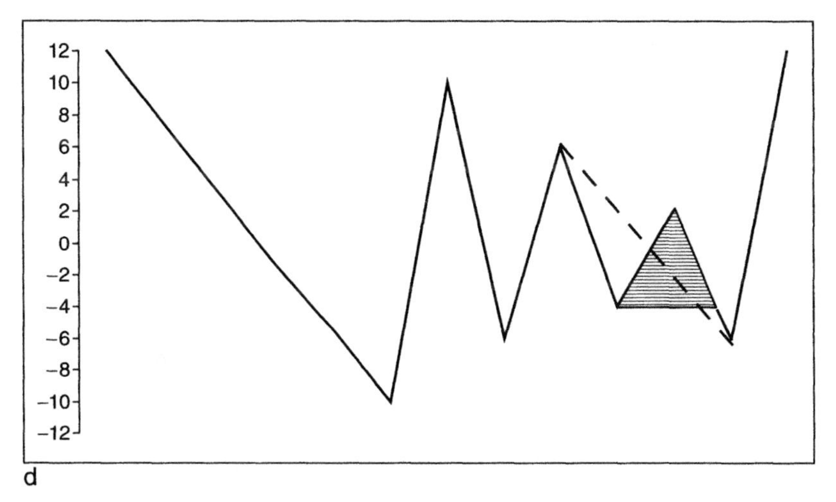

FIGURE 3.7 *Cont'd* (c) Extracted cycle: from −10 to 2; (d) extracted cycle: from −4 to 2

Continued

The rainflow cycle counting results are tabulated in Table 3.7. They can be alternatively described by the from–to rainflow matrix in Table 3.8 or the range–mean matrix in Table 3.9. In the form of a rainflow matrix, the number of cycles is stored in the cell (i, j) that defines each constant-amplitude loading starting from i to j in the from–to matrix or the combination of range i and mean j in the range–mean matrix. The from–to format is useful for reconstructing a loading sequence because of the directional characteristic, whereas

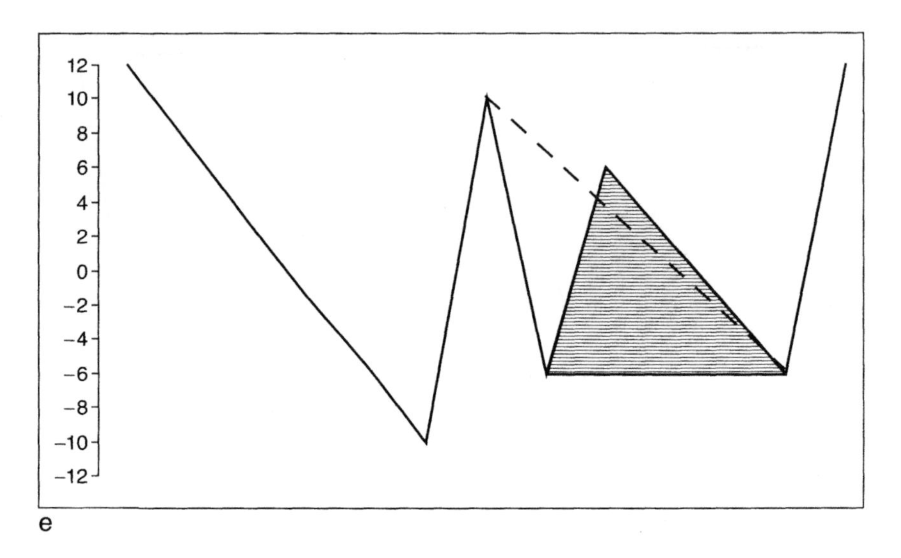

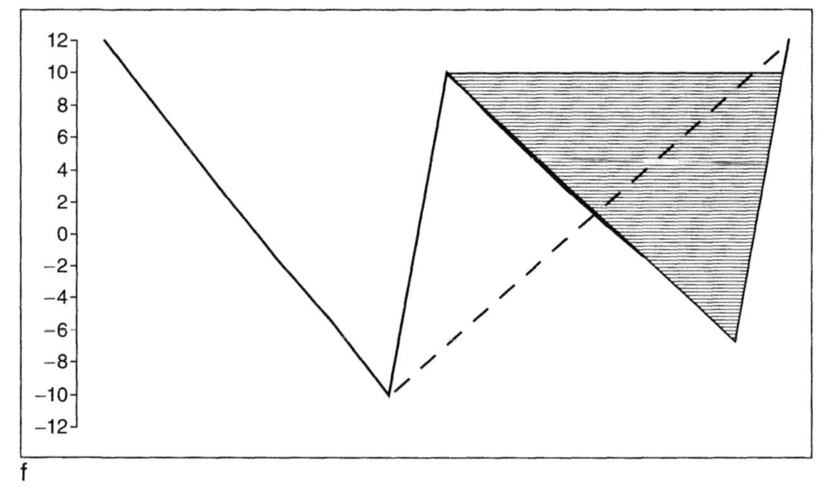

FIGURE 3.7 *Cont'd* (e) Extracted cycle: from −6 to 6; (f) extracted cycle: from 10 to −6.

the range–mean one is often used in the U.S. automotive industry to generate a load-history profile.

Example 3.2. Use the range-pair cycle counting method to determine the number of cycles in the load-time history given in Figure 3.6.

Solution. There is no need to rearrange the loading-time history for the range-pair cycle counting method. The cycle extraction technique used in the three-point rainflow cycle counting method is applied to every three consecutive load points in the original load-time history. The first loading

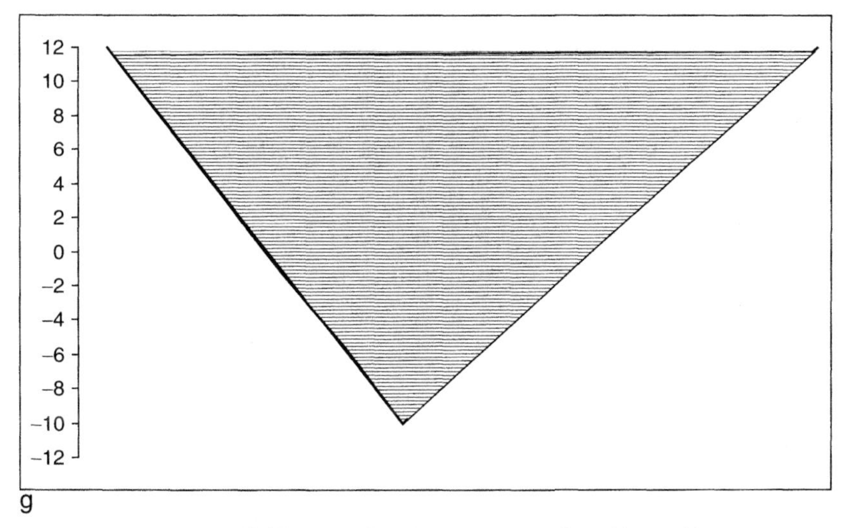

FIGURE 3.7 *Cont'd* (g) Extracted cycle: from 12 to −10.

TABLE 3.7 Summary of the Three-Point Rainflow Cycle Counting Results

No. of Cycles	From	To	Range	Mean
1	−2	6	8	2
1	−10	2	12	−4
1	−4	2	6	−1
1	−6	6	12	0
1	10	−6	16	2
1	12	−10	22	1

points from 2 to −6 are counted as one cycle and the two data points are discarded. A new load-time history is generated by connecting the point before 2 and the point after −6 to each other. This process is illustrated in Figure 3.8(a). The same process is carried out until all the cycles are identified, as shown in Figures 3.8(b)–(d). The remaining data points that are not qualified to form the cycles shown in Figure 3.8(e) can be recounted by starting at the end of the load-time history and counting backward. In this example, one cycle from 6 to −6 and one half-cycle from 10 to −10 are extracted. The rainflow cycle counting results are tabulated in Table 3.10.

3.3.2 FOUR-POINT CYCLE COUNTING METHOD

Similar to the three-point cycle counting method, the four-point cycle counting rule uses four consecutive points to extract a cycle. Figure 3.9 illustrates the principles for two possible cycles counted in a nominal

TABLE 3.8 The From–To Rainflow Cycle Counting Matrix: Three-Point Cycle Counting

		To											
		−10	−8	−6	−4	−2	0	2	4	6	8	10	12
From	−10							1					
	−8												
	−6									1			
	−4							1					
	−2									1			
	0												
	2												
	4												
	6												
	8												
	10			1									
	12	1											

TABLE 3.9 The Range-Mean Rainflow Cycle Counting Matrix

		Mean						
		−4	−3	−2	−1	0	1	2
Range	6				1			
	7							
	8							1
	9							
	10							
	11							
	12	1				1		
	13							
	14							
	15							
	16							1
	17							
	18							
	19							
	20							
	21							
	22						1	

stress-time history and the corresponding local stress–strain response. Two possible cycles are defined: one with a hanging cycle in (a) and another with a standing cycle in (b). The four consecutive stress points (S_1, S_2, S_3, S_4) define the inner ($\Delta S_1 = |S_2 - S_3|$) and the outer stress range ($\Delta S_0 = |S_1 - S_4|$). If the inner stress range is less than or equal to the outer stress range ($\Delta S_I \leq \Delta S_0$) and the points comprising the inner stress range are bounded by (between) the points of the outer stress range, the inner cycle from S_2 to S_3 is extracted,

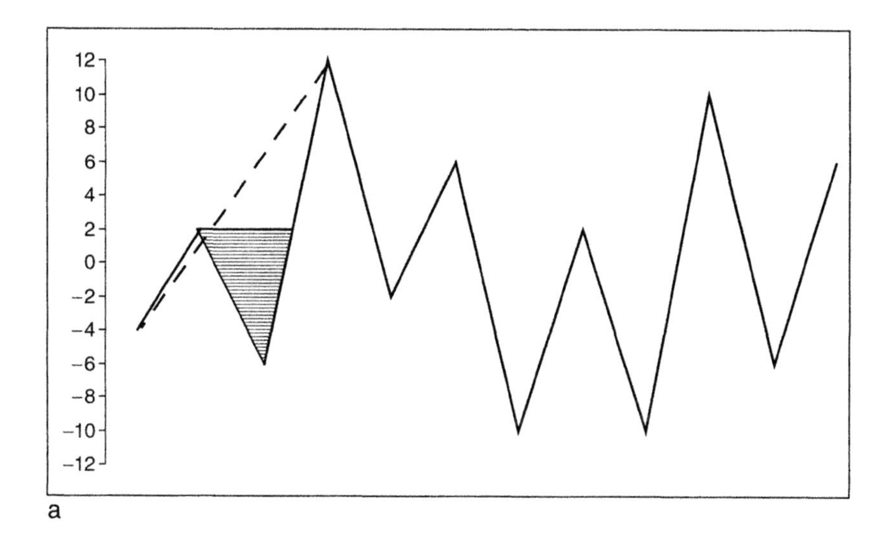

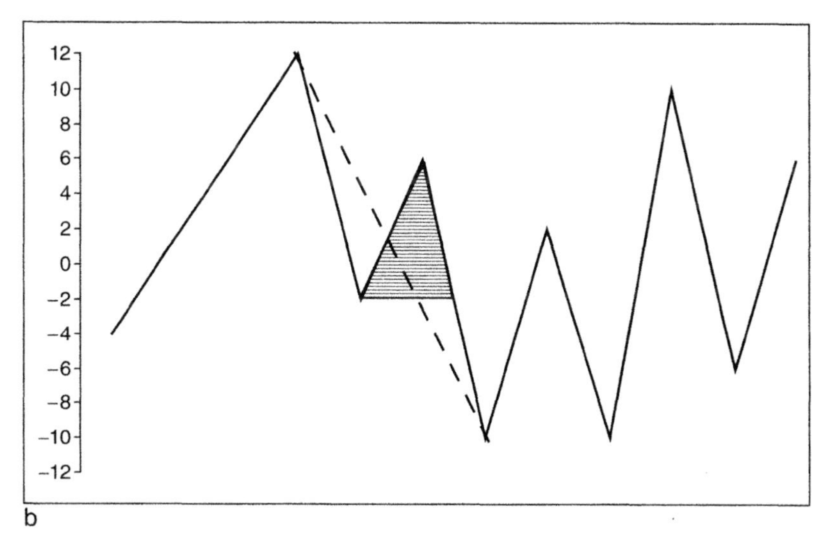

FIGURE 3.8 (a) Extracted cycle: from 2 to −6. (b) extracted cycle: from −2 to 6.

Continued

the two inner points discarded, and the two outer points (S_1 and S_4) connected to each other. Otherwise, no cycle is counted, and the same check is done for the next four consecutive stress points (S_2, S_3, S_4, S_5) until no data remain.

Unlike the three-point rainflow cycle counting method, this technique does not guarantee that all the data points will form closed cycles. The remaining data points that cannot constitute a cycle are called the *residue*. With

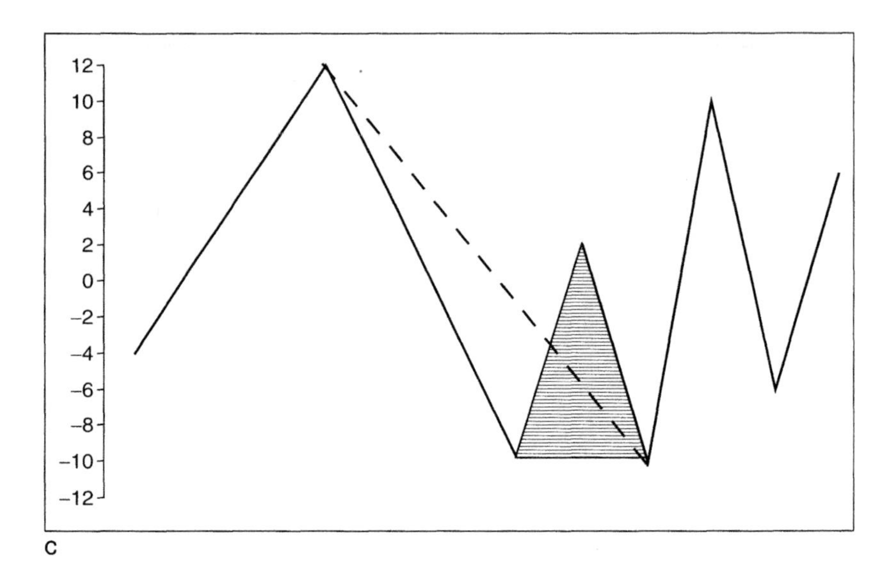

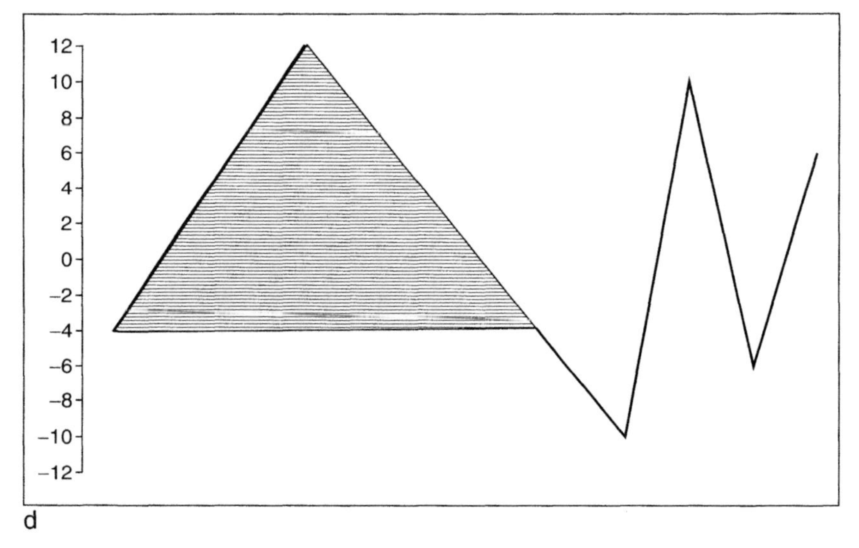

FIGURE 3.8 *Cont'd* (c) Extracted cycle: from −10 to 2; (d) extracted cycle: from −4 to 12.

this difference, the three-point rainflow counts can still be derived from the four-point counting as follows:

1. Extract the cycles and the residue, based on the four-point counting method.
2. Duplicate the residue to form a sequence of [residue + residue].
3. Apply the same rainflow technique to the sequence of [residue + residue].

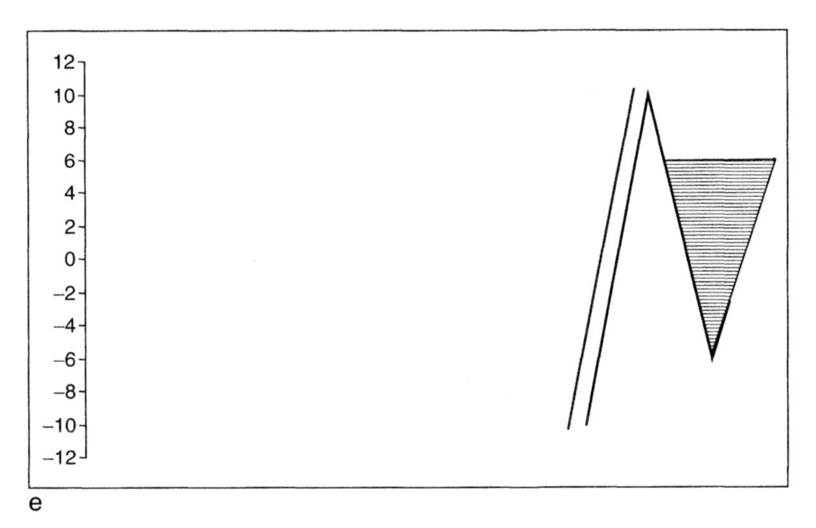

FIGURE 3.8 (e) Extracted remaining cycle: from 6 to −6.

4. Add the newly extracted cycles to the original cycles.

Both rainflow cycle counting methods lead to the identical range–mean rainflow matrix.

There are some unique features of the four-point cycle counting method. First, this tecnique is very easy to use in conjunction with as-recorded data acquisition and data reduction, because it does not require rearrangement of the load-time history. Second, this method can be easily implemented for cycle extrapolation and load-time history reconstruction. Finally, this cycle counting method is very generic, because the three-point rainflow matrix can be deduced from the four-point rainflow matrix and its residue.

TABLE 3.10 Summary of the Range-Pair Cycle Counting Results

No. of Cycles	From	To	Range	Mean
1	−2	−6	4	−4
1	−2	6	8	2
1	−10	2	12	−4
1	−4	12	16	4
1	6	−6	12	0
0.5	10	−10	20	0

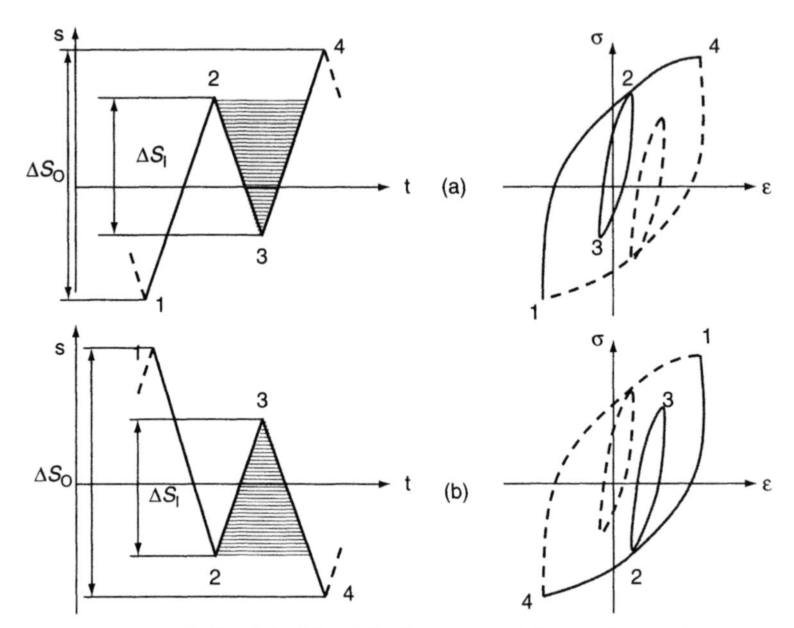

FIGURE 3.9 Principle of the four-point rainflow cycle counting.

Example 3.3. Use the four-point rainflow cycle counting method to determine the number of cycles in the load-time history given in Figure 3.6.

Solution. Double-check the load-time history to ensure that it contains only the peaks and valleys. Apply the four-point rainflow cycle counting rule to every four consecutive load points in the history. The first cycle from −2 to 6 and the two data points are extracted. A new load-time history is generated by connecting the point before −2 and the point after 6 to each other. This is illustrated in Figure 3.10(a). The same process is carried out until the second cycle from −10 to 2 is identified as shown in Figure 3.10(b). Finally, the four-point rainflow technique results in two cycles extracted and the residue with eight remaining points shown in Figure 3.10(c).

To arrive at a three-point rainflow count, Figure 3.11(a) shows the new sequence because of duplication of the residues, and Figure 3.11(b)–(f) illustrates the extraction of the cycles from this new time history and its residue. It is interesting to note that the residue remains the same with and without the process of counting the residue into cycles. The exactly identical range–mean rainflow matrix as in Table 3.3 is formed by adding the six counted cycles into the cells. However, the resulting from–to rainflow matrix in Table 3.10 appears slightly different than the one by the three-point rainflow method in Table 3.8, in which the last largest cycle (a hanging cycle from 12 to −10) extracted by the three-point counting method is

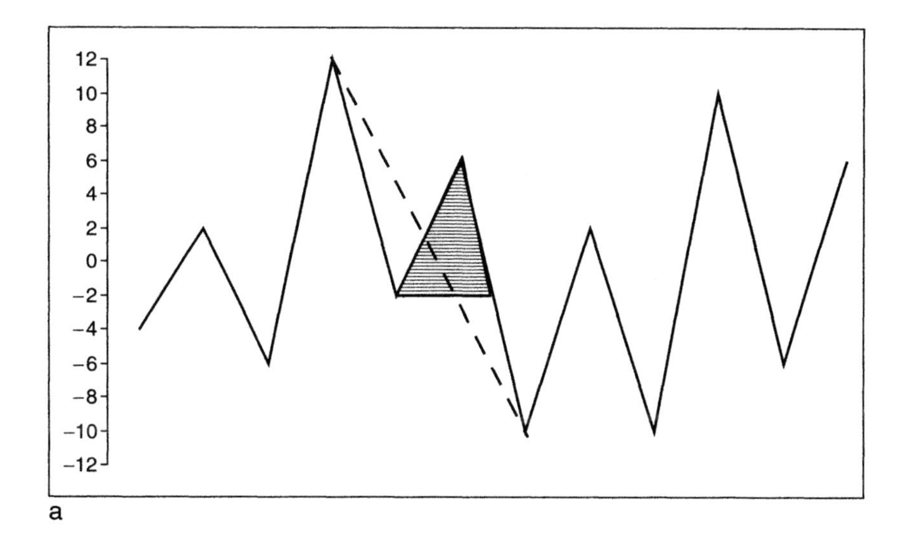

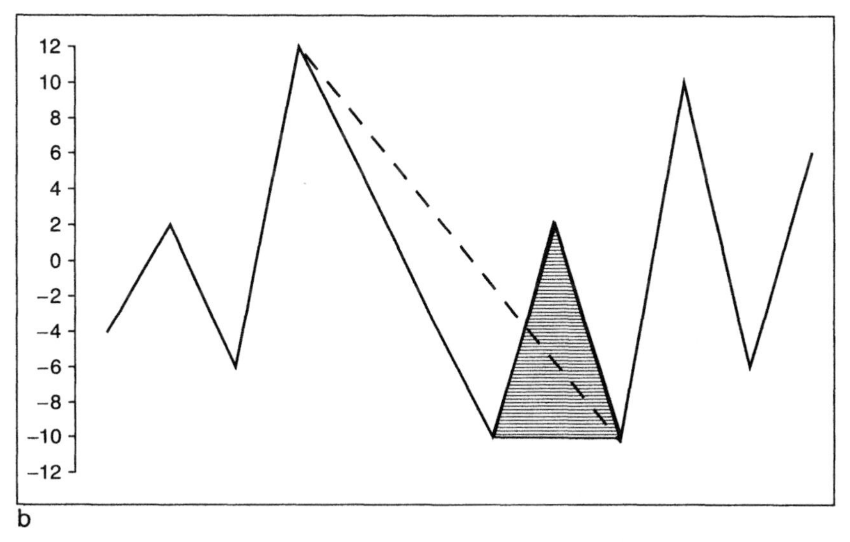

FIGURE 3.10 (a) Extracted cycle from −2 to 6; (b) Extracted cycle from −10 to 2.

Continued

opposite to the one (a standing cycle from −10 to 12) by the four-point counting method.

3.4 RECONSTRUCTION OF A LOAD-TIME HISTORY

Often, it is necessary to take the results from cycle counting and reconstruct a load-time history as an input to fatigue testing or simulation

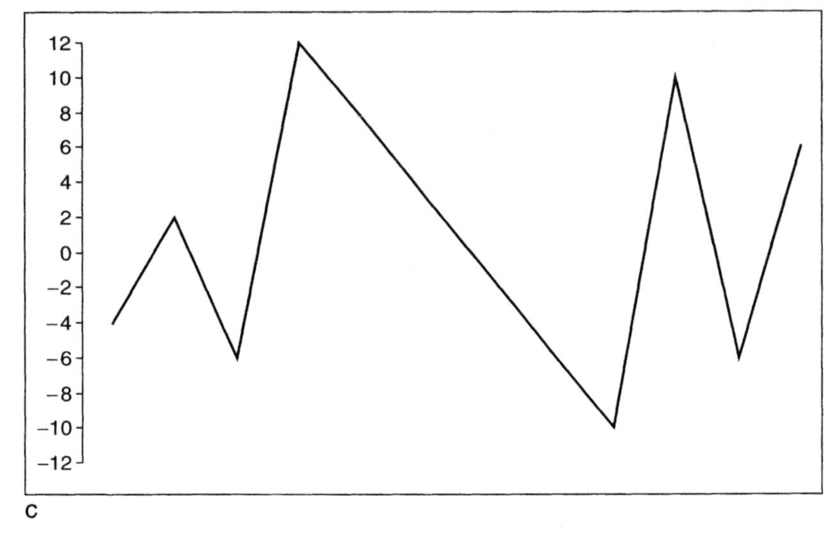

FIGURE 3.10 *Cont'd* (c) Residues: (−4)-(2)-(−6)-(12)-(−10)-(10)-(−6)-(6).

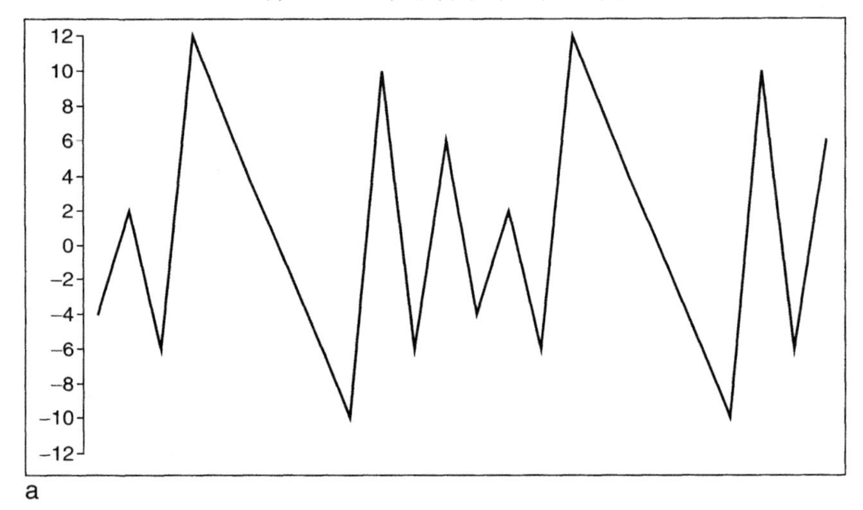

FIGURE 3.11 (a) Duplication of residues: [residue] + [residue].

Continued

(Khosrovaneh and Dowling, 1990; Amzallag et al., 1994). A method to reconstruct a new load-time history based on the four-point counting matrix and its residue is given in this section. The procedure for reconstruction is the reverse process of the four-point cycle extraction. The old and the reconstructed loading sequences are considered equivalent in terms of the same rainflow counting results and different in the exact loading sequence that is lost during the reconstruction process. The load-time reconstruction process follows the technique developed (Amzallag et al., 1994).

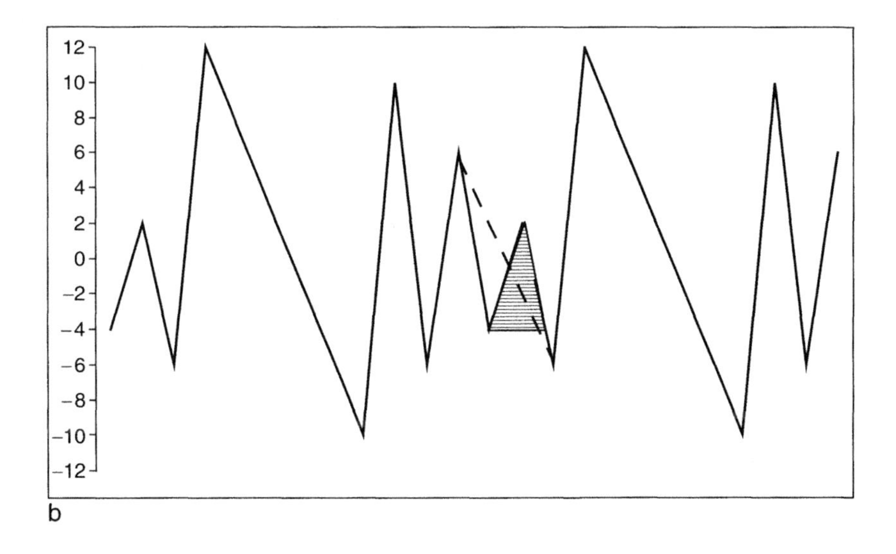

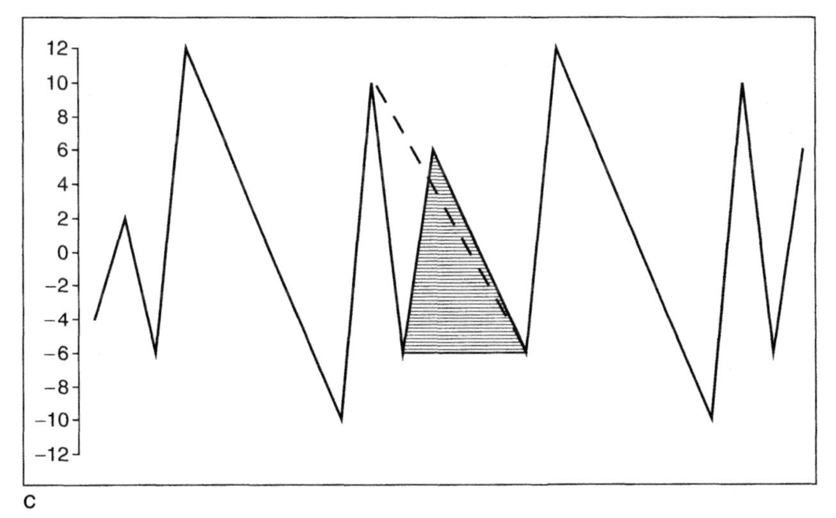

FIGURE 3.11 *Cont'd* (b) extracted cycle: −4 to 2; (c) Extracted cycle: −4 to 2.

Continued

The procedure for reconstruction is illustrated in Figure 3.12. A hanging cycle can be inserted in a positive-sloped reversal, and a standing cycle can be inserted in a negative-sloped reversal. According to this rule, a cycle may be inserted in many different reversals where the two principles of inserting a cycle are satisfied. Therefore, the reconstructed load-time history is not unique. This can be demonstrated in Figure 3.13, in which a cycle to be inserted is denoted by a-b-a and a residue is formed by A-B-C-D-E-F. In this case, cycle a-b-a may be inserted into the three possible reversals: A-B, C-D,

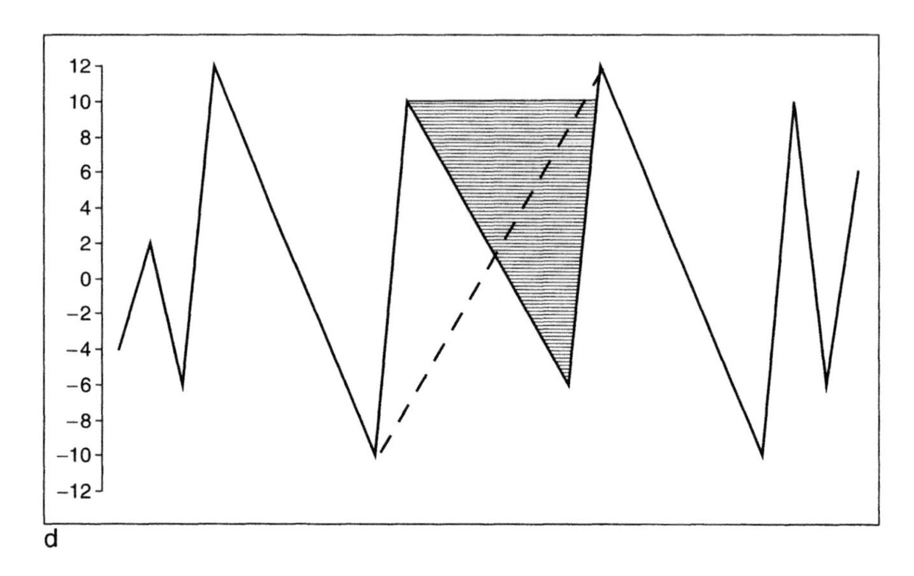

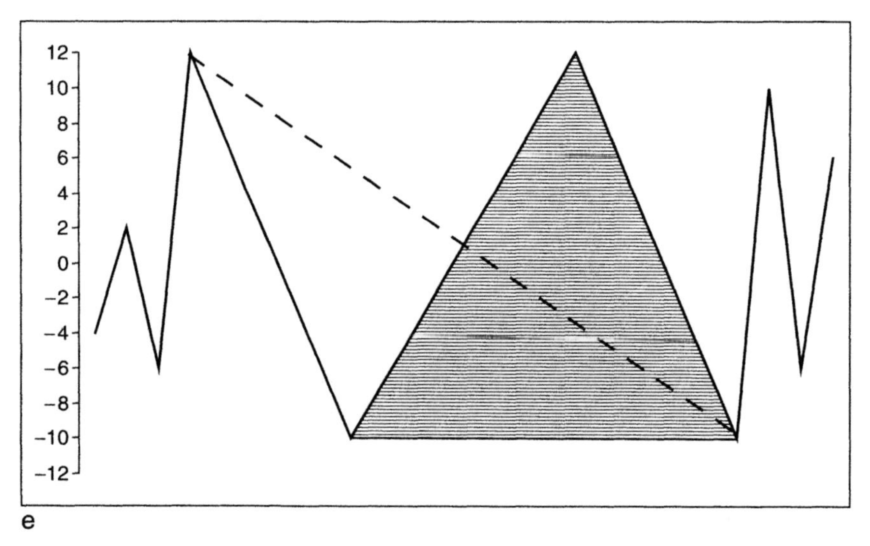

FIGURE 3.11 *Cont'd* (d) Extracted cycle: 10 to −6; (e) extracted cycle: −10 to 12.

and E-F. The possible reconstructed load-time histories including this cycle are shown in Figure 3.14(a)–(c). The choice of one of the three possible reversals is a random process. Moreover, the biggest cycle must be treated first in order to insert all the cycles to the residue.

Example 3.4. Reconstruct a new load-time history based on the cycle extracted and the residue obtained in Example 3.3 (Figure 3.15).

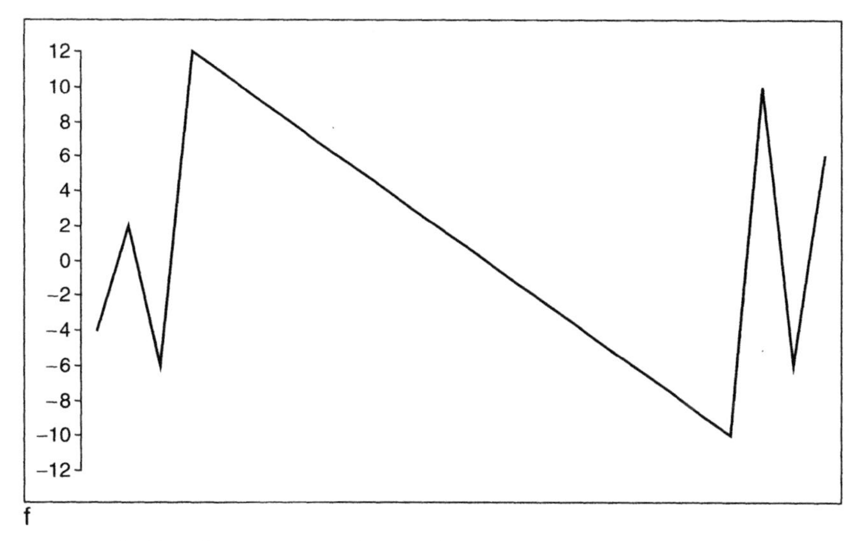

FIGURE 3.11 (f) residues: (−4)-(2)-(−6)-(12)-(−10)-(10)-(−6)-(6).

Solution. Figure 3.15 shows the two cycles extracted (a-b-a and c-d-c) and the residue (A-H) from Example 3.3 by the four-point rainflow cycle counting method. First, the larger cycle c-d-c is identified and is inserted into the only reversal (D-E) that includes cycle c-d-c, as shown in Figure 3.16(a). It is found next that the remaining cycle a-b-a can be inserted into the two possible reversals (D-C and F-G). From the random selection process, the cycle a-b-a is inserted into the reversal F-G as shown in Figure 3.16(b). This completes the new load-time reconstruction procedure.

TABLE 3.11 The From–To Rainflow Cycle Counting Matrix: Four-Point Cycle Counting

		To											
		−10	−8	−6	−4	−2	0	2	4	6	8	10	12
	−10							1					1
	−8												
	−6									1			
	−4							1					
	−2									1			
	0												
From	2												
	4												
	6												
	8												
	10			1									
	12												

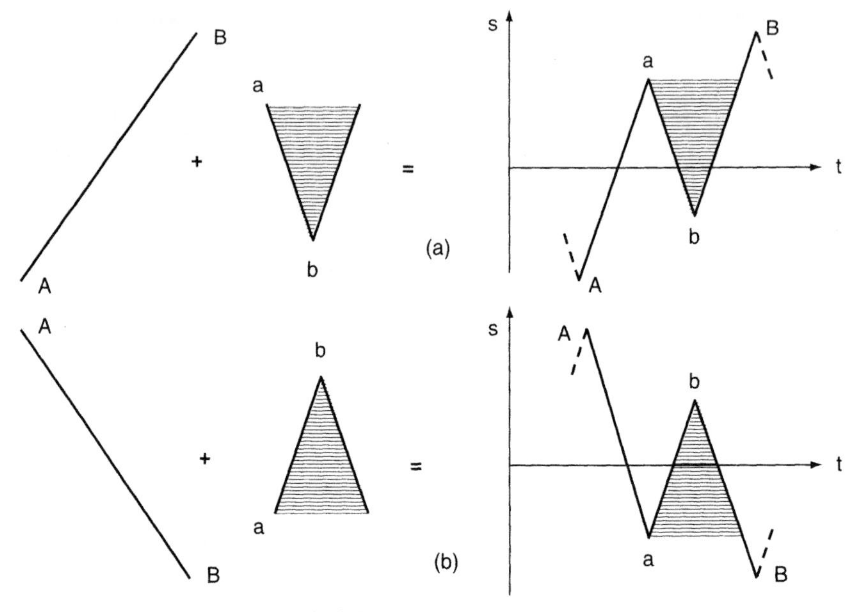

FIGURE 3.12 Principle of inserting a cycle.

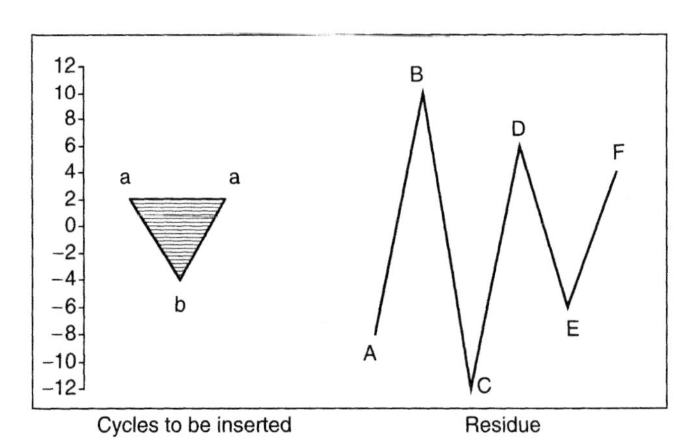

FIGURE 3.13 Example of reconstructed load-time history.

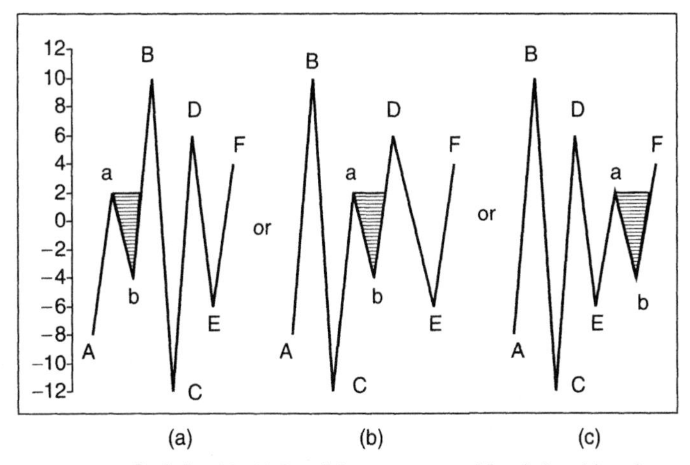

FIGURE 3.14 (a)–(c) Possible reconstructed load-time histories.

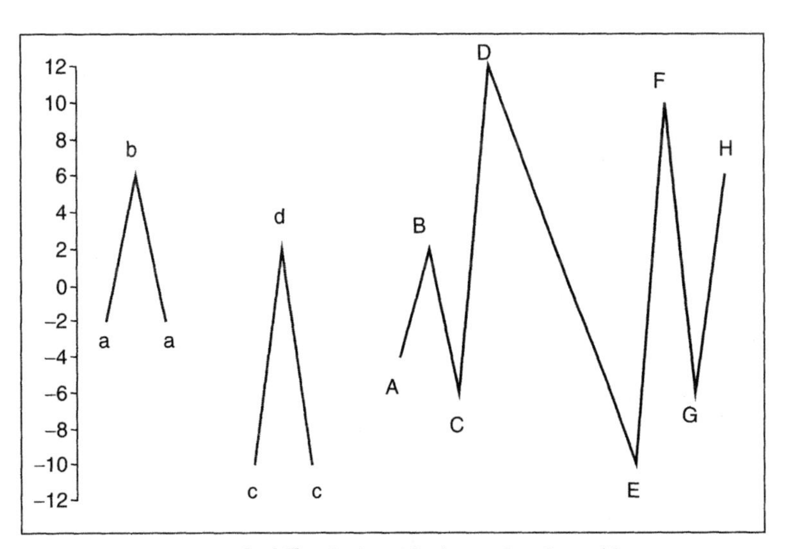

FIGURE 3.15 Cycles to be inserted and a residue.

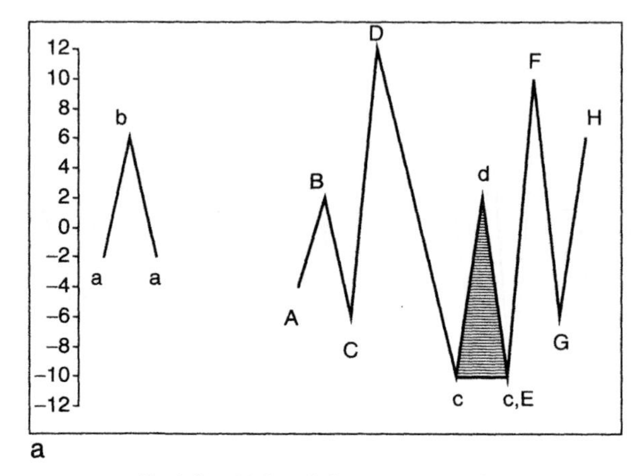

FIGURE 3.16 (a) Load-time reconstruction process 1.

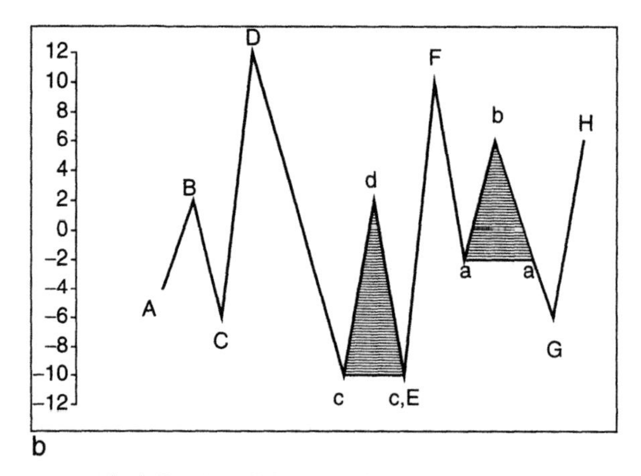

FIGURE 3.16 *Cont'd* (b) Load-time reconstruction process 2.

REFERENCES

Amzallag, C., Gerey, J. P., Robert J. L., and Bahuaud, J., Standardization of the rainflow counting method for fatigue analysis, *International Journal of Fatigue*, Vol. 16, 1994, pp. 287–293.

ASTM Standard E1049, Standard Practices for Cycle Counting in Fatigue Analysis, Philadelphia, PA, 1985.

Dowling, N. E., Fatigue at notches and the local strain and fracture mechanics approaches, *Fracture Mechanics*, ASTM STP 677, American Society of Testing and Materials, 1979.

Khosrovaneh, A. K., and Dowling, N. E., Fatigue loading history reconstruction based on the rainflow technique, *International Journal of Fatigue*, Vol. 12, No. 2, 1990, pp. 99–106.

Matsuishi, M. and Endo, T., Fatigue of metals subjected to varying stress. Presented to the Japan Society of Mechanical Engineers, Fukuoka, Japan, 1968.

Rice, R. C., et al. (Eds.), *Fatigue Design Handbook*, 3rd ed., Society of Automotive Engineers, Warrendale, PA, 1997.

4

STRESS-BASED FATIGUE ANALYSIS AND DESIGN

YUNG-LI LEE
DAIMLERCHRYSLER
DARRYL TAYLOR
DAIMLERCHRYSLER

4.1 INTRODUCTION

The fatigue damage theories in Chapter 2 indicate that fatigue damage is strongly associated with the cycle ratio $(n_i/N_{i,f})$, where n_i and $N_{i,f}$ are, respectively, the number of applied stress cycles and the fatigue life at a combination of stress amplitude and mean stress levels. The cycle counting techniques for the determination of n_i have been addressed in Chapter 3. This chapter focuses on how to establish baseline fatigue data, determine the fatigue life based on a stress-based damage parameter, and demonstrate engineering applications of this methodology.

Since the mid-1800s, a standard method of fatigue analysis and design has been the stress-based approach. This method is also referred to as the stress–life or the *S–N* approach and is distinguished from other fatigue analysis and design techniques by several features:

- Cyclic stresses are the governing parameter for fatigue failure
- High-cycle fatigue conditions are present
  - High number of cycles to failure
  - Little plastic deformation due to cyclic loading

During fatigue testing, the test specimen is subjected to alternating loads until failure. The loads applied to the specimen are defined by either a constant stress range (S_r) or a constant stress amplitude (S_a). The stress range is defined as the algebraic difference between the maximum stress (S_{max}) and minimum stress (S_{min}) in a cycle:

$$S_r = S_{max} - S_{min} \tag{4.1.1}$$

The stress amplitude is equal to one-half of the stress range:

$$S_a = \frac{S_r}{2} = \frac{(S_{max} - S_{min})}{2} \tag{4.1.2}$$

Typically, for fatigue analysts, it is a convention to consider tensile stresses positive and compressive stresses negative. The magnitude of the stress range or amplitude is the controlled (independent) variable and the number of cycles to failure is the response (dependent) variable. The number of cycles to failure is the fatigue life (N_f), and each cycle is equal to two reversals ($2N_f$). The symbols of stresses and cycles mentioned previously are illustrated in Figure 4.1.

Most of the time, S–N fatigue testing is conducted using fully reversed loading. *Fully reversed* indicates that loading is alternating about a zero mean stress. The mean stress (S_m) is defined as

$$S_m = \frac{(S_{max} + S_{min})}{2} \tag{4.1.3}$$

Exceptions exist when stress-life testing is performed for specimens for which this type of loading is physically not possible or is unlikely. One example is the fatigue testing of spot welded specimens. Cyclic loading varying from

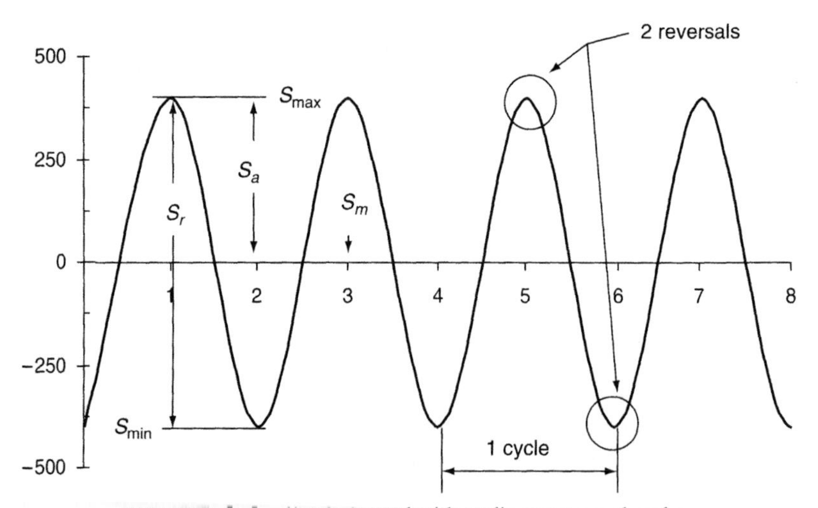

FIGURE 4.1 Symbols used with cyclic stresses and cycles.

zero to tension is used in fatigue testing on specimens with a single spot weld, because compression may cause local buckling of the thin sheet metal.

Actual structural components are usually subjected to alternating loads with a mean stress. Two parameters, the stress ratio (R) and the amplitude ratio (A), are often used as representations of the mean stress applied to an object. The stress ratio is defined as the ratio of minimum stress to maximum stress:

$$R = \frac{S_{\min}}{S_{\max}} \tag{4.1.4}$$

The amplitude ratio is the ratio of the stress amplitude to mean stress:

$$A = \frac{S_a}{S_m} = \frac{1 - R}{1 + R} \tag{4.1.5}$$

To generate data useful for fatigue designs using the stress-life approach, stress-life fatigue tests are usually carried out on several specimens at different fully reversed stress amplitudes over a range of fatigue lives for identically prepared specimens. The fatigue test data are often plotted on either semilog or log–log coordinates. Figure 4.2 shows the bending fatigue data of steel plotted on semilog coordinates. In this figure, the single curve that represents the data is called the S–N curve or the Wöhler curve. When plotted on log–log scales, the curve becomes linear. The portion of the curve or the line with a negative slope is called the finite life region, and the horizontal line is the infinite life region. The point of this S–N curve at which the curve changes from a negative slope to a horizontal line is called the *knee* of the S–N curve and represents the fatigue limit or the endurance limit. The fatigue limit is associated with the phenomenon that crack nucleation is arrested by the first

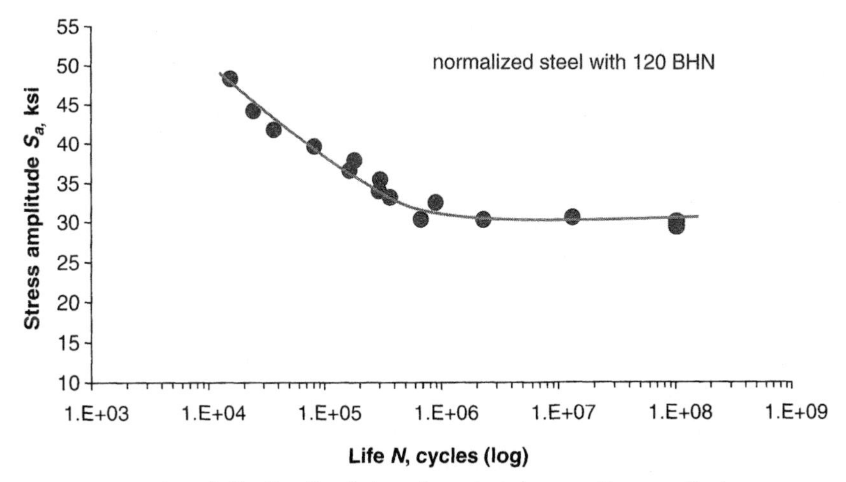

FIGURE 4.2 Bending fatigue data plotted on semilog coordinates.

grain boundary or a dominant microstructural barrier. Also, it can be overcome by the application of a few overloads, in a corrosive environment, etc.

When generating log–log graphs of applied stress versus fatigue life from S–N fatigue tests, the y-coordinate is expressed in terms of the stress amplitude or the stress range, and the x-coordinate is expressed in terms of the number of reversals to failure or the number of cycles to failure. Here, the fatigue life (cycles or reversals) refers to the life required to nucleate and grow a small crack to visible crack length.

Equation 4.1.6 represents the typical S–N curve:

$$S_a = S'_f(2N_f)^b \tag{4.1.6}$$

where b is the fatigue strength exponent, and S'_f is the fatigue strength coefficient. This expression developed from log–log S–N graphs is the most widely used equation (known as the Basquin relation) in the stress-based approach to fatigue analysis and design. There are also other S–N expressions, such as the following:

$$S_a = A(N_f)^B = (S'_f \times 2^b)(N_f)^b \tag{4.1.7}$$

or

$$S_r = C(2N_f)^D = (S'_f \times 2)(2N_f)^b \tag{4.1.8}$$

or

$$S_r = E(N_f)^F = (S'_f \times 2^{b+1})(N_f)^b \tag{4.1.9}$$

In this chapter, the stress-life approach refers to the use of the cyclic nominal stresses (S) versus fatigue life. Determination of the nominal stress depends on the loading and configuration of the specimen. The most common loading modes are bending (M), axial force (P), or torque (T). These loads can be related to nominal stresses by using traditional elastic stress formulas. For example, these familiar equations are

$$S = \frac{Mc}{I} \quad \text{for bending } M \tag{4.1.10}$$

$$S = \frac{P}{A} \quad \text{for axial load } P \tag{4.1.11}$$

$$S = \frac{Tr}{J} \quad \text{for torsion } T \tag{4.1.12}$$

where S can be the nominal normal or shear stress, depending on the equation used, A is the cross-sectional area, I is the moment of inertia, J is the polar moment of inertia, c is the distance from the neutral axis to the point of interest, and r is the distance from the center of a cross section to the point of interest.

An *S–N* curve can be generated for standard smooth material specimens, for individual manufactured structural components, for subassemblies, or for complete structures. Standard smooth specimens can be flat or cylindrical unnotched precision-machined coupons with polished surfaces so as to minimize surface roughness effects. The material *S–N* curve provides the baseline fatigue data on a given geometry, loading condition, and material processing for use in subsequent fatigue life and strength analyses. This baseline data can be adjusted to account for realistic component conditions such as notches, size, surface finish, surface treatments, temperature, and various types of loading. Other than from testing, there is no rational basis for determining these correction factors. The *S–N* curve for real components, subassemblies, or structures represents the true fatigue behavior of production parts/structures including all the aforementioned variables. However, if a design has changed, it is necessary to regenerate the *S–N* curve to incorporate the change effect. This adds cost and time to the fatigue design process. In general, a limitation of the *S–N* method is that it cannot predict local plasticity and mean stress effect.

4.2 THE STRESS-LIFE (S–N) AND FATIGUE LIMIT TESTING

Between 1852 and 1870, August Wöhler, a German railway engineer, conducted the first systematic fatigue investigations. In honor of his contributions to the study of fatigue, *S–N* fatigue tests in which constant-amplitude stress cycles with a specific mean stress level are applied to test specimens are sometimes called classical Wöhler tests. These tests are the most common type of fatigue testing. From these tests, it is possible to develop *S–N* curves that represent the fatigue life behavior of a component or of a material test specimen. Regardless of the type of test sample used, these *S–N* fatigue tests provide valuable information to an engineer during the design process. When conducting *S–N* fatigue tests, engineers do not have an unlimited amount of time or an unlimited number of test samples. Thus, it is necessary that the requirements, limitations, and approaches to the construction of an *S–N* curve be understood in order to effectively plan fatigue life tests. These are the major topics discussed in this section.

Fatigue life data exhibit widely scattered results because of inherent microstructural inhomogeneity in the materials properties, differences in the surface and the test conditions of each specimen, and other factors. In general, the variance of log life increases as the stress level decreases. It has been observed that once grains nucleate cracks in a material at high-stress levels, these cracks have a better chance of overcoming the surrounding microstructure. Most of the grains can successfully nucleate cracks at low stress levels, but only very few of them can overcome the surrounding obstacles (such as orientation, size, and microstructure) to grow a crack. As a result of the unavoidable variation in fatigue data, median *S–N* fatigue life curves are not sufficient for fatigue analysis and design. The statistical nature of fatigue must be considered.

There is a need for statistical *S–N* testing to predict fatigue life at various stress amplitude and mean stress combinations. The *S–N* test methods presented by the Japan Society of Mechanical Engineers (1981), Nakazawa and Kodama (1987), ASTM (1998), Shen (1994), Wirshing (1983), and Kececioglu (2003) are widely used by researchers for *S–N* testing and fatigue life predictions.

The median *S–N* test method with a small sample size as described elsewhere (Japan Society of Mechanical Engineers, 1981; Nakazawa and Kodama, 1987) can be used as a guideline to determine an *S–N* curve with a reliability of 50% and a minimum sample size. This method requires 14 specimens. Eight specimens are used to determine the finite fatigue life region, and six specimens are used to find the fatigue limit. The curve for the finite life region is determined by testing two samples at each of four different levels of stress amplitude and the fatigue limit is tested by the staircase method (introduced in Section 4.2.3) with six specimens. The recommended test sequence is shown in Figure 4.3, in which the number next to the data point represents the order in which the specimen is to be tested. The finite life region data is assumed linear in the log–log coordinates, and the data are analyzed by the least-squares method. The fatigue limit is determined by taking the average of the stress levels in a staircase test.

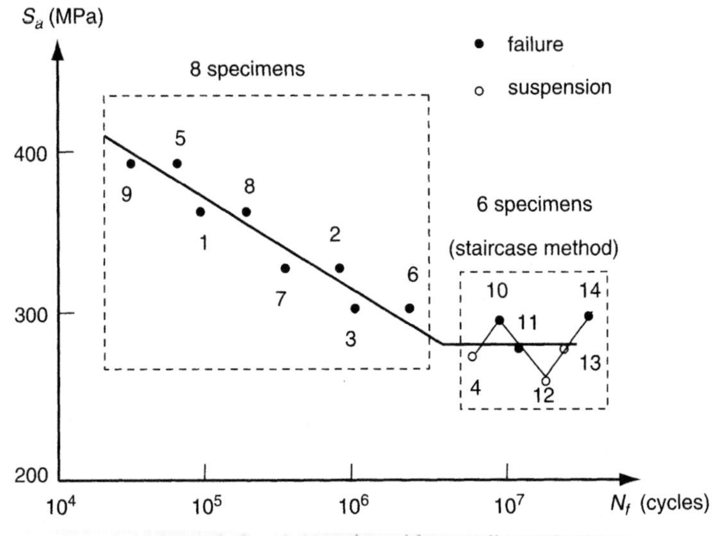

FIGURE 4.3 *S–N* testing with a small sample size.

The guidelines for the generation of statistical S–N curves are documented elsewhere (Wirshing, 1983; Shen, 1994; ASTM, 1998). It is recommended that more than one specimen be tested at each stress level. Tests with more than one test sample at a given stress amplitude level are called tests with replicate data. Replicate tests are required to estimate the variability and the statistical distribution of fatigue life. Depending on the intended purpose of the S–N curve, the recommended number of samples and number of replicated tests vary. The recommended sample size for the number of test samples used to generated the S–N curve is

- 6–12 for preliminary and research and development tests
- 12–24 for design allowable and reliability tests

The percent replication (PR), based on the number of stress levels (L) and a sample size (n_s), is defined as follows:

$$PR = 100(1 - L/n_s) \tag{4.2.1}$$

The percent replication value indicates the portion of the total number of specimens tested that may be used for determining an estimate of the variability of replicate tests. These guidelines recommend that the percent replication for various tests be

- 17–33 for preliminary and exploratory tests
- 33–50 for research and development tests
- 50–75 for design allowable data tests
- 75–88 for reliability data tests

Example 4.1. As a test engineer, you are requested to present a test plan (number of samples and number of stress levels required) to generate an S–N curve for reliability design. Provide justification for your proposal.

Solution. The guideline states that for reliability test, the sample size should be between 12 and 24. It also states that for reliability test, the percent replication should be between 75 and 88. To demonstrate the application of Equation 4.2.1, the numbers of stress levels corresponding to the two extreme sample sizes are tabulated in Table 4.1. Because any of the rows fits the aforementioned criteria, the engineering decision is made on how many samples are available and cost.

4.2.1 ANALYSIS OF FATIGUE DATA IN THE FINITE LIFE REGION

Once fatigue life data in the finite life region has been collected from S–N tests, the least-squares method for generating a line of best fit from the data is recommended. For the statistical analysis of the fatigue data, this method of

TABLE 4.1 Summary of Sample Size, Stress Levels, and PR for Sample Sizes 12 and 24

Sample Size (n_s)	Stress Levels (L)	PR
12	2	83.3
12	3	75.0
24	3	87.5
24	4	83.3
24	5	79.2
24	6	75.0

generating a line of best fit is feasible because the data can be represented as a straight line on a log–log plot of stress amplitude versus reversals to failure. It is assumed that the fatigue life resulting at a given stress amplitude level follows the lognormal distribution and that the variance of log life is constant over the tested range. The assumption of constant variance for all stress levels is referred to in statistics as the assumption of homoscedasticity. The least-squares regression model is

$$Y = A + BX + \varepsilon \tag{4.2.2}$$

where ε is a random variable of error. The regression line is

$$\hat{Y} = \hat{A} + \hat{B}X \tag{4.2.3}$$

where the estimated values of $\hat{A}$ and $\hat{B}$ are obtained by minimizing the sum of the square of the deviations of the observed values of Y from those predicted, i.e.,

$$\Delta^2 = \sum_{i=1}^{n_s} (Y_i - \hat{Y}_i)^2 = \sum_{i=1}^{n_s} (Y_i - \hat{A} - \hat{B}X_i)^2 \tag{4.2.4}$$

where the number of test samples is represented by the variable n_s. This leads to the following:

$$\frac{\partial \Delta^2}{\partial \hat{A}} = \sum_{i=1}^{n_s} 2(Y_i - \hat{A} - \hat{B}X_i)(-1) = 0 \tag{4.2.5}$$

$$\frac{\partial \Delta^2}{\partial \hat{B}} = \sum_{i=1}^{n_s} 2(Y_i - \hat{A} - \hat{B}X_i)(-X_i) = 0 \tag{4.2.6}$$

From this, the least-squares method estimates $\hat{B}$ and $\hat{A}$ as follows:

$$\hat{B} = \frac{\sum_{i=1}^{n_s} (X_i - \overline{X})(Y_i - \overline{Y})}{\sum_{i=1}^{n} (X_i - \overline{X})^2} \tag{4.2.7}$$

$$\hat{A} = \overline{Y} - \hat{B}\overline{X} \tag{4.2.8}$$

where $\overline{X}$ and $\overline{Y}$ are the average values of X and Y (for example, $\overline{X} = \sum X_i/n_s$ and $\overline{Y} = \sum Y_i/n_s$).

The measure of deviation about Y_i on X_i (i.e., the residual mean square) is assumed to be constant within the range of X_i of interest and is described as follows:

$$s^2 = \frac{1}{n_s - 2}\sum_{i=1}^{n_s}\left[Y_i - \left(\hat{A} + \hat{B}X_i\right)\right]^2 \tag{4.2.9}$$

where s is the sample standard error of Y_i estimates on X_i.

Taking logarithms (either the log base10 or the natural log) of both sides of the S–N equation (Equation 4.1.6) and rearranging yield the following equation:

$$\log(2N_f) = -\frac{1}{b}\log(S'_f) + \frac{1}{b}\log(S_a) \tag{4.2.10}$$

In this equation, the $\log_{10}$ is used. By comparing Equation 4.2.10 to the least-squares equation, $Y = \hat{A} + \hat{B}X$, it can be determined that X($= \log(S_a)$) is the independent variable; and Y ($= \log(2N_f)$) is the dependent variable. Similarly, the coefficients from the least squares equation take the forms $\hat{A} = (-1/b)\log(S'_f)$ and $\hat{B} = 1/b$.

In summary, the fatigue strength exponent b can be related to the linear regression constant $\hat{B}$ by the following expression:

$$b = 1/\hat{B} \tag{4.2.11}$$

The fatigue strength coefficient can be calculated as follows:

For the case using the base-10 logarithm,

$$S'_f = 10^{\left(-\hat{A}\times b\right)} \tag{4.2.12}$$

For the case using the natural logarithm,

$$S'_f = \exp(-\hat{A} \times b) \tag{4.2.13}$$

For probabilistic design and analysis, the following statistical fatigue properties were derived (Wirshing, 1983; Lee et al., 1999). It is assumed that the slope of the S–N curve b and the variance of Y estimate on X are constant. Based on the given s (the sample standard error of Y_i estimates on X_i), the coefficient of variation of S'_f ($C_{S'_f}$) is defined as the ratio of the standard deviation to the mean and is estimated as follows:

For the case of base-e logarithm,

$$C_{S'_f} = \sqrt{\exp(b^2 s^2) - 1} \tag{4.2.14}$$

For the case of base-10 logarithm:

$$C_{S_f'} = \sqrt{10^{b^2 s^2} - 1} \tag{4.2.15}$$

Example 4.2. Based on the engineering decision in Example 4.1, rotating bending fatigue testing under $R = -1$ loading was conducted on 12 structural components for reliability development. Table 4.2 shows the fatigue test data in the finite life region.

- Determine the S–N curve using the least-squares method and plot the fatigue data on the log–log coordinates.
- Compare the test data to the predicted S–N curve. Also determine the statistical fatigue properties.
- State the limitations of the S–N equation.

Solution. The independent and dependent variables (X and Y) are defined in the third and fourth columns in Table 4.2 by taking the base-10 logarithm of the stress amplitude and reversals in the first and second columns. Using the least-squares analysis method, the regression coefficients are

$$\hat{A} = 65.564$$

$$\hat{B} = -26.536$$

$$\mathrm{s} = 0.48850$$

According to Equations 4.2.11, 4.2.12, and 4.2.15,

$$b = 1/\hat{B} = -0.0377$$

$$S_f' = 10^{(-\hat{A} \times b)} = 296\text{MPa}$$

TABLE 4.2 Summary of the Component S–N Fatigue Test Data

Stress Amplitude S_a (MPa)	Fatigue Life $2N_f$ (Reversals)	X Log(S_a)	Y Log($2N_f$)
200	9.8E + 03	2.30103	3.99123
200	1.2E + 04	2.30103	4.07918
200	4.1E + 04	2.30103	4.61278
200	2.4E + 04	2.30103	4.38021
175	7.7E + 06	2.24304	6.88649
175	5.6E + 05	2.24304	5.74819
175	4.0E + 06	2.24304	6.60206
175	5.2E + 06	2.24304	6.71600
150	2.5E + 07	2.17609	7.39794
150	9.0E + 07	2.17609	7.95424
150	4.2E + 07	2.17609	7.62325
150	3.0E + 07	2.17609	7.47712

$$C_{S'_f} = \sqrt{10^{b^2 s^2} - 1} = \sqrt{10^{(-0.0377)^2 (0.4885)^2} - 1} = 0.0280$$

Thus, the equation for the *S–N* curve is

$$S_a = S'_f (2N_f)^b = 296 \times (2N_f)^{-0.0377}$$

Figure 4.4 shows the comparison between the predicted median *S–N* curve and the experimental fatigue data in a semilog scale. The limitations of the predicted *S–N* equation are as follows:

1) Only valid for the test data presented (i.e., effective life range from 9.8×10^3 to 9.0×10^7 reversals).
2) Only valid for the same failure mode due to rotating bending testing.
3) Representative of the median fatigue data.
4) Lognormal distribution of N_f is assumed.
5) Variance of N_f is considered constant for all N_f.

4.2.2 DESIGN S–N CURVES IN THE FINITE LIFE REGION

This section presents a practical method for constructing a design *S–N* curve that characterizes the minimum fatigue life at a given fatigue strength level so that majority of the fatigue data fall above the minimum or the lower-bound value. A schematic representation of the design *S–N* curve is given in Figure 4.5. The choice of the lower-bound *S–N* curve (a so-called design *S–N* curve) is fairly arbitrary and dependent on materials cost, safety policy, and industry standards. For example, if a value of R95C90 is used for component designs, this particular value ensures that there is a 95% possibility of survival (reliability) with a 90% of confidence level for a fatigue life at a specified stress

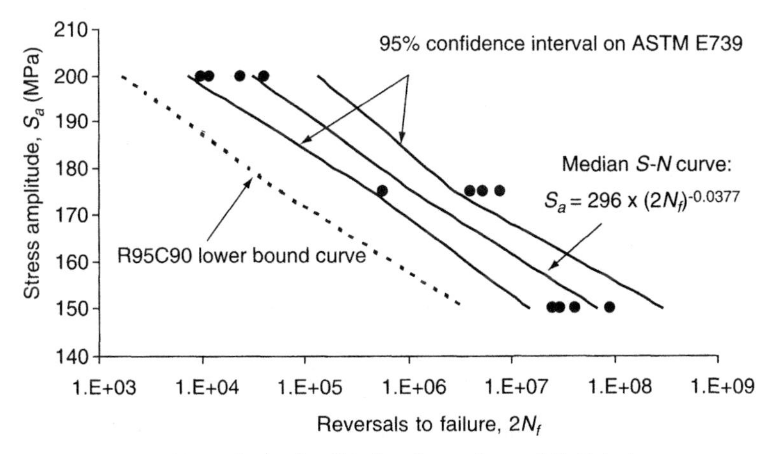

FIGURE 4.4 Predicted and experimental *S–N* data.

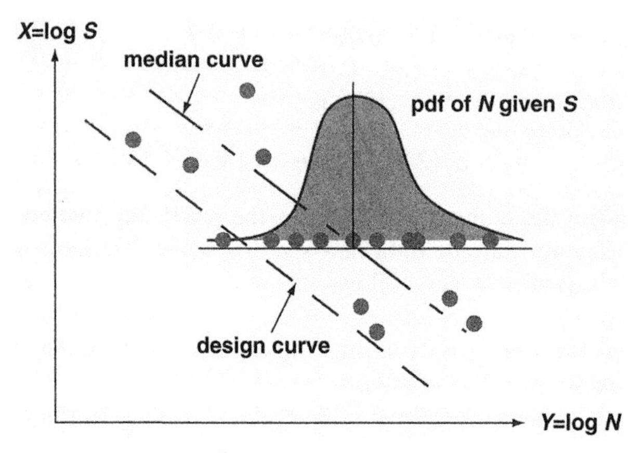

FIGURE 4.5 Concept of a design S–N curve.

level. Because a sample size is usually limited, the 90% confidence level is introduced to ensure that there is a 90% of possibility that the actual 95% of reliability value may be expected to fall above this lower bound.

Traditionally the engineering community prefers to use the lower 2-sigma or 3-sigma design curve in which the design curve can be derived by shifting the median S–N curve in logarithm coordinates to the left by two or three times the sample standard deviation. That means the design S–N curve can be expressed as follows:

$$Y_L(X_i) = \hat{Y}(X_i) - K \times s \tag{4.2.16}$$

where s is the sample standard error of Y on X_i, K is a multiplier equal to 2 or 3, Y_L is defined as the lower limit of $Y(= \ln 2N_f$ or $\ln N_f)$ at a given $X_i(= \ln S_{a,i})$. An example of this application can be found in Kececioglu (2003). However, these methods fail to account for the statistical distribution of fatigue life due to the sample size and the reliability level of interest.

Another design curve methodology, the one-side lower-bound tolerance limit method (Lieberman, 1958), has been proposed to overcome the previously mentioned shortcomings. The design S–N curve is obtained by moving the median to the left by a factor of K dependent on the sample size of Y_i at X_i. This method is valid when samples are tested at a fixed stress amplitude and the only variable (e.g., the log life) follows a normal distribution. When performing regression analyses in which both X and Y are variables, this method appears incorrect because it accounts for only the variable Y.

The double-sided confidence intervals approach (ASTM, 1998) and the approximate Owen one-side tolerance limit (Shen et al., 1996) have been proposed to account for uncertainty in regression analyses. Though more widely used and accepted at this time than the latter method, the ASTM standard only provides an exact confidence band for the entire median *S–N* curve, i.e., the two-sided confidence level for an R50 value. On the other hand, the one-sided tolerance limit, derived based on the approximate Owen tolerance limit method, has the flexibility of obtaining the lower-bound confidence level with any reliability value.

The double-sided confidence intervals approach the ASTM method, includes the effect of the additional variable X in the K value and is defined as

$$K_{\text{ASTM}} = \mp\sqrt{2F_{p,(2,n_s-2)}} \times \sqrt{\frac{1}{n_s} + \frac{(X_i - \overline{X})^2}{\sum_{k=1}^{n_s}(X_i - \overline{X})^2}} \tag{4.2.17}$$

where $F_{p,(2,n_s-2)}$ is the F-distribution value with the desired confidence interval p for $(2, n_s-2)$ degrees of freedom in Appendix 9.3. Because the ASTM standard is used to generate a double-sided confidence band, there must be two values of K_{ASTM}. K_{ASTM} takes the negative value for the upper confidence band whereas the lower confidence band takes the positive value when K_{ASTM} is used in Equation 4.2.16. Note that n_s is the test sample size.

For the approximate Owen one-sided tolerance limit, the K value was derived in the following expression:

$$K_{\text{owen}} = K_D \times R_{\text{owen}} \tag{4.2.18}$$

where

$$K_D = c_1 K_R + K_C\sqrt{c_3 K_R^2 + c_2 a} \tag{4.2.19}$$

$$R_{\text{owen}} = b_1 + \frac{b_2}{f^{b_3}} + b_4 \exp(-f) \tag{4.2.20}$$

in which

$$K_R = \varphi^{-1}(R) \tag{4.2.21}$$

$$K_C = \varphi^{-1}(C) \tag{4.2.22}$$

$$f = n_s - 2 \tag{4.2.23}$$

$$a = \frac{1.85}{n_s} \tag{4.2.24}$$

where the subscripts R and C denote the reliability and confidence levels, and $\phi(.)$ is the standard normal cumulative distribution function. Coefficients for empirical forms of K_{owen} are shown in Tables 4.3 and 4.4. A schematic illustration of this method is depicted in Figure 4.6.

Example 4.3. Determine the K_{owen} with the sample size of 10, a 90% probability of survival (R90), and a 95% confidence level (C95).

Solution.

$$f = n_s - 2 = 10 - 2 = 8$$

$$a = \frac{1.85}{n_s} = 0.185$$

$$K_R = \phi^{-1}(0.90) = 1.282$$

$$K_C = \phi^{-1}(0.95) = 1.645$$

$$c_1 = 1 + \frac{3}{4(f - 1.042)} = 1.1078$$

TABLE 4.3 Empirical Coefficients $b_i (i = 1, 2, 3, 4)$ for K_{owen}

Confidence Level (C)	b_1	b_2	b_3	b_4
0.95	0.9968	0.1596	0.60	−2.636
0.90	1.0030	−6.0160	3.00	1.099
0.85	1.0010	−0.7212	1.50	−1.486
0.80	1.0010	−0.6370	1.25	−1.554

Source: From Shen et al., 1996.

TABLE 4.4 Empirical Coefficients $c_i (i = 1, 2, 3)$ for K_{owen}

	c_1	c_2	c_3
$f < 2$	1	1	$\frac{1}{2f}$
$f \geq 2$	$1 + \frac{3}{4(f - 1.042)}$	$\frac{f}{f - 2}$	$c_2 - c_1^2$

Source: From Shen et al., 1996.

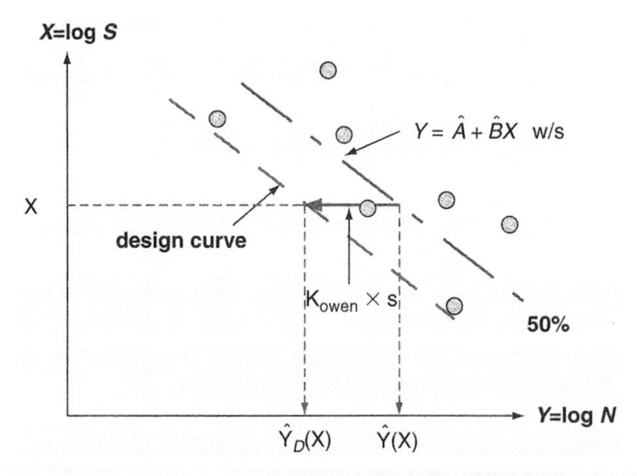

FIGURE 4.6 Design S–N curve based on the approximate Owen tolerance limit method.

$$c_2 = \frac{f}{f-2} = 1.3333$$

$$c_3 = c_2 - c_1^2 = 0.1061$$

$$K_D = c_1 K_R + K_C \sqrt{c_3 K_R^2 + c_2 a} = 2.488$$

Therefore, based on the empirical coefficient b_1, b_2, b_3, b_4 corresponding to the 95% confidence level in Table 4.3, R and K are calculated as follows:

$$R_{\text{owen}} = b_1 + \frac{b_2}{f^{b_3}} + b_4 \exp(-f) = 1.0417$$

$$K = K_D R_{\text{owen}} = 2.488 \times 1.0417 = 2.592$$

The K_{owen} factors for various sample sizes, reliability levels, and confidence levels generated by Williams et al. (2003) are tabulated in Table 4.5.

TABLE 4.5 K_{owen} Factors for the Approximate Owen Tolerance Limits

n_S	$C = 0.90$			$C = 0.95$		
	$R = 0.90$	$R = 0.95$	$R = 0.99$	$R = 0.90$	$R = 0.95$	$R = 0.99$
6	2.862	3.504	4.750	3.560	4.331	5.837
7	2.608	3.190	4.319	3.167	3.846	5.173
8	2.441	2.987	4.043	2.910	3.534	4.751
9	2.323	2.843	3.851	2.728	3.314	4.455
10	2.253	2.736	3.707	2.592	3.151	4.237
11	2.162	2.651	3.595	2.485	3.024	4.069
12	2.105	2.583	3.505	2.400	2.923	3.936
13	2.057	2.526	3.430	2.331	2.840	3.827
14	2.016	2.478	3.367	2.272	2.771	3.737
15	1.980	2.436	3.313	2.222	2.712	3.660
16	1.949	2.400	3.266	2.178	2.661	3.594
17	1.922	2.369	3.225	2.140	2.617	3.536
18	1.898	2.340	3.189	2.106	2.577	3.485
19	1.876	2.315	3.156	2.076	2.542	3.440
20	1.857	2.292	3.127	2.048	2.510	3.399
21	1.839	2.272	3.100	2.024	2.482	3.363
22	1.822	2.252	3.076	2.001	2.456	3.329
23	1.807	2.235	3.054	1.981	2.432	3.300
24	1.793	2.219	3.034	1.961	2.410	3.271
25	1.781	2.204	3.015	1.947	2.389	3.247
26	1.769	2.191	2.997	1.927	2.370	3.221
27	1.757	2.178	2.981	1.912	2.353	3.199
28	1.747	2.166	2.966	1.898	2.337	3.178
29	1.737	2.155	2.952	1.885	2.321	3.158
30	1.728	2.144	2.939	1.872	2.307	3.140

Source: From Williams et al., 2003.

Example 4.4. According to Example 4.2, an S–N curve with 50% reliability ($S_f' = 296$ MPa and $b = -0.0377$) is determined to be

$$S_a = 296 \times (2N_f)^{-0.0377}$$

where S_a is the nominal stress amplitude. (1) Given that the sample size is 12 and the sample standard error is 0.4885, determine the lower-bound S–N curve with 95% probability of survival (R95) and 90% confidence level (C90) based on the approximate Owen tolerance limit method. (2) Establish the 95% confidence intervals for the median S–N curve, based on the ASTM Standard E739.

Solution. (1) A stress amplitude of $S_a = 171$ MPa is determined by the input of $2N_f = 2 \times 10^6$ reversals to the previous equation. Note that this point ($S_a = 171$, $2N_f = 2 \times 10^6$) on the median S–N curve is arbitrarily chosen. The new design curve with R95C90 in the log–log scales can be constructed by a line with the same fatigue strength exponent b of -0.0377 that is horizontally shifted to the left from the median baseline with a distance of $K_{owen} \times s$. In Table 4.5, the K_{owen} factor with the sample size ($n_s = 12$), the 95% possibility of survival, and the 90% confidence level is 2.583. Therefore, the logarithm of the R95C90 fatigue life in reversals can be determined by subtracting from the logarithm of the median fatigue life by a value of $K_{owen} \times s$, i.e., $\log(2N_{f,R95C90}) = \log(2 \times 10^6) - 2.583 \times 0.4885 = 5.039$ or

$$2N_{f,R95,C90} = 10^{(5.039)} = 109{,}400 \text{ reversals}$$

The S–N equation with R95C90 can be determined by the identical slope, b, and the pair of $S_a = 171$ and $2N_{f,R95C90} = 109{,}400$. The fatigue strength coefficient $S'_{f,R90C95}$ is then obtained as follows:

$$S'_{f,R95C90} = \frac{S_a}{(2N_{f,R95C90})^b} = \frac{171}{(109{,}400)^{-0.0377}} = 265 \text{ MPa}$$

Thus, the new S–N equation with R95C90 shown in Figure 4.4 has the following form:

$$S_a = S'_{f,R95C90}(2N_f)^b = 265 \times (2N_f)^{-0.0377}$$

(2) Based on the desired confidence level $p = 0.95$ and the test sample size $n_s = 12$, look up $F_{0.95}$ value for $(2, n_s - 2)$ degrees of freedom in Appendix 9.3. It is found that $F_{0.95,(2,10)} = 4.10$. Also, $\bar{X} = 2.24005$ and $\sum_{k=1}^{12}(X - \bar{X})^2 = 0.031274$ can be obtained from Table 4.2. Choose a number of X values within the range of the data at which to compute points for drawing the confidence level. For example, let X be 2.30103, 2.24304, and 2.17609 at the three given stress amplitudes. At each selected X value, compute $\hat{Y}$ based on the least-squares regression line:

$$\hat{Y} = \hat{A} + \hat{B}X = 65.5638 + (-26.536) \times X$$

and K_{ASTM} on Equation 4.2.17. A 95% confidence band for the whole line can then be obtained by

$$\hat{Y} \pm K_{ASTM} \times s = \hat{Y} \pm K_{ASTM} \times 0.4885$$

Finally, as illustrated in Figure 4.4, the 95% confidence band can be constructed by connecting the upper series of three points and the lower series of three points via smooth curves. The upper bound and the lower bound represent the curves with R50C2.5 and R50C97.5, respectively.

4.2.3 FATIGUE STRENGTH TESTING

The objective of the fatigue strength test (also called the fatigue limit test, the strength test, or the response test) is to estimate a statistical distribution of the fatigue strength at a specific high-cycle fatigue life. Among many fatigue strength test methods, the staircase method (often referred to as the up-and-down method) is the most popular one that has been adopted by many standards (e.g., the British Standard Institution, 1966; Japanese Society of Mechanical Engineers, 1981, L'association Francaise de Normalisation, 1991; MPIF, 2000) to assess statistical properties of a fatigue limit.

In this test, the mean fatigue limit has to be first estimated, and a fatigue life test is then conducted at a stress level a little higher than the estimated mean. If the specimen fails prior to the life of interest, the next specimen has to be tested at a lower stress level. If the specimen does not fail within this life of interest, a new test has to be conducted at a higher stress level. Therefore, each test is dependent on the previous test results, and the test continues with a stress level increased or decreased. This process is illustrated in Figure 4.7. Collins (1993) recommends that the test be run with at least 15 specimens. The stress increments are usually taken to be less than about 5% of the initial estimate of the mean fatigue limit.

Two typical data reduction techniques, the Dixon-Mood (1948) and the Zhang-Kececioglu (1998) methods, are used to determine the statistical parameters of the test results. Discussion of the two methods can be found elsewhere (Lin et al., 2001). The Zhang-Kececioglu method proposed to use either the maximum likelihood estimation method or the suspended-items analysis method for statistical analysis. The Dixon-Mood method is derived on the maximum likelihood estimation method and assumes a normal distribution best fit for the fatigue limit. The former method has the flexibility of fitting the test data to a statistical distribution other than the normal distribution and can be used for variable stress steps. The latter method is easy to use and often provides conservative results. Thus, it is the focus of this chapter.

The Dixon-Mood method provides approximate formulas to calculate the mean (μ_S) and the standard deviation (σ_S) of a fatigue limit (S_e). The method assumes that the fatigue limit follows a normal distribution. It requires that the two statistical properties be determined by using the data of the less frequent event (i.e., either only the failures or only the survivals). The stress levels S_i spaced equally with a chosen increment d are numbered i, where $i = 0$ for the lowest stress level S_o. The stress increment should be in the range of half to twice the standard deviation of the fatigue limit.

Denoting by $n_{DM,\,i}$ the number of the less frequent event at the numbered stress level i, two quantities A_{DM} and B_{DM} can be calculated:

$$A_{DM} = \sum i \times n_{DM,i} \tag{4.2.25}$$

$$B_{DM} = \sum i^2 \times n_{DM,i} \tag{4.2.26}$$

The estimate of the mean is then

$$\mu_S = S_o + d \times \left(\frac{A_{DM}}{\sum n_{DM,i}} \pm \frac{1}{2} \right) \tag{4.2.27}$$

where the plus sign (+) is used if the less frequent event is survival and the minus sign (−) is used if the less frequent event is failure.

The standard deviation is estimated by

$$\sigma_S = 1.62 \times d \times \left[\frac{B_{BM} \sum n_{DM,i} - A_{DM}^2}{\left(\sum n_{DM,i}\right)^2} + 0.029 \right]$$
$$\text{if} \quad \frac{B_{DM} \sum n_{DM,i} - A_{DM}^2}{\left(\sum n_{DM,i}\right)^2} \geq 0.3 \tag{4.2.28}$$

or

$$\sigma_S = 0.53 \times d \quad \text{if} \quad \frac{B_{DM} \sum n_{DM,i} - A_{DM}^2}{\left(\sum n_{DM,i}\right)^2} < 0.3 \tag{4.2.29}$$

With the mean μ_S and standard deviation σ_S determined for a normal distribution, the lower-bound value associated with reliability and confidence levels can be determined by the one-side lower-bound tolerance limit K factor (Lieberman, 1958) as follows:

$$S_{e,\,R,\,C} = \mu_S - K \times \sigma_s \tag{4.2.30}$$

where the K factors for one-side tolerance limit for a normal distribution are given in Tables 4.6 and 4.7. The lower-bound fatigue limit $S_{e,\,R,\,C}$ means that with a confidence level of $C\%$, $R\%$ of the tested fatigue limit might be expected to exceed the $S_{e,\,R,\,C}$ level.

Example 4.5. Determine the statistical properties of the fatigue limit and its R95C90 value based on the staircase test results shown in Figure 4.7, where

TABLE 4.6 K-Factor for One-Side Lower-Bound Tolerance Limit for a Normal Distribution

C	0.75					0.90				
R	0.75	0.90	0.95	0.99	0.999	0.75	0.90	0.95	0.99	0.999
n										
3	1.464	2.501	3.152	4.396	5.805	2.602	4.258	5.310	7.340	9.651
4	1.256	2.134	2.680	3.726	4.910	1.972	3.187	3.967	5.437	7.128
5	1.152	1.961	2.463	3.421	4.507	1.698	2.742	3.400	4.666	6.112
6	1.087	1.860	2.336	3.243	4.273	1.540	2.494	3.091	4.242	5.556
7	1.043	1.791	2.250	3.126	4.118	1.435	2.333	2.894	3.972	5.201
8	1.010	1.740	2.190	3.042	4.008	1.360	2.219	2.755	3.783	4.955
9	0.984	1.702	2.141	2.977	3.924	1.302	2.133	2.649	3.641	4.772
10	0.964	1.671	2.103	2.927	3.858	1.257	2.065	2.586	3.532	4.629
11	0.947	1.646	2.073	2.885	3.804	1.219	2.012	2.503	3.444	4.515
12	0.933	1.624	2.048	2.851	3.760	1.188	1.966	2.448	3.371	4.420
13	0.919	1.606	2.026	2.822	3.722	1.162	1.928	2.403	3.310	4.341
14	0.909	1.591	2.007	2.796	3.690	1.139	1.895	2.363	3.257	4.274
15	0.899	1.577	1.991	2.776	3.661	1.119	1.866	2.329	3.212	4.215
16	0.891	1.566	1.977	2.756	3.637	1.101	1.842	2.299	3.172	4.164
17	0.883	1.554	1.964	2.739	3.615	1.085	1.820	2.272	3.136	4.118
18	0.876	1.544	1.951	2.723	3.595	1.071	1.800	2.249	3.106	4.078
19	0.870	1.536	1.942	2.710	3.577	1.058	1.781	2.228	3.078	4.041
20	0.865	1.528	1.933	2.697	3.561	1.046	1.765	2.208	3.052	4.009
21	0.859	1.520	1.923	2.686	3.545	1.035	1.750	2.190	3.028	3.979
22	0.854	1.514	1.916	2.675	3.532	1.025	1.736	2.174	3.007	3.952
23	0.849	1.508	1.907	2.665	3.520	1.016	1.724	2.159	2.987	3.927
24	0.845	1.502	1.901	2.656	3.509	1.007	1.712	2.145	2.969	3.904
25	0.842	1.496	1.895	2.647	3.497	0.999	1.702	2.132	2.952	3.882
30	0.825	1.475	1.869	2.613	3.545	0.966	1.657	2.080	2.884	3.794
35	0.812	1.458	1.849	2.588	3.421	0.942	1.623	2.041	2.833	3.730
40	0.803	1.445	1.834	2.568	3.395	0.923	1.598	2.010	2.793	3.679
45	0.795	1.435	1.821	2.552	3.375	0.908	1.577	1.986	2.762	3.638
50	0.788	1.426	1.811	2.538	3.358	0.894	1.560	1.965	2.735	3.604

Source: From Lieberman, 1958.

the black dots and the open circles represent test data failed prior to 10^7 cycles and suspended at 10^7 cycles, respectively.

Solution. The survivals are used in this analysis because they are the less frequent event for these results. Thus, for

$$i = 0 \quad S_o = 360\,\text{MPa} \quad n_{DM,0} = 1$$
$$i = 1 \quad S_1 = 380\,\text{MPa} \quad n_{DM,1} = 4$$
$$i = 2 \quad S_2 = 400\,\text{MPa} \quad n_{DM,2} = 6$$
$$i = 3 \quad S_3 = 420\,\text{MPa} \quad n_{DM,3} = 1$$
$$i = 4 \quad S_4 = 440\,\text{MPa} \quad n_{DM,4} = 0$$

TABLE 4.7 *K* Factors for One-Side Lower-Bound Tolerance Limit for a Normal Distribution

C	0.95					0.99				
R	0.75	0.90	0.95	0.99	0.999	0.75	0.90	0.95	0.99	0.999
n										
3	3.804	6.158	7.655	10.55	13.86	—	—	—	—	—
4	2.619	4.163	5.145	7.042	9.215	—	—	—	—	—
5	2.149	3.407	4.202	5.741	7.501	—	—	—	—	—
6	1.895	3.006	3.707	5.062	6.612	2.849	4.408	5.409	7.334	9.550
7	1.732	2.755	3.399	4.641	6.061	2.490	3.856	4.730	6.411	8.348
8	1.617	2.582	3.188	4.353	5.686	2.252	3.496	4.287	5.811	7.566
9	1.532	2.454	3.031	4.143	5.414	2.085	3.242	3.971	5.389	7.014
10	1.465	2.355	2.911	3.981	5.203	1.954	3.048	3.739	5.075	6.603
11	1.411	2.275	2.815	3.852	5.036	1.854	2.897	3.557	4.828	6.284
12	1.366	2.210	2.736	3.747	4.900	1.771	2.773	3.410	4.633	6.032
13	1.329	2.155	2.670	3.659	4.787	1.702	2.677	3.290	4.472	5.826
14	1.296	2.108	2.614	3.585	4.690	1.645	2.592	3.189	4.336	5.561
15	1.268	2.068	2.566	3.520	4.607	1.596	2.521	3.102	4.224	5.507
16	1.242	2.032	2.523	3.463	4.534	1.553	2.458	3.028	4.124	5.374
17	1.220	2.001	2.486	3.415	4.471	1.514	2.405	2.962	4.038	5.268
18	1.200	1.974	2.453	3.370	4.415	1.481	2.357	2.906	3.961	5.167
19	1.183	1.949	2.423	3.331	4.364	1.450	2.315	2.855	3.893	5.078
20	1.167	1.926	2.396	3.295	4.319	1.424	2.275	2.807	3.832	5.003
21	1.152	1.905	2.371	3.262	4.276	1.397	2.241	2.768	3.776	4.932
22	1.138	1.887	2.350	3.233	4.238	1.376	2.208	2.729	3.727	4.866
23	1.126	1.869	2.329	3.206	4.204	1.355	2.179	2.693	3.680	4.806
24	1.114	1.853	2.309	3.181	4.171	1.336	2.154	2.663	3.638	4.755
25	1.103	1.838	2.292	3.158	4.143	1.319	2.129	2.632	3.601	4.706
30	1.059	1.778	2.220	3.064	4.022	1.249	2.029	2.516	3.446	4.508
35	1.025	1.732	2.166	2.994	3.934	1.195	1.957	2.431	3.334	4.364
40	0.999	1.697	2.126	2.941	3.866	1.154	1.902	2.365	3.250	4.255
45	0.978	1.669	2.092	2.897	3.811	1.122	1.857	2.313	3.181	4.168
50	0.961	1.646	2.065	2.863	3.766	1.096	1.821	2.269	3.124	4.096

Source: From Lieberman, 1958.

$\sum n_{DM} = 12$, $d =$ equal stress increment $= 20$ MPa.
From Equations 4.2.25 and 4.2.26, one obtains

$$A_{DM} = \sum_{i=0}^{4} i \times n_{DM,i} = 0 + (1 \times 4) + (2 \times 6) + (3 \times 1) + 0 = 19$$

$$B_{DM} = \sum_{i=0}^{4} i^2 \times n_{DM,i} = 0 + (1^2 \times 4) + (2^2 \times 6) + (3^2 \times 1) + 0 = 37$$

From Equation 4.2.27,

$$\mu_S = S_o + d \times \left(\frac{A_{DM}}{\sum n_{DM}} + \frac{1}{2} \right) = 360 + 20\left(\frac{19}{12} + \frac{1}{2} \right) = 401.7 \text{ MPa}$$

Since

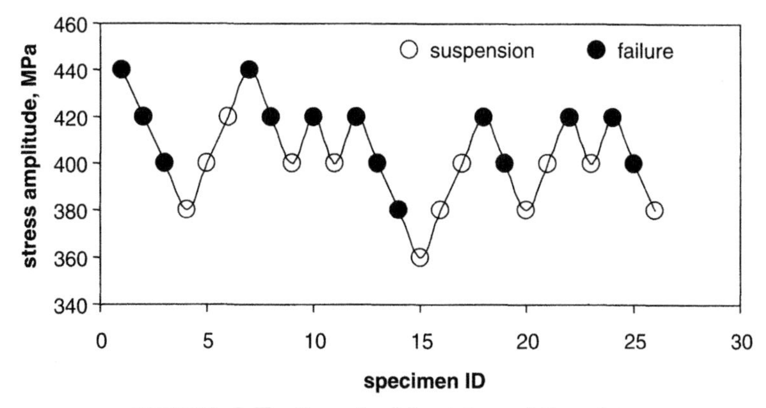

FIGURE 4.7 Example of the staircase fatigue data.

$$\frac{B_{DM}\sum n_{DM} - A_{DM}^2}{(\sum n_{DM})^2} = 0.58 > 0.3,$$

the standard deviation can be estimated from Equation 4.2.28 as follows:

$$\sigma_S = 1.62d\left(\frac{B_{DM}\sum n_{DM} - A_{DM}^2}{(\sum n_{DM})^2} + 0.029\right) = 1.62(20)(0.58 + 0.029) = 19.7\,\text{MPa}$$

Hence, the estimated mean and standard deviation of the fatigue endurance strength are 401.7 MPa and 19.7 MPa, respectively.

Assuming that the fatigue limit follows a normal distribution, the R95C90 fatigue limit value can be found from the one-side lower-bound tolerance limit method:

$$S_{e,0.95,0.9} = \mu_{S_e} - K \times \sigma_{S_e}$$

where K is obtained as 2.448 from Table 4.6 with $n_S = 12$. Thus,

$$S_{e,0.95,0.9} = \mu_{S_e} - K \times \sigma_{S_e} = 401.7 - 2.448 \times 19.7 = 353.5\,\text{MPa}$$

Example 4.6 (Application). An automotive intake-valve cam with translating follower, long push rod, rocker arm, valve, and valve spring is shown in Figure 4.8. The automotive intake-valve rocker arms are made of SAE 1008 hot rolled steels with a surface finish of 1.1 μm. The Rockwell hardness (RB) of the material varies from 50 to 70. Check the rocker arm for the durability criterion. This means that the R90C90 fatigue strength at 10^7 cycles should exceed the maximum operating load by a safety factor. It is assumed that the safety factor of 1.3 is used in this example.

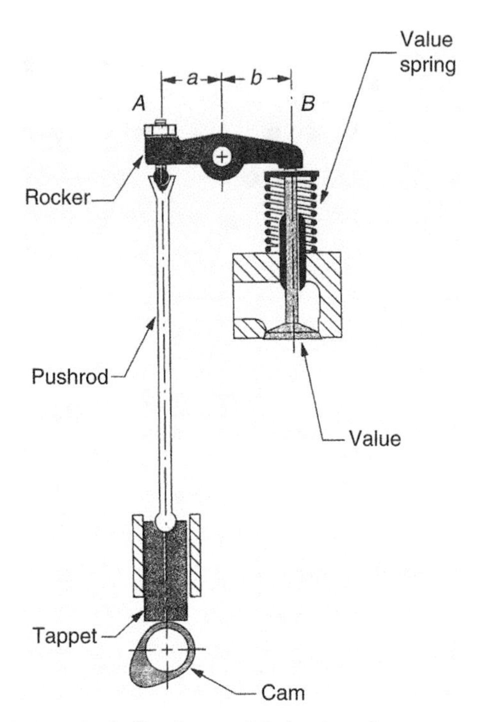

FIGURE 4.8 Automobile intake-valve system.

The staircase (or up-and-down) fatigue tests for fatigue strength of the rocker arm at 10^7 cycles were performed by cycling loading with $R = 0$. The test fixture was designed to sustain the cyclic load from the valve tip and to constrain the pivot shaft and push rod in three translations. The cyclic valve tip load varies from zero to a specified maximum load. Figure 4.9 shows the staircase results and Figure 4.10 illustrates the calculated valve tip load in Newtons versus engine speed (rpm).

(1) Use the Dixon-Mood method to analyze the staircase fatigue tests for statistical fatigue strength at 10^7 cycles.
(2) Use the one-side lower-bound tolerance limit method to determine the R90C90 fatigue strength at 10^7 cycles.
(3) Calculate the factor of safety to examine if the rocker arm design meets the durability target.

Solution. Counting the survival specimens since the number survivals are less than the number failures:

$$i = 0 \quad P_0 = 14,500\,N \quad n_{DM,0} = 3$$

$$i = 1 \quad P_1 = 14,900\,N \quad n_{DM,1} = 4$$

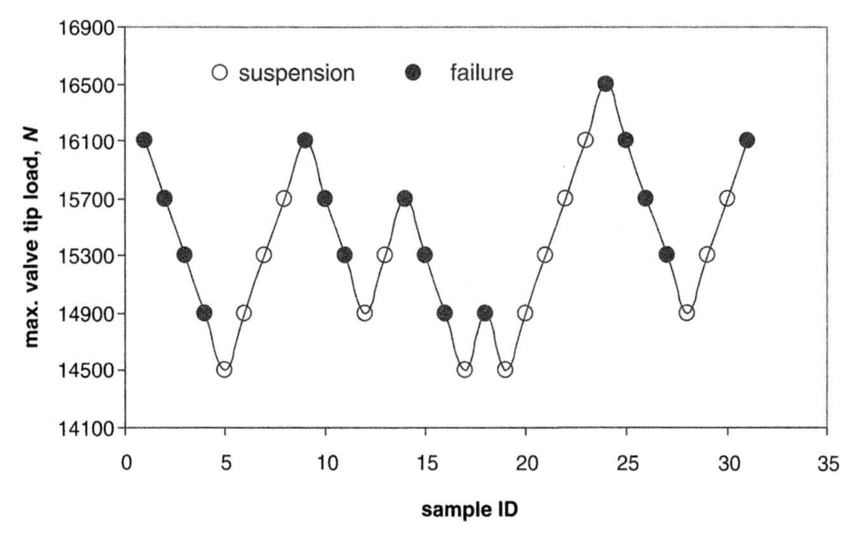

FIGURE 4.9 Staircase fatigue data for the rocker arm.

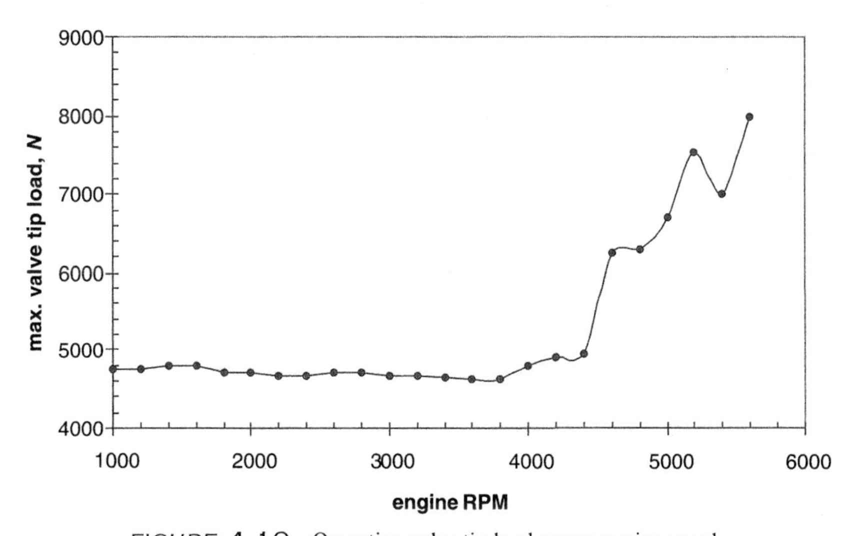

FIGURE 4.10 Operating valve tip load versus engine speed.

$$i = 2 \quad P_2 = 15,300\,N \quad n_{DM,2} = 4$$
$$i = 3 \quad P_3 = 15,700\,N \quad n_{DM,3} = 3$$
$$i = 4 \quad P_4 = 16,100\,N \quad n_{DM,4} = 1$$
$$i = 5 \quad P_5 = 16,500\,N \quad n_{DM,5} = 0$$
$$\sum n_{DM,i} = 15$$

(1) Solving for mean of the fatigue limit,

$$\mu_s = P_o + P_d\left(\frac{A_{DM}}{\sum n_{DM,i}} \pm \frac{1}{2}\right)$$

where $A_{DM} = \sum(i)(n_i)$, μ_s = mean fatigue strength at a prescribed life; P_o = lowest load level at which the less frequent event occurred; P_d = load increment; and $n_{DM,i}$ = number of the less frequent event at the numbered stress level, i.

Note that the plus sign is used if the less frequent event is survival, and the minus sign is used if the less frequent event is failure.

$$A_{DM} = \sum(i)(n_{DM,i}) = (0)(3) + (1)(4) + (2)(4) + (3)(3) + (4)(1) + (5)(0) = 25$$

$$\mu_s = P_o + P_d\left(\frac{A}{\sum n_{DM,i}} + \frac{1}{2}\right) = 14,500 + 400\left(\frac{25}{15} + \frac{1}{2}\right) = 15,400$$

(2) Solving for standard deviation of the fatigue limit,

$$B_{DM} = \sum(i^2)(n_{DM,i})$$

$$B_{DM} = (0^2)(3) + (1^2)(4) + (2^2)(4) + (3^2)(3) + (4^2)(1) + (5^2)(0) = 63$$

$$\frac{B_{DM}\sum n_{DM,i} - A_{DM}{}^2}{(\sum n_{DM,i})^2} = \frac{(63)(15) - 25^2}{15^2} = 1.42$$

$$\text{As } 1.42 > 0.3,\ \sigma_s = 1.62(P_d)\left(\frac{B_{DM}\sum n_{DM,i} - A_{DM}{}^2}{(\sum n_{DM,i})^2} + 0.029\right)$$

$$\sigma_s = 1.62(400)\left(\frac{(63)(15) - 25^2}{(15)^2} + 0.029\right) = 940$$

(3) Use the one-sided lower-bound tolerance limit method and $n = 15$ in Table 4.6 to determine the R90C90 fatigue strength at 10^7 cycles:

$$P_{R90C90} = \mu_s - K_{R90C90}\sigma_s = 15,400 - (1.866)(940) = 13,646$$

(4) Calculate the factor of safety to examine whether the rocker arm design meets the target:

$$\text{With } P_{\max} = 8000N$$

$$\text{Design factor} = \frac{P_{R90C90}}{P_{\max}} = \frac{13,646}{8,000} = 1.71 > \text{S.F.} = 1.3$$

4.3 ESTIMATED *S–N* CURVE OF A COMPONENT BASED ON ULTIMATE TENSILE STRENGTH

In the event that experimental *S–N* data are not available, methods for estimating the *S–N* behavior of a component becomes useful and crucial for the design process. This section focuses on the technique for estimating an *S–N* curve based on limited information. Materials presented here have drawn heavily from numerous references (Juvinall, 1967; Bannantine et al., 1990; Dowling, 1998; Stephens et al., 2001). Large amounts of *S–N* data have been historically generated based on fully reversed rotating bending testing on standard specimens. The standard test specimen is a round hourglass-shaped bar with a diameter of 7.6 mm (0.3 in.) and a mirror-polished surface. Two types of rotating bending test set-up are illustrated in Figure 4.11.

The *S–N* curve derived on the standard specimens under fully reversed bending loads can be constructed as a piecewise-continuous curve consisting of three distinct linear regions when plotted on log–log coordinates. As shown schematically in Figure 4.12, there are two inclined linear segments and one horizontal segment in a typical log–log *S–N* curve. The two inclined linear segments represent the low-cycle fatigue (LCF) and high-cycle fatigue (HCF) regions, and the horizontal asymptote represents the bending fatigue limit. The boundary between low- and high-cycle fatigue cannot be defined by a specific number of cycles. A rational approach to the difference is discussed in detail in Chapter 5.

For specimens made of steels, the fatigue strength values at 1, 10^3, 10^6 cycles define an *S–N* curve. These fatigue strength values will be referred to as S'_f, S_{1000}, and S_{be}, respectively. The slope of the *S–N* curve in the high-cycle fatigue region is denoted as b and expressed as follows:

$$b = \frac{\log S_{1000} - \log S_{be}}{\log 10^3 - \log 10^6} = -\frac{1}{3}\log\left(\frac{S_{1000}}{S_{be}}\right) \tag{4.3.1}$$

The inverse slope of the curve (slope factor) is denoted as k:

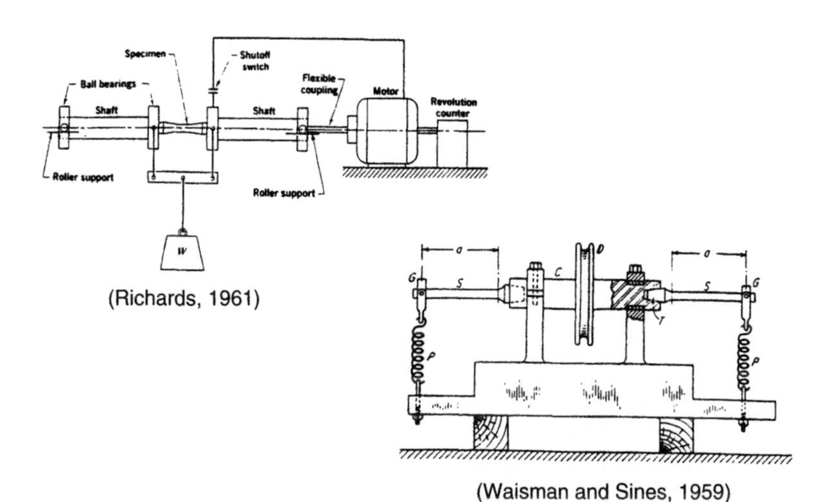

FIGURE 4.11 Rotating beam and rotating cantilever fatigue testing machines. Reproduced from Richards (1961) and Sines and Waisman (1959) with permission of the author.

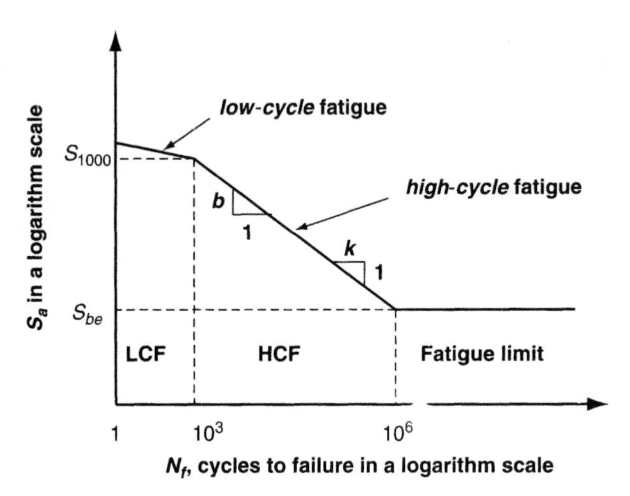

FIGURE 4.12 Schematic of an S–N curve for steels.

$$k = -\frac{1}{b} \tag{4.3.2}$$

With a given S–N data point (S_1, N_1), the slope b or the inverse slope k can be used to determine the number of equivalent damage cycles (N_2) for a stress amplitude (S_2) as follows:

$$N_2 = N_1 \left(\frac{S_1}{S_2}\right)^{-1/b} \quad \text{or} \quad N_2 = N_1 \left(\frac{S_1}{S_2}\right)^{k} \tag{4.3.3}$$

In general, if fatigue behavior is dominated by the crack propagation mechanism (e.g., welded joints or sharp notched components, etc.), the S–N curve often has a steep slope ($b \approx -0.3$; $k \approx 3$). If fatigue behavior is controlled by the crack initiation mode (e.g., smooth and blunt notched components), the S–N curve has a flatter slope ($b \approx -0.15$; $k \approx 7$). Equation 4.3.3 indicates that in the case of $b = -0.15$, an increase of the stress amplitude by 10% decreases the fatigue life by 53%.

Through many years of experience and testing, empirical relationships that relate fatigue strength data and the ultimate tensile strength, S_u, have been developed. These relationships are not scientifically based but are simple and useful engineering tools to estimate the fatigue lives of components in the high-cycle fatigue region. Thus, the estimated fatigue strength corresponding to the two fatigue lives (e.g., 10^3 cycles and 10^6 cycles) for the baseline test specimens are discussed. If a component or loading condition deviates from the standard test setup used to generate the aforementioned S–N data, it is necessary to modify the baseline S–N curve to account for the differences. Such adjustments are accomplished through the use of modifying factors. Figure 4.13 illustrates the effects of modifying factors on a baseline S–N curve.

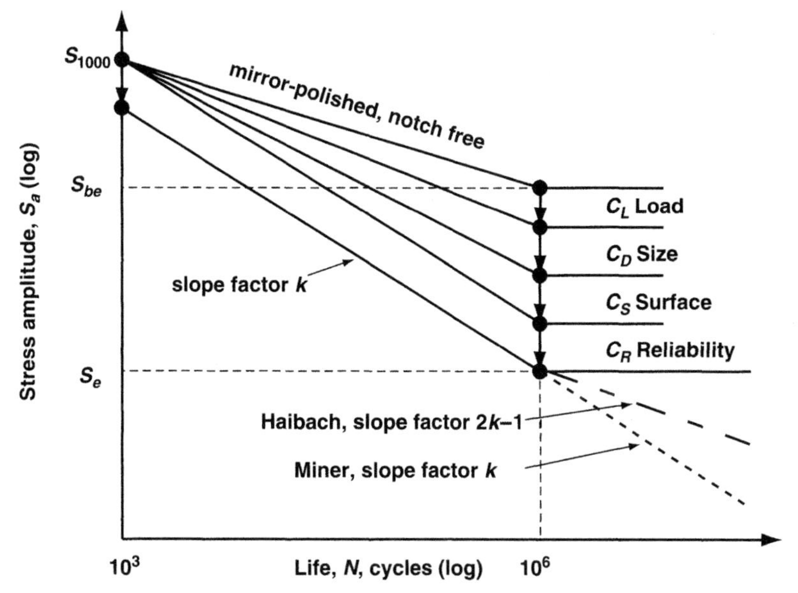

FIGURE 4.13 Modified S–N curves for smooth components made of steels.

4.3.1 METHODOLOGY

To generate a baseline S–N curve for a component, the ultimate strength of that given material must first be known or determined by some means. Once determined, the fatigue strengths at 10^3 cycles and at the fatigue limit life can be approximated using the ultimate strength. These strength values can be later modified to include the effects of other factors such as loading, surface finish, size, and reliability on the S–N curve.

4.3.2 ESTIMATED ULTIMATE TENSILE STRENGTH OF A STEEL

For low and medium strength of steels with low Brinell hardness (<500 BHN), the ultimate tensile strength of a material can be linearly approximated as

$$S_u(\text{MPa}) = 3.45\text{BHN} \tag{4.3.4}$$

or

$$S_u(\text{ksi}) = 0.5\text{BHN} \tag{4.3.5}$$

The ultimate tensile strength for cast iron can be estimated (Krause, 1969) as

$$S_u(\text{MPa}) = 1.58\text{BHN} - 86 \tag{4.3.6}$$

or

$$S_u(\text{ksi}) = 0.23\text{BHN} - 12.5 \tag{4.3.7}$$

4.3.3 ESTIMATED FATIGUE STRENGTH AT 10^3 CYCLES

The fatigue strength at 10^3 cycles (S_{1000}) depends on the reliability level and the type of loading. For example,

$$S_{1000,R} = S_{1000} \times C_R \tag{4.3.8}$$

where S_{1000} is the fatigue strength at 10^3 cycles based on the standard test specimens under different loading conditions and C_R is the modifying factor at a specified reliability level.

4.3.3.1 Baseline Fatigue Strength at 10^3 Cycles for Various Types of Loading

Empirically, the S_{1000} value is observed to be approximately 90% of ultimate tensile strength, S_u, for bending and 75% of the ultimate tensile strength for axial loading. For torsional loading, the S_{1000} value is estimated as 90% of the ultimate shear strength, S_{us}. Note that the ultimate shear strength is material dependent and is roughly approximated as 80%, 70%, and 130% of the ultimate tensile strength for steel, nonferrous metals, and cast irons, respectively. Estimates of S_{1000} are summarized in Table 4.8.

4.3.3.2 Modifying Factor for Reliability, C_R, at the Fatigue Strength at 10^3 Cycles

If the statistical scatter of fatigue data is considered, the fatigue strength of a component would be modified from the median baseline *S–N* data for a specified reliability level. If fatigue test data are not available, a rigorous statistical analysis cannot be performed to account for the random nature of fatigue life. In the absence of a rigorous statistical analysis, suggested values of modifying factors C_R for various reliability levels are given in Table 4.9. The tabulated values for C_R are derived in Appendix 4-A. Although presented as a modifying factor for fatigue strengths at 10^3 cycles, the reliability factors listed in Table 4.9 can be used for estimating the fatigue strength at the fatigue limit as well.

TABLE 4.8 Estimates of S_{1000}

Type of Material	Type of Loading	S_{1000}
All	Bending	$0.9 \times S_u$
All	Axial	$0.75 \times S_u$
Steel	Torsion	$0.9 \times S_{us} = 0.72 \times S_u$
Nonferrous	Torsion	$0.9 \times S_{us} = 0.63 \times S_u$
Cast iron	Torsion	$0.9 \times S_{us} = 1.17 \times S_u$

TABLE 4.9 Reliability Factors, C_R

Reliability	C_R
0.50	1.000
0.90	0.897
0.95	0.868
0.99	0.814
0.999	0.753
0.9999	0.702
0.99999	0.659
0.999999	0.620

4.3.4 ESTIMATED FATIGUE LIMIT

In view of the asymptotic character of the *S–N* curve, the fatigue limit can be defined as the constant stress amplitude at which the fatigue life becomes infinite or fatigue failures do not occur. However, in some cases where fatigue testing must be terminated at a specific large number of cycles (e.g., 10^7 cycles), this nonfailure stress amplitude is often referred to as the endurance limit. Even the endurance limit needs not be a fatigue limit, but from the engineering point of view, the term of a fatigue limit is commonly used to refer to either case.

A fatigue limit can be interpreted from the physical perspective of the fatigue damage phenomenon under constant amplitude loading. Because of operating cyclic stresses, a microcrack will nucleate within a grain of material and grow to the size of about the order of a grain until a grain boundary barrier impedes its growth. If the grain barrier is not strong enough, the microcrack will eventually propagate to a macrocrack and may lead to final failure. If the grain barrier is very strong, the microcrack will be arrested and become a nonpropagating crack. The minimum stress amplitude to overcome the crack growth barrier for further crack propagation is referred to as the fatigue limit (McGreevy and Socie, 1999; Murakami and Nagata, 2002).

The fatigue limit might be negatively influenced by other factors such as periodic overloads, elevated temperatures, or corrosion. When the Miner rule is applied in variable-amplitude loading, the stress cycles with amplitudes below the fatigue limit could become damaging if some of the subsequent stress amplitudes exceed the original fatigue limit. It is believed that the increase in crack driving force due to the periodic overloads will overcome the original grain barrier and help the crack propagate until failure. As shown in Figure 4.13, two methods, the Miner rule and the Miner–Haibach model (Haibach, 1970), were proposed to include the effect of periodic overloads on the stress cycle behavior below the original fatigue limit. The Miner rule extends the *S–N* curve with the same slope factor k to approach zero-stress amplitude, whereas the Miner–Haibach rule extends the original

S–N curve below the fatigue limit to the zero-stress amplitude with a flatter slope factor $2k - 1$. Stanzl et al. (1986) found good agreement between measured and calculated results when using the Miner–Haibach model.

Historically, engineers have relied on macroscopic properties such as hardness and tensile strength to estimate fatigue limits. A general relationship between the fatigue limit and the ultimate strength of a material has been observed. As shown in Figure 4.14, the bending fatigue limit S_{be} at 10^6 cycles for wrought steels can be estimated as 0.5 times the ultimate tensile strength S_u for ultimate tensile strengths less than 1400 MPa and as 700 MPa for ultimate tensile strengths greater than 1400 MPa. The value for the ratio of the bending fatigue limit and the ultimate tensile strength in the linear range varies from 0.25 to 0.6, depending on the microstructure of the steel (e.g., grain size, inclusions, porosity, and graphite nodules). Based on Juvinall (1967), Table 4.10 shows the effect of microstructure of steels on the baseline bending fatigue limit.

The relationship between the fatigue limit and ultimate strength of steels can be examined from the microscopic perspective. For mild- and high-strength steels in which the flaws and the inhomogeneity (e.g., inclusions, carbides, graphite nodules, and porosity) are approximately smaller than the size of a grain, cracks nucleate within a grain and are arrested by the grain barrier that is approximately proportional to the hardness of a material. Thus, the fatigue limit initially increases in a nearly linear fashion with ultimate strength. Once a critical ultimate strength value is reached, the fatigue limit either remains constant or decreases with ultimate strength. This phenomenon is attributed to flaws in the material due to processing, e.g., grinding cracks, surface finish, and severe quench cracks. For example,

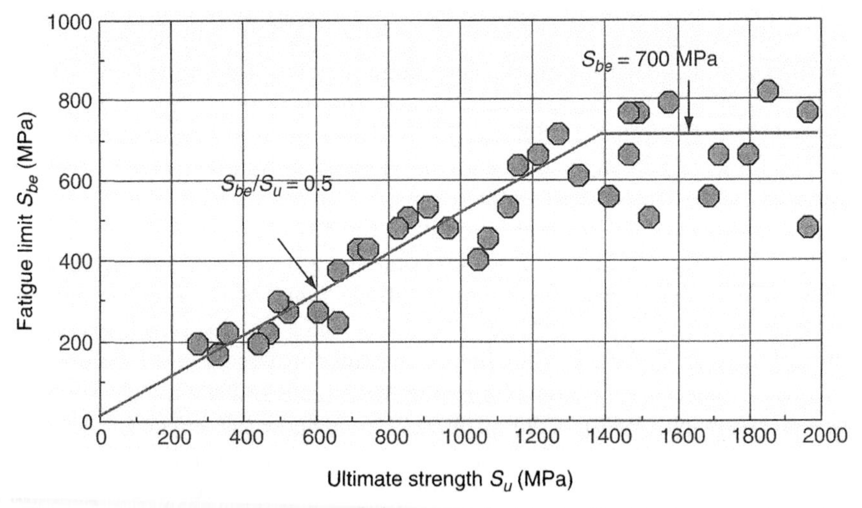

FIGURE 4.14 Schematic of the fatigue strength data at 10^6 cycles for wrought steels.

quenching of high-strength steel to produce superior wear performance can introduce microcracking and promote intergranular crack nucleation that are detrimental to fatigue resistance.

Expanding this concept to materials other than steel, the bending fatigue limit of cast iron occurs at 5×10^7 cycles and is estimated to be 0.4 times the ultimate tensile strength. Even though aluminum alloys do not have a true fatigue limit, a common practice is to take the fatigue strength at 5×10^8 cycles as the endurance limit, the pseudo fatigue limit value. As shown in Figure 4.15, the bending fatigue limit of wrought aluminum is 0.4 times the ultimate tensile strength with the ultimate tensile strength less than 336 MPa and is 130 MPa otherwise. Also, bending fatigue limits for permanent mold (PM) cast aluminum and sand cast aluminum are 80 MPa and 55 MPa, respectively. Table 4.10 summarizes the estimates of baseline bending fatigue limit for various materials.

The fatigue limit S_e can be estimated by modifying the bending fatigue limit (S_{be}) with the four factors for the type of loading (C_L), surface finish (C_S), size (C_D), and reliability level (C_R):

$$S_{e,R} = S_{be} \times C_L \times C_S \times C_D \times C_R \tag{4.3.9}$$

Note that these modifying factors are empirically based and usually range from 0.0 to 1.0. The modifying factors are introduced in the following sections.

TABLE 4.10 Estimates of Baseline Bending Fatigue Limits for Various Materials (From Juvinall, 1967)

	S_{be}	@ Cycles	Comments
Type of Material			
Microstructure of Steels			
Steel - Ferrite	$0.58 \times S_u$	10^6	
Steel – Ferrite + Pearlite	$0.38 \times S_u$	10^6	
Steel - Pearlite	$0.38 \times S_u$	10^6	
Steel – Untempered martensite	$0.26 \times S_u$	10^6	
Steel – Highly tempered Martensite	$0.55 \times S_u$	10^6	
Steel – Highly Tempered Martensite + Tempered Bainite	$0.5 \times S_u$	10^6	
Steel – Tempered Bainite	$0.5 \times S_u$	10^6	
Steel - Austenite	$0.37 \times S_u$	10^6	
Type of Material			
Wrought Steels	$0.5 \times S_u$	10^6	$S_u < 1400$ MPa
Wrought Steels	700 MPa	10^6	$S_u \geq 1400$ MPa
Cast iron	$0.4 \times S_u$	5×10^7	-
Aluminum alloys	$0.4 \times S_u$	5×10^8	$S_u < 336$ MPa
Aluminum alloys	130 MPa	5×10^8	$S_u \geq 336$ MPa
PM cast aluminum	80 MPa	5×10^8	-
Sand cast aluminum	55 MPa	5×10^8	-

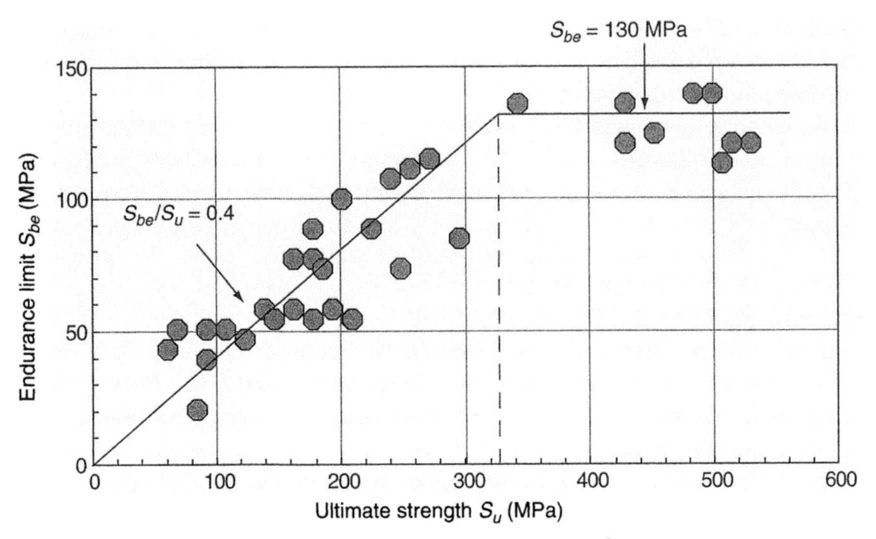

FIGURE 4.15 Schematic of fatigue strength data at 5×10^8 cycles for wrought aluminum alloys.

4.3.4.1 Modifying Factor for the Type of Loading (C_L) at the Fatigue Limit

Historically, baseline *S–N* data have been generated from fully reversed bending stresses. However, real components (e.g., powertrain and driveline shafts in automobiles) are often subjected to other loading conditions. This requires that the fatigue limit for bending be modified for loading conditions other than bending. In general, the loading factor is determined by considering the effects of the stress gradient and the type of stresses (normal or shear stresses).

Test results show that C_L varies from 0.7 to 0.9 for unnotched components under axial loading. The difference between axial loading and bending is the result of the different stress gradients for each loading condition. Under the circumstance of identical maximum nominal stresses, the fatigue strength of a component in axial loading is smaller than that in bending. In axial loading, a large percentage of the cross-sectional material is loaded with a uniform high stress level and it is more likely to initiate a crack in a correspondingly larger volume of the material. The C_L value is recommended as 0.9 for pure axial loading without bending and 0.7 for loads with slight bending induced by misalignment.

Experimental results for ductile steels indicate C_L ranges from 0.5 to 0.6 for an unnotched component under torsional loading. Based on the von Mises theory, a C_L of 0.58 is recommended for ductile materials. However, a C_L of 0.8 is suggested for use in cast iron. The effects of C_L are summarized in Table 4.11. Generalized *S–N* curves that incorporate these modifying factors for steels under different loading conditions are illustrated in Figure 4.16.

TABLE 4.11 Load Factors, C_L

Type of Loading	C_L	Comments
Pure axial loading	0.9	
Axial loading (with slight bending)	0.7	
Bending	1.0	
Torsional	0.58	For steels
Torsional	0.8	For cast iron

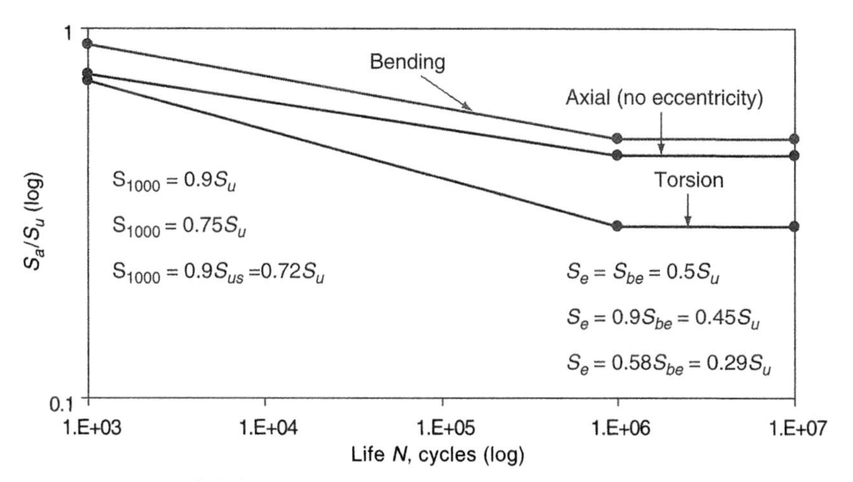

FIGURE 4.16 Generalized S–N curves for different loading—steels.

4.3.4.2 Modifying Factor for the Surface Finishing Factor (C_S) at the Fatigue Limit

Because fatigue cracks initiate predominantly at the free surface of a material, the surface condition of a test sample becomes critical. The surface condition can be characterized by considering two factors: (1) notch-like surface irregularities or roughness and (2) residual stress in the surface layer. Irregularities along the surface act as stress concentrations and result in crack initiation at the surface. Figure 4.17 is an empirical chart used to determine the surface finishing modifying factor for steels when measurement of the surface roughness is known.

Surface treatments are usually used to induce residual surface stresses. Some operations such as chemical process (decarburization, carburization, and nitriding), electroplating, and thermal processing can alter the physical properties of the surface layer of the materials. Decarburization processes (e.g., forging and hot rolling) deplete carbon from the surface layer and induce tensile residual stresses. These tensile stresses result in fatigue strength reduction. Carburization and nitriding processes force carbon and nitrogen

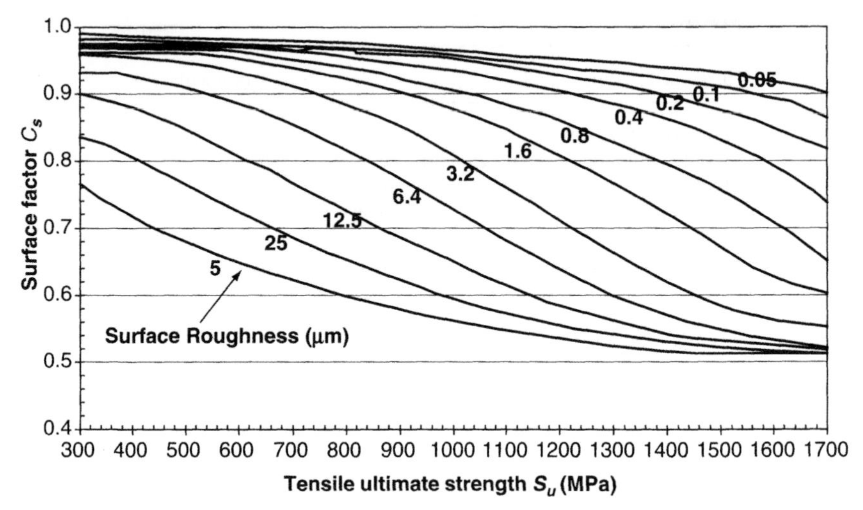

FIGURE 4.17 Qualitative description of C_S surface finish factor. Adapted from Johnson (1973) with permission of Penton Media, Inc.

into the surface layer and strengthen the surface layer and produce compressive residual stress. These compressive residual stresses increase the fatigue strength of materials.

Mechanical processes can introduce residual stresses into a material. For example, tensile residual stresses can be generated from machining and compressive residual stresses may be introduced in a material by peening and cold working processing. Figure 4.18 is used to estimate the surface-modifying factors based on manufacturing process for steels. In general, surface finish is more critical for high-strength steel and at high-cycle fatigue lives where crack initiation dominates the fatigue life. At short lives where crack propagation dominates the fatigue life, the effect of surface finish on fatigue strength is minimal.

4.3.4.3 Modifying Factor for the Specimen Size (C_D) at the Fatigue Limit

The size effect on fatigue strength can be explained by the critical volume theory of Kuguel (1961), according to which fatigue damage can be related to the volume of the material subjected to the critical stress range taken as 95 to 100% of the maximum applied stress. In the same maximum bending stress, as the diameter of a round specimen increases, a larger volume of the material is subjected to the critical stress range. This results in higher damage and smaller fatigue strength than those of a round specimen with a smaller diameter. The quantitative description of the size factor for a round component under bending and torsion is illustrated in Figure 4.19 and expressed in the following forms:

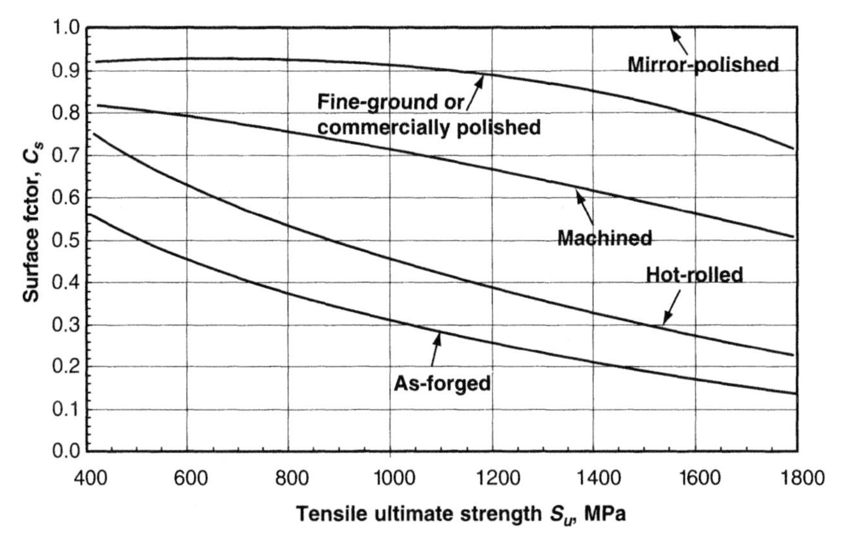

FIGURE 4.18 Qualitative description of C_s surface finish factor. Adapted from Juvinall and Marshek (2000) Copyright © 2000 John Wiley & Sons, Inc., and used with permission of John Wiley & Sons, Inc.

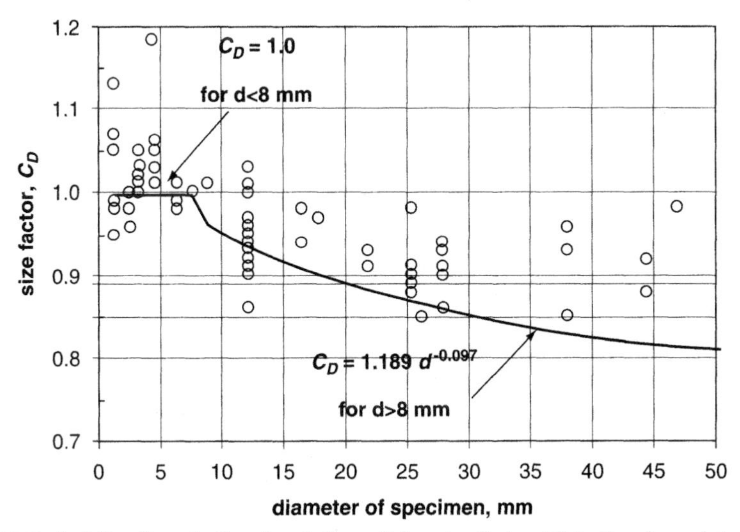

FIGURE 4.19 Quantitative description of the size factor (C_D). Based on data from Heywood (1962) and Horger (1965).

$$C_D = 1.0 \quad \text{for } d < 8\,\text{mm } (= 0.3\,\text{in.}) \tag{4.3.10}$$

$$C_D = 1.189 \times d^{-0.097} \quad \text{for } 8\,\text{mm} < d < 250\,\text{mm} \tag{4.3.11}$$

or

$$C_D = 0.869 \times d^{-0.097} \quad \text{for } 0.3\,\text{in.} < d < 10\,\text{in.} \tag{4.3.12}$$

where d is the diameter of the component.

For a component subjected to axial loading, the macroscopic stress gradient does not exist, and the critical axial stress is the same throughout the cross-section of the material, regardless the size of the diameters. It may be supposed that the fatigue strength for a larger component is lower than that of a smaller one because larger components have more volume of material loaded with the same critical axial stress. However, fatigue tests have indicated that there is little size effect for axially loaded components. Thus, a modifying factor ($C_D = 1.0$) is recommended.

For situations where specimens do not have a round cross-section, an effective diameter d_e can be estimated by equating the area of material stressed at and above 95% of the maximum stress to the same area in the rotating beam specimen. The effective stressed area for a round cross section under rotating bending is

$$A_{0.95S} = \frac{\pi}{4}\left[d_e^2 - (0.95d_e)^2\right] = 0.0766d_e^2 \tag{4.3.13}$$

A rectangular section of dimensions $w \times t$ under fully reversed bending has the effective stressed area

$$A_{0.95S} = 0.05w \times t \tag{4.3.14}$$

Setting Equations 4.3.13 and 4.3.14 equal to each other leads to the solution for the effective diameter:

$$d_e = \sqrt{0.65w \times t} \tag{4.3.15}$$

Example 4.7. Given a fine-ground unnotched steel bar with a diameter of 14 mm and an ultimate strength of 800 MPa, estimate the allowable axial stress amplitude for lives of 10^3, 10^4, 10^5, and 10^6 cycles.

Solution. For axial loading, fatigue strength at 1000 cycles is estimated to be

$$S_{1000} = 0.75S_u = 0.75(800) = 600\,\text{MPa}$$

For axial loading, the fatigue limit of the bar is estimated to be

$$S_e = S_{be}C_LC_DC_SC_R$$

where the bending fatigue limit for wrought steels with $S_u = 800\,\text{MPa} < 1400\,\text{MPa}$ is

$$S_{be} = 0.5S_u = 0.5(800) = 400\,\text{MPa}$$

The load factor for pure axial loading without bending is $C_L = 0.9$. Because tests indicate little size effect for axially loaded parts, the size factor is $C_D = 1.0$. Using the graph of surface finish factor versus ultimate strength, the curve representing fine ground materials gives for a surface finish factor of $C_S = 0.9$ for steels with $S_u = 800$ MPa. Because no reliability requirement is stated, it is assumed that the median S–N curve is adequate. Therefore, the reliability factor is taken as $C_R = 1.000$.

The resulting fatigue limit at 10^6 cycles is

$$S_e = (400\,\text{MPa})(0.9)(1.0)(0.9)(1.000) = 324\,\text{MPa}$$

After both S_{1000} and S_e have been determined, they can be graphed to estimate the S–N curve as shown in Figure 4.20. Also, the fatigue strength at a specific fatigue life can be determined by the following equation:

$$\frac{\log(10^3) - \log(10^6)}{\log(S_{1000}) - \log(S_e)} = \frac{\log(10^3) - \log(N_f)}{\log(S_{1000}) - \log(S_N)}$$

This leads to fatigue strengths at 10^4 and 10^5 cycles of 489 and 403 MPa, respectively.

Example 4.8. Generate the S–N curve with 90% of reliability for a forged steel shaft under torsional loading. The shaft has a diameter of 20 mm and an ultimate strength S_u of 1000 MPa.

Solution. For torsional loading, the fatigue strength at 1000 cycles is estimated to be

$$S_{1000} = 0.9 S_{us} C_R$$

where $S_{us} = 0.8\,S_u$ for steel and $C_R = 0.897$ for 90% reliability.

$$S_{1000} = 0.9(0.8)(0.897)(1000) = 646\,\text{MPa}$$

For torsional stress the fatigue limit is estimated to be

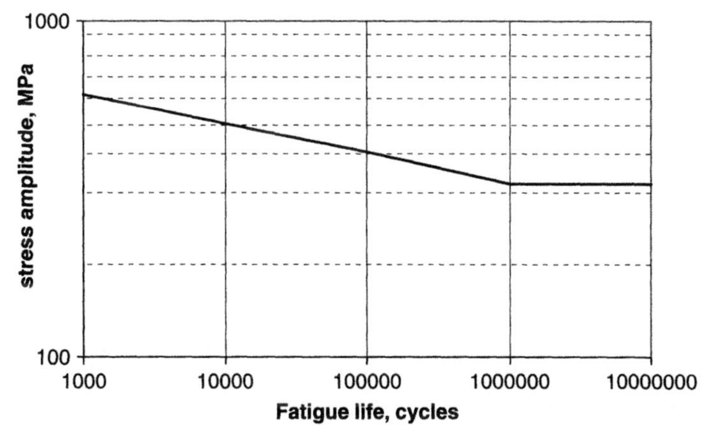

FIGURE 4.20 Estimated S–N curve for fine-ground unnotched steel bar under axial loading.

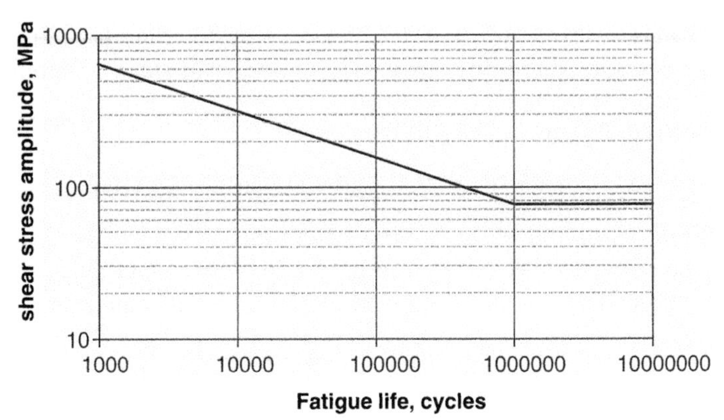

FIGURE 4.21 Estimated *S–N* curve with 90% reliability for forged unnotched steel shafts under torsional loading.

$$S_e = S_{be} C_L C_D C_S C_R$$

where the bending fatigue limit for wrought steels with $S_u = 1000\,\text{MPa} < 1400\,\text{MPa}$ is

$$S_{be} = 0.5\,S_u = 0.5(1000) = 500\,\text{MPa}$$

The load factor is $C_L = 0.58$ for ductile steels in torsion.

The size factor for $d = 20\,\text{mm} > 8.0\,\text{mm}$ is

$$C_D = 1.189 \times (d)^{-0.097} = 1.189 \times (20)^{-0.097} = 0.89$$

Using the curve on the surface finish factor graph for forged steel with $S_u = 1000\,\text{MPa}$, the surface factor is $C_S = 0.33$. The reliability factor with 90% reliability is $C_R = 0.897$.

Therefore, the fatigue limit of the shaft under torsional loading is

$$S_e = (500\,\text{MPa})(0.58)(0.89)(0.33)(0.897) = 76.4\,\text{MPa}$$

After both S_{1000} and S_e have been determined, they can be graphed to estimate the *S–N* curve as shown in Figure 4.21.

4.4 NOTCH EFFECT

Fatigue failure of a component typically occurs at a notch on a surface where the stress level increases due to the stress concentration effect. The term *notch* is defined as a geometric discontinuity that may be introduced either by design, such as a hole, or by the manufacturing process in the form of material and fabrication defects such as inclusions, weld defects, casting defects, or machining marks. For a component with a surface notch, the maximum elastic notch stress (σ^e) can be determined by the product of a nominal stress (S) and the elastic stress concentration factor (K_t):

$$\sigma^e = S \times K_t \qquad (4.4.1)$$

The maximum elastic notch stress can be calculated from an elastic finite element analysis and is sometimes referred to as the pseudo-stress if the material at a notch is actually inelastic. Because notch stresses and strains are controlled by net section material behavior, the nominal stress for determination of K_t is defined by an engineering stress formula based on basic elasticity theory and the net section properties that do not consider the presence of the notch.

The elastic stress concentration factor is a function of the notch geometry and the type of loading. For the cases in which the component geometry and loading conditions are relatively simple and the nominal stress can be easily defined, elastic stress concentration factors are often available in the references (Peterson, 1974; Pilkey, 1997). However, because of complexities of the geometry and loads in most real components, the value of σ^e can be directly obtained from the elastic finite element analysis.

4.4.1 NOTCH EFFECT AT THE FATIGUE LIMIT

Theoretically speaking, based on the same high-cycle fatigue life (i.e., the same maximum stress creating a microcrack in the notched and the unnotched members), the nominal strength of a smooth component should be higher than that of a notched component by a factor of K_t. However, tests indicate that at the fatigue limit, the presence of a notch on a component under cycling nominal stresses reduces the fatigue strength of the smooth component by a factor K_f and not the factor K_t. Tryon and Dey (2003) presented a study revealing the effect of fatigue strength reduction for Ti-6Al-4V in the high-cycle fatigue region in Figure 4.22. The K_f factor is called the fatigue strength reduction factor or the fatigue notch factor, and is usually defined as the ratio of the nominal fatigue limits for smooth and notched test samples or components, i.e.,

$$K_f = \frac{\text{unnotched fatigue limit}}{\text{notched fatigue limit}} \qquad (4.4.2)$$

In general, K_f is equal to or less than K_t. The difference between K_t and K_f increases with a decrease in both the notch root radius and ultimate tensile strength. This difference can be explained by either the local cyclic yielding behavior or the theory of the stress field intensity (Yao, 1993; Qylafku et al., 1999; Adib and Pluvinage, 2003).

It is believed that cyclic yielding at a notch root in materials reduces the peak stress from the value predicted by K_t. Also, based on the concept of the stress field intensity, it indicates that the fatigue strength of a notched component depends on the average stress in a local damage zone, rather than the peak notch stress. The average stress is associated with the stress

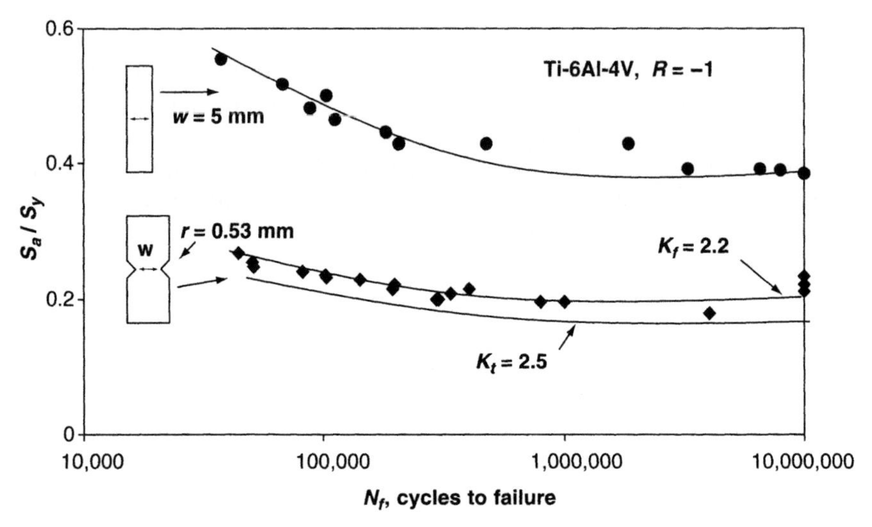

FIGURE 4.22 Effect of a notch on *S–N* behavior. Adapted from Tryon and Dey (2003).

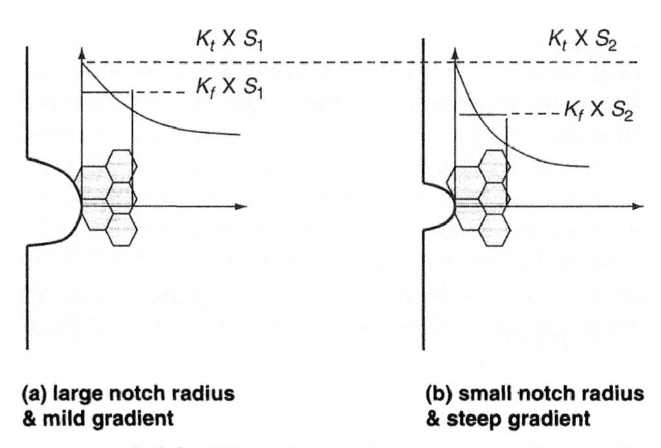

FIGURE 4.23 Effect of notch size and stress gradient on K_f.

distribution and the local damage volume at the notch. Figure 4.23 schematically shows two notched components with the same peak stress and material for which the damage zones are identical. As the notch radius decreases, the stress gradient becomes steeper, resulting in a lower average stress level and K_f. Figure 4.24 illustrates another example of the same notched components made of steel with different ultimate strength values. Because the damage zone for high-strength steel is usually smaller than that for mild-strength steel, a larger damage zone decreases the average stress value. Thus, the value of K_f is lower for the material with lower ultimate strength.

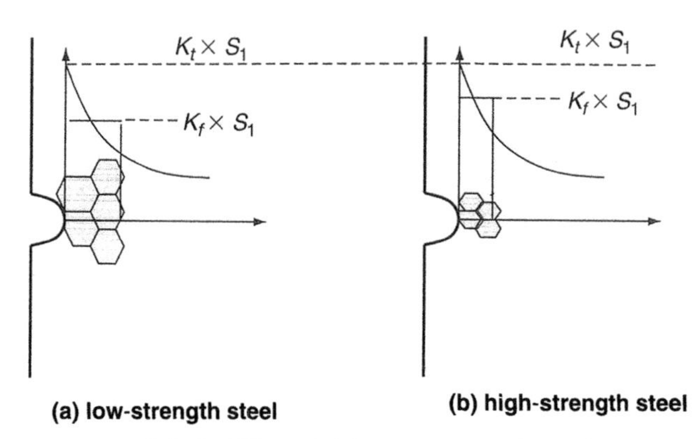

FIGURE 4.24 Effect of ultimate strength of materials on K_f.

For engineering applications, the fatigue strength reduction factor can be empirically related to the elastic stress concentration factor by a notch sensitivity factor q:

$$q = \frac{K_f - 1}{K_t - 1} \quad 0 \leq q \leq 1 \tag{4.4.3}$$

Equation 4.4.3 can be written in the following practical formula for K_f:

$$K_f = 1 + (K_t - 1) \times q \tag{4.4.4}$$

If $q = 1$, $K_f = K_t$ and the material is considered to be fully notch sensitivity; on the other hand, if $q = 0$, $K_f = 1.0$ and the material is not considered to be notch sensitivity (the so-called notch blunting effect).

Peterson (1959) assumed that fatigue damage occurs when a point stress at a critical distance (a_p) away from the notch root is equal to the fatigue strength of a smooth component. Assuming the stress near the notch drops linearly, Peterson obtained the following empirical equation for q:

$$q = \frac{1}{1 + \frac{a_P}{r}} \tag{4.4.5}$$

where r is the notch root radius and a_P is a material constant related to grain size and loading. A plot by Peterson (1959) is provided in Figure 4.25 to determine the notch sensitivity for high- and low-strength steels. For relatively high-strength steels ($S_u > 560\,\text{MPa}$), a_P can be approximately related to the ultimate tensile strength S_u by the following expressions:

$$a_p(\text{mm}) = 0.0254 \times \left(\frac{2079}{S_u(\text{MPa})}\right)^{1.8} \quad \text{for axial and bending} \tag{4.4.6}$$

$$a_p(\text{mm}) = 0.01524 \times \left(\frac{2079}{S_u(\text{MPa})}\right)^{1.8} \quad \text{for torsion} \tag{4.4.7}$$

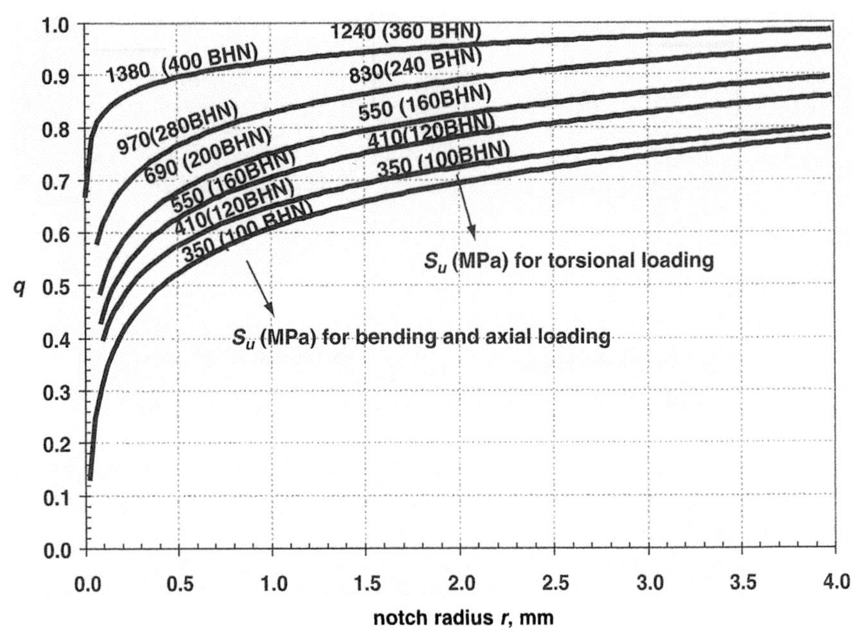

FIGURE 4.25 Peterson's notch sensitivity curves for steels. Adapted from Sines and Waisman (1959) with permission of the author.

or

$$a_p(inches) = 0.001 \times \left(\frac{300}{S_u(\text{ksi})}\right)^{1.8} \quad \text{for axial and bending} \tag{4.4.8}$$

$$a_p(inches) = 0.0006 \times \left(\frac{300}{S_u(\text{ksi})}\right)^{1.8} \quad \text{for torsion} \tag{4.4.9}$$

Assuming that the fatigue failure occurs if the average stress over a length (a_N) from the notch root is equal to the fatigue limit of a smooth component, Neuber (1946) proposed the following empirical equation for q:

$$q = \frac{1}{1 + \sqrt{\frac{a_N}{r}}} \tag{4.4.10}$$

where a_N is the Neuber's material constant related to the grain size. Figure 4.26 shows the plots of aluminum alloys for determining Neuber's material constant $\sqrt{a_N}$ in $\sqrt{\text{mm}}$ versus the ultimate tensile strength.

A uniform fine-grained material is very sensitive to the presence of a notch, whereas gray cast iron has a low sensitivity factor because the graphite flakes act as internal notches and significantly reduce the effect of external notches. Based on intrinsic defects, Heywood (1962) obtained the following empirical equation of estimating K_f for cast irons:

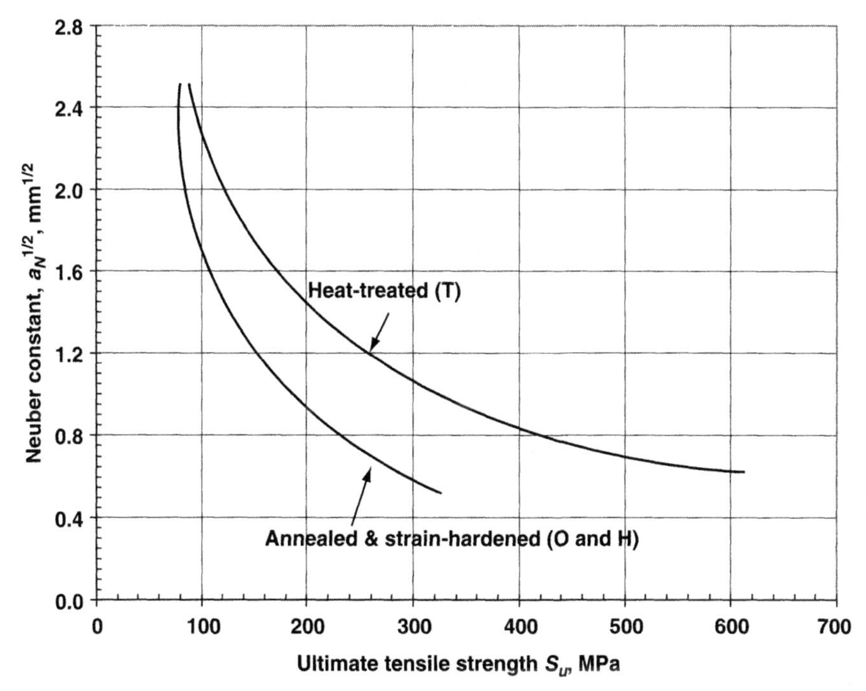

FIGURE 4.26 Neuber's notch sensitivity curves for aluminum alloys. Adapted from Kuhn and Figge (1962).

$$K_f = \frac{K_t}{1 + 2\frac{\sqrt{a'}}{\sqrt{r}}\left(\frac{K_t - 1}{K_t}\right)} \tag{4.4.11}$$

where a' corresponds to the length of equivalent material flaws. Note that for cast iron with flake graphite, $\sqrt{a'} = 0.605$, for cast iron with spheroid graphite, $\sqrt{a'} = 173.6/S_u$, and for magnesium alloys, $\sqrt{a'} = 0.0756$, where a' and r are in mm and S_u in MPa.

Instead of using the notch root radius to account for the stress gradient effects on the fatigue strength reduction, Siebel and Stieler (1955) introduced a new parameter, the relative stress gradient (RSG), defined by

$$\mathrm{RSG} = \frac{1}{\sigma^e(x)}\left(\frac{d\sigma^e(x)}{dx}\right)_{x=0} \tag{4.4.12}$$

where x is the normal distance from the notch root and $\sigma^e(x)$ is the theoretically calculated elastic stress distribution shown in Figure 4.27. By testing the fatigue strength of smooth and notched components at 2×10^7 cycles, they generated a series of empirical curves relating the values of K_t/K_f to RSG for various materials. These curves are shown in Figure 4.28 and can be expressed by the following formula:

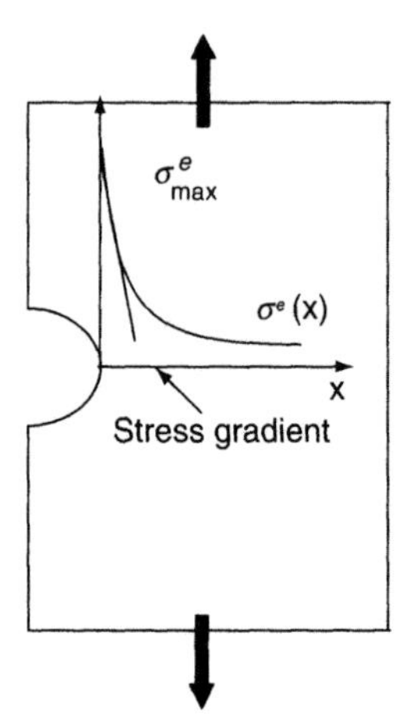

FIGURE 4.27 Elastic stress distribution and the stress gradient at a notch root.

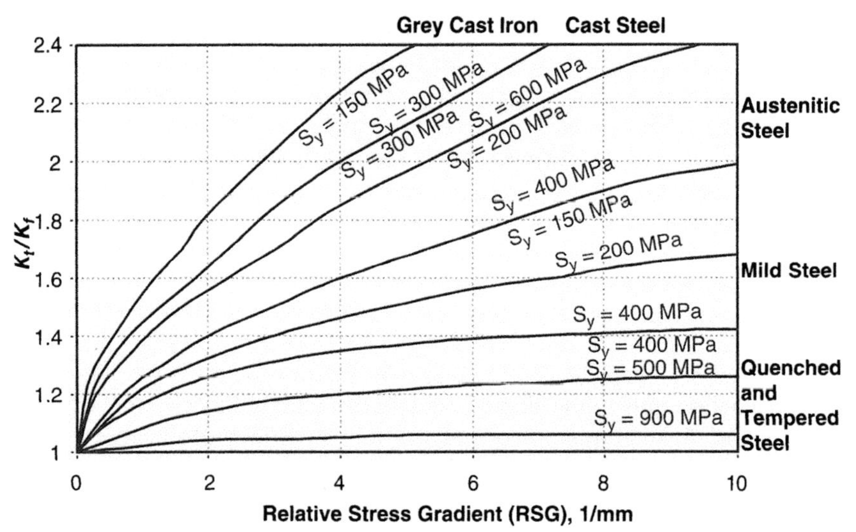

FIGURE 4.28 Relative stress gradient effect on K_t/K_f. Adapted from Siebel and Stieler (1995).

$$\frac{K_t}{K_f} = 1 + \sqrt{C_{SS} \times \text{RSG}} \tag{4.4.13}$$

where C_{SS} is a material constant dependent on the yield stress of a material (S_y).

RSG values was calculated by the theory of elasticity and presented by Siebel and Stieler for various notched members:

1. A notch plate with a notch radius (r) and a net cross-sectional width (w):
   a) Axial loading:

$$\mathrm{RSG} = \frac{2}{r} \tag{4.4.14}$$

b) Bending:

$$\mathrm{RSG} = \frac{2}{r} + \frac{2}{w} \tag{4.4.15}$$

2. A grooved shaft with a notch radius (r) and a minimum diameter (d)
   a) Axial loading:

$$\mathrm{RSG} = \frac{2}{r} \tag{4.4.16}$$

b) Bending:

$$\mathrm{RSG} = \frac{2}{r} + \frac{2}{d} \tag{4.4.17}$$

c) Torsion:

$$\mathrm{RSG} = \frac{1}{r} + \frac{2}{D+d} \tag{4.4.18}$$

3. A shoulder shaft with a notch radius (r) and two different diameters (D and d):
   a) Axial loading:

$$\mathrm{RSG} = \frac{2}{r} \tag{4.4.19}$$

b) Bending:

$$\mathrm{RSG} = \frac{2}{r} + \frac{4}{D+d} \tag{4.4.20}$$

c) Torsion:

$$\mathrm{RSG} = \frac{1}{r} + \frac{4}{D+d} \tag{4.4.21}$$

Despite being basically empirical, these formulas for K_f have found wide application and are still recommended in many textbooks and handbooks of fatigue design.

4.4.2 NOTCH EFFECT AT INTERMEDIATE AND SHORT LIVES

In the low and intermediate fatigue life regions at which larger local yielding occurs at a notch, the notch may be less sensitive to that predicted by K_f at the fatigue limit. Figure 4.29 shows the general trends of the fatigue strength reduction factor versus the cycles to failure. The fatigue notch sensitivity factor at 1000 cycles is empirically defined as follows:

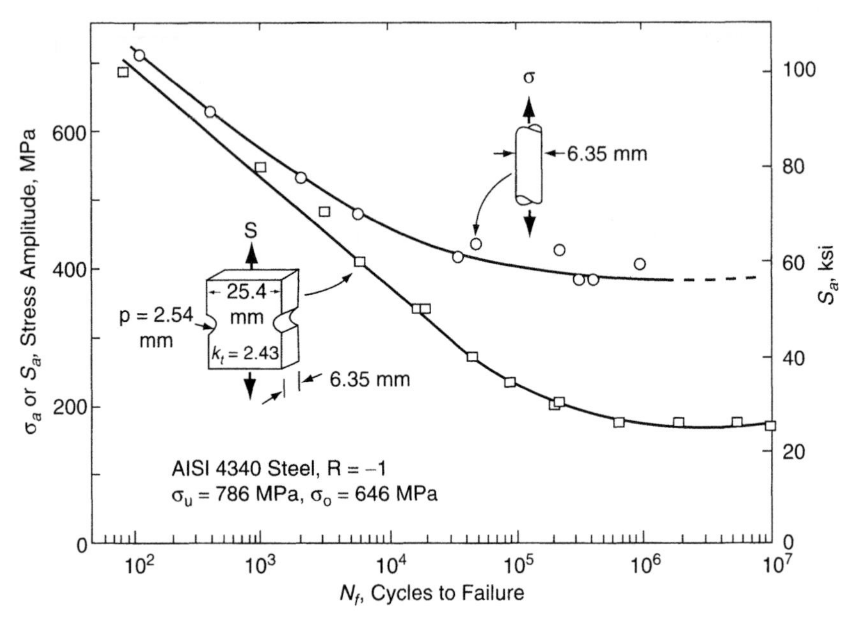

FIGURE 4.29 Notch effects at intermediate and short lives. Reprinted by permission of Pearson Education, Inc., Upper Saddle River, NJ, from *Mechanical Behavior of Materials*, 2nd edition by Dowling (1998).

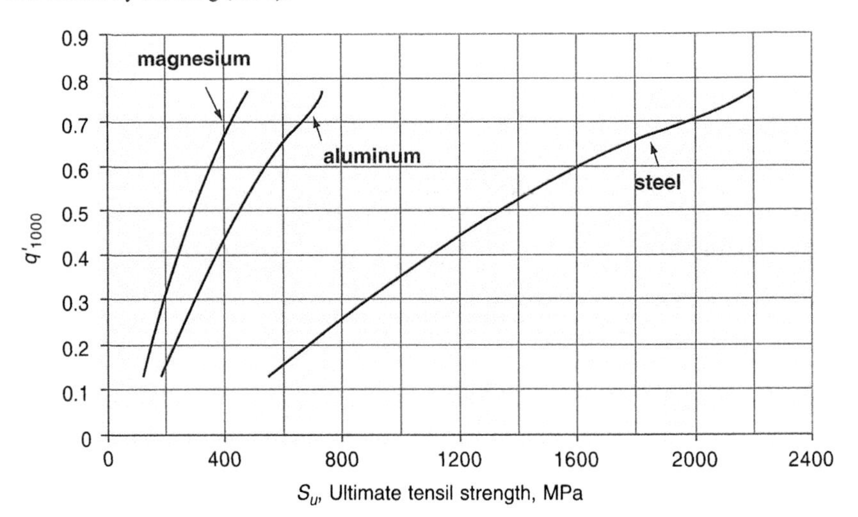

FIGURE 4.30 Fatigue notch factor K_f' at $N_f = 1000$ cycles. Based on data from Heywood, 1962.

$$q'_{1000} = \frac{K_f' - 1}{K_f - 1} \tag{4.4.22}$$

where K_f' is the fatigue strength reduction factor at 1000 cycles, and K_f is the fatigue strength reduction factor at the fatigue limit. Figure 4.30 shows the empirical curves between q'_{1000} and S_u for materials such as steel, aluminum, and magnesium.

4.4.3 ESTIMATE OF FATIGUE LIFE FOR A NOTCHED COMPONENT

There are two ways of predicting fatigue life for a notched component. One is to calculate the notch stress by multiplying the nominal stress with an assumed fatigue strength reduction factor and to enter the calculated notch stress to the *S–N* curve of a smooth component for life estimates. Based on the relationship between K'_f at 10^3 cycles and K_f at 10^6 cycles, this process requires numerous iterations for adjustment of the fatigue strength reduction factor to the life solution. For example, if one is provided with the fatigue strength reduction factors K'_f and K_f for a component at 1000 cycles and the fatigue limit, the fatigue strength reduction factor $K_{f,N}$ at a life N, where $10^3 \leq N \leq 10^6$, can be determined as follows:

$$\frac{\log(10^3) - \log(10^6)}{\log(K'_f) - \log(K_f)} = \frac{\log(10^3) - \log(N)}{\log(K'_f) - \log(K_{f,N})} \tag{4.4.23}$$

This approach has been widely adopted in multiaxial states of stresses and is discussed in Section 4.6.

Another approach is to modify the *S–N* curve of a smooth member for notch effects so that the fatigue life can be determined for any given nominal stress. This approach appears simpler and widely used for uniaxial states of stress. Two versions of the modified *S–N* curves for the notch effect, one by Heywood (1962) and the other by Collins (1993), are illustrated in Figures 4.31 (a) and (b), respectively. Based on Heywood's model, the *S–N* curve for a notched component takes into account the specific effects of a notch in the high-cycle and low-cycle fatigue regions. Based on the Collins model, the *S–N* curve for the notched component is established by a straight line connecting the corrected fatigue limit for a notch and the fatigue strength at one cycle. In general, the Collins model has the benefit of being easier to employ and can be used if the fatigue properties (*S–N* curve) are given, whereas the Heywood model would be a better choice when the fatigue properties are estimated based on the ultimate strength of a material.

Example 4.9. A grooved shaft made of high-grade alloy steel is heat-treated to 400 BHN and finished with a careful grinding operation. The geometry of the shaft is shown in Figure 4.32, with the dimensions of $D = 28.0$ mm, $d = 25.5$ mm, and the notch radius $r = 1.28$ mm. The elastic stress concentration factors due to bending, axial loading, and torsional loading are calculated as $K_{t,\text{bending}} = 2.26$, $K_{t,\text{tension}} = 2.50$, and $K_{t,\text{torsion}} = 1.64$. Based on the Heywood approach, estimate the fatigue limit and the fatigue strength at 1000 cycles of the notched shaft with a reliability of 0.999 for the following loading conditions:

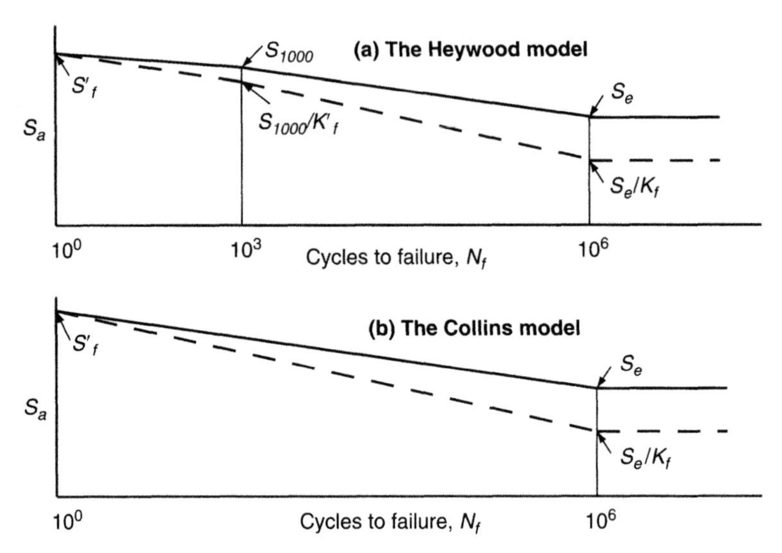

FIGURES 4.31 (a) and (b) Models for the effect of notches.

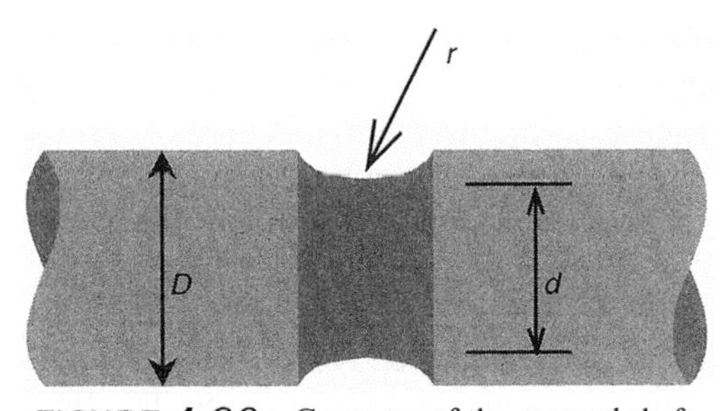

FIGURE 4.32 Geometry of the grooved shaft.

(1) Reversed bending loading
(2) Reversed axial loading assuming negligible intermediate bending
(3) Reversed torsional loading

Solution. The general formulas to determine S_e and S_{1000} for the notched shaft made of a steel are

$$S_e = \frac{S_{be} C_L C_S C_D C_R}{K_f} \text{ and } S'_{1000} = \frac{S_{1000} C_R}{K'_f}$$

where S_{1000} depends on the loading condition, i.e., $0.90S_u$ for bending, $0.75S_u$ for axial loading, or $0.72S_u$ for torsion. For a steel with the Brinell hardness of 400 BHN, $S_u = 3.5\text{BHN} = 3.5(400) = 1400\,\text{MPa}$.

$$S_{be} = 700\,\text{MPa for a steel where } S_u \geq 1400\,\text{MPa}.$$

The following correction factors apply to all three loading conditions:

$C_R = 0.753$ for the reliability of 99.9%.

$C_S = 0.85$ is obtained by entering the surface finish curve for the grinding operation for steels with $S_u = 1400$ MPa

For bending and torsion,

$$C_D = 1.189(d)^{-0.097} = 1.189(25.5)^{-0.097} = 0.868, \text{since } d = 25.5\,\text{mm} > 8.0\,\text{mm}$$

For tension, $C_D = 1.0$.

C_L depends on the loading condition. For example, $C_L = 1.0$ for bending, $C_L = 0.58$ for torsion, and because of the presence of similar steep stress gradient as seen in bending stress load cases, $C_L = 1.0$ for a notched shaft under tension.

For a steel for which $S_u = 1400\,\text{MPa} > 560\,\text{MPa}$, Peterson's formula for fatigue strength reduction factor at the fatigue limit life can be estimated as follows:

$$K_f = 1 + (K_t - 1)q$$

where for bending and tension:

$$a = 0.0254 \times \left(\frac{2079}{S_u}\right)^{1.8} = 0.0254 \times \left(\frac{2079}{1400}\right)^{1.8} = 0.0518\,\text{mm}$$

$$q = \frac{1}{1 + \frac{a}{r}} = \frac{1}{1 + \frac{0.0518}{1.28}} = 0.961$$

Thus,

$$K_{f,\text{ bending}} = 1 + (K_{t,\text{ bending}} - 1)q = 1 + (2.26 - 1) \times 0.961 = 2.21$$
$$K_{f,\text{ tension}} = 1 + (K_{t,\text{ tension}} - 1)q = 1 + (2.50 - 1) \times 0.961 = 2.45$$

For torsion,

$$a = 0.01524 \times \left(\frac{2079}{S_u}\right)^{1.8} = 0.01524 \times \left(\frac{2079}{1400}\right)^{1.8} = 0.0311$$

$$q = \frac{1}{1 + \frac{a}{r}} = \frac{1}{1 + \frac{0.0311}{1.28}} = 0.976$$

$$K_{f,\text{ torsion}} = 1 + (K_{t,\text{ torsion}} - 1)q = 1 + (1.64 - 1) \times 0.961 = 1.62$$

Next, it is necessary to determine the fatigue strength reduction factor at 1000 cycles by the empirical relation between K_f and K_f' as follows:

$$\frac{K_f' - 1}{K_f - 1} = q_{1000}'$$

where $q_{1000}' = 0.5$ is obtained from the empirical curve for steels with $S_u = 1400$ MPa. Therefore,

$$K'_{f,\,\text{bending}} = 1 + (K_{f,\,\text{bending}} - 1)q'_{1000} = 1 + (2.21 - 1) \times 0.5 = 1.61$$

$$K'_{f,\,\text{tension}} = 1 + (K_{f,\,\text{tension}} - 1)q'_{1000} = 1 + (2.45 - 1) \times 0.5 = 1.73$$

$$K'_{f,\,\text{torsion}} = 1 + (K_{f,\,\text{torsion}} - 1)q'_{1000} = 1 + (1.62 - 1) \times 0.5 = 1.31$$

In summary, the fatigue limits of the notched shaft for bending, tension, and torsion are

$$S_{e,\,\text{bending}} = \frac{S_{be}C_L C_S C_D C_R}{K_{f,\,\text{bending}}} = \frac{700(1)(0.85)(0.868)(0.753)}{2.21} = 176\,\text{MPa}$$

$$S_{e,\,\text{tension}} = \frac{S_{be}C_L C_S C_D C_R}{K_{f,\,\text{tension}}} = \frac{700(1)(0.85)(1)(0.753)}{2.45} = 183\,\text{MPa}$$

$$S_{e,\,\text{torsion}} = \frac{S_{be}C_L C_S C_D C_R}{K_{f,\,\text{torsion}}} = \frac{700(0.58)(0.85)(0.868)(0.753)}{1.62} = 139\,\text{MPa}$$

And the fatigue strengths of the notched shaft at $N = 1000$ cycles for bending, tension, and torsion are

$$S'_{1000,\,\text{bending}} = \frac{S_{1000,\,\text{bending}}C_R}{K'_{f,\,\text{bending}}} = \frac{(0.9 \times 1400)(0.753)}{1.61} = 589\,\text{MPa}$$

$$S'_{1000,\,\text{tension}} = \frac{S_{1000,\,\text{tension}}C_R}{K'_{f,\,\text{tension}}} = \frac{(0.75 \times 1400)(0.753)}{1.73} = 457\,\text{MPa}$$

$$S'_{1000,\,\text{torsion}} = \frac{S_{1000,\,\text{torsion}}C_R}{K'_{f,\,\text{torsion}}} = \frac{(0.72 \times 1400)(0.753)}{1.31} = 579\,\text{MPa}$$

4.5 MEAN STRESS EFFECT

From the perspective of applied cyclic stresses, fatigue damage of a component correlates strongly with the applied stress amplitude or applied stress range and is also influenced by the mean stress (a secondary factor). In the high-cycle fatigue region, normal mean stresses have a significant effect on fatigue behavior of components. Normal mean stresses are responsible for the opening and closing state of microcracks. Because the opening of microcracks accelerates the rate of crack propagation and the closing of microcracks retards the growth of cracks, tensile normal mean stresses are detrimental and compressive normal mean stresses are beneficial in terms of fatigue strength. The shear mean stress does not influence the opening and closing state of microcracks, and, not surprisingly, has little effect on crack propagation. There is very little or no effect of mean stress on fatigue strength in the low-cycle fatigue region in which the large amounts of plastic deformation erase any beneficial or detrimental effect of a mean stress.

Early empirical models by Gerber (1874), Goodman (1899), Haigh (1917), and Soderberg (1930) were proposed to compensate for the tensile normal mean stress effects on high-cycle fatigue strength. These empirical models can be plotted as constant life diagrams. The most useful graphical representations of experimental fatigue data are constant life plots of S_{max} versus S_{min} or S_a versus S_m. As illustrated in Figure 4.33, these constant-life models can be determined experimentally from a family of *S–N* curves generated with specific values of S_a and S_m.

In 1874, Gerber proposed a parabolic representation of Wohler's fatigue limit data on a plot of S_{max}/S_u versus S_{min}/S_u as shown in Figure 4.34. In 1899, Goodman introduced a theoretical line representing the available fatigue data (i.e., the two straight lines in Figure 4.34), based on an impact criterion for bridge designs. Goodman justified the use of the impact criterion on the basis that it was easy, simple to use, and provided a good fit to the data. In 1917, Haigh first plotted fatigue data for brasses on a S_a versus S_m plot. Figure 4.35 illustrates the Haigh plot of the Gerber and the Goodman mean stress corrections. The ordinate of the Haigh plot is the normalized fatigue limit, and the maximum mean stress is limited to the ultimate strength S_u. The curve connecting these two points on the two axes represents combinations of stress amplitudes and means stresses given at the fatigue limit life.

Mathematically, the Gerber parabola and the Goodman line in Haigh's coordinates can be expressed as the following expressions:

- Gerber's mean stress correction

$$S_e = \frac{S_a}{1 - \left(\frac{S_m}{S_u}\right)^2} \tag{4.5.1}$$

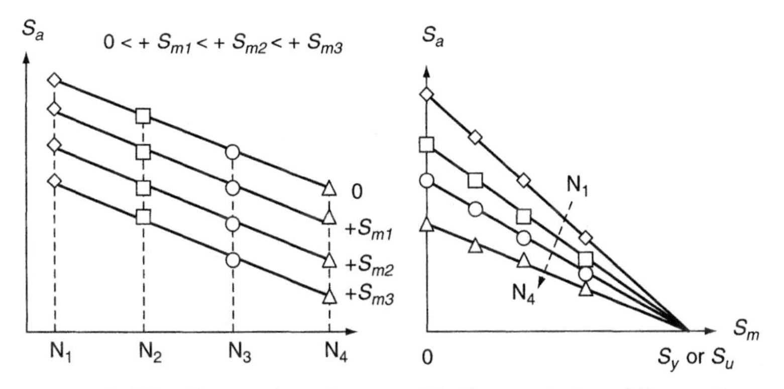

FIGURE 4.33 Construction of constant life diagrams in S_a and S_m coordinates.

- Goodman's mean stress correction

$$S_e = \frac{S_a}{1 - \dfrac{S_m}{S_u}} \tag{4.5.2}$$

where S_e is the fatigue limit for fully reversed loading that is equivalent to the load case with a stress amplitude S_a and a mean stress S_m. In 1930, Soderberg

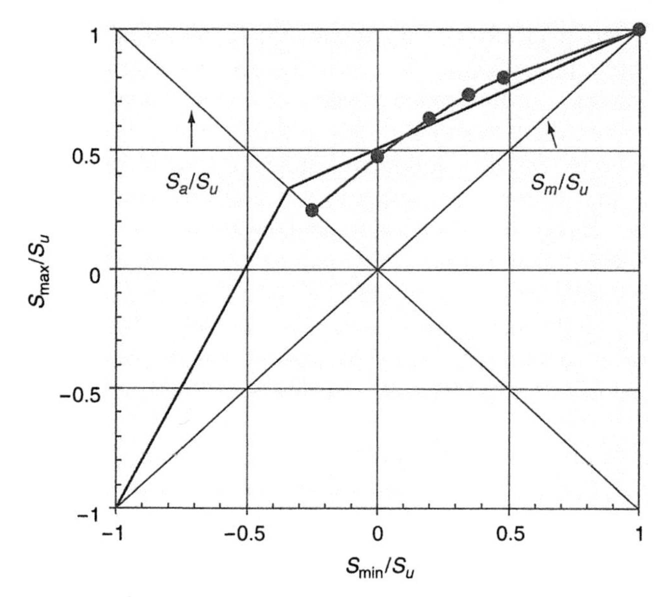

FIGURE 4.34 Gerber's and Goodman's diagrams for Wohler's data.

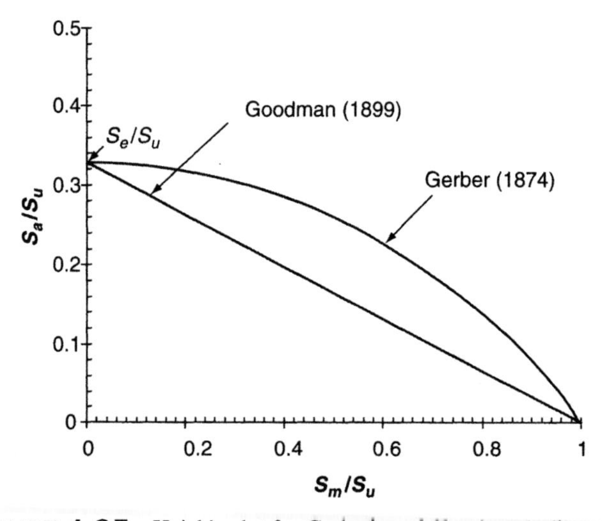

FIGURE 4.35 Haigh's plot for Gerber's and Goodman's diagrams.

suggested that the maximum normal mean stress should be limited to the yield strength S_y and this mean stress correction is represented by

$$S_e = \frac{S_a}{1 - \dfrac{S_m}{S_y}} \tag{4.5.3}$$

These equations are later extended to account for the normal mean stress effect by simply replacing the fatigue limit S_e with a fully reversed stress amplitude S_{ar}, corresponding to a specific life in the high-cycle fatigue region.

Among these early empirical equations, the Goodman formula is simple, attractive, and works reasonably well for tensile normal mean stress situations at the fatigue limit. For ductile materials, the ultimate strength of a notched member is approximately the same as that for a smooth member. Therefore, the Goodman equation can be summarized as follows for notched and smooth members of ductile materials:

$$S_{ar} = \frac{S_a}{1 - \dfrac{S_m}{S_u}} \tag{4.5.4}$$

It is conservative to assume that for most ductile materials, the compressive normal mean stress does not benefit fatigue strength. This means that the fully reversed stress amplitude is the same as the stress amplitude if the normal mean stress is negative. A modified Goodman diagram for both tensile and compressive normal mean stresses is schematically illustrated in the Haigh plot at the fatigue limit as shown in Figure 4.36. Wilson and Haigh (1923) introduced the line of constant yield strength as an additional constraint for ductile materials on the safe design stress region, named the safe design region for fatigue limit and yield strength, shown in Figure 4.36.

Since 1960, some models for the mean stress effect have been proposed as improvements on past models. Fatigue test data indicate that a tensile normal

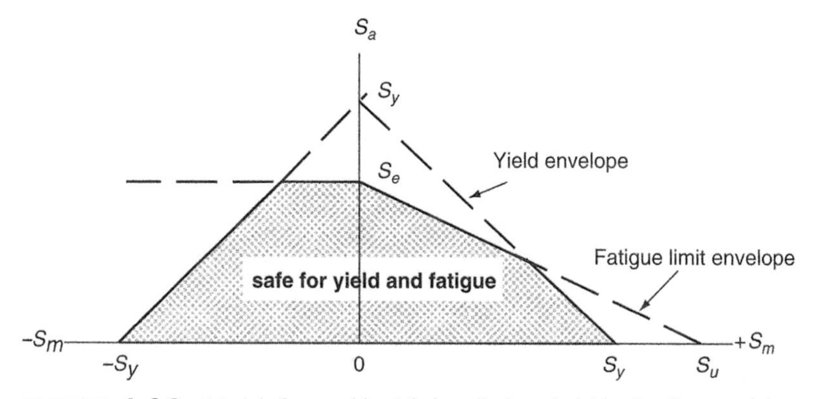

FIGURE 4.36 Models for combined fatigue limit and yield—ductile materials.

mean stress would reduce the fatigue strength coefficient and a compressive normal mean stress would increase it. Also, considering that the monotonic yield and ultimate strength are not appropriate for describing the fatigue behavior of a material, Morrow (1968) suggested that the stress amplitude plus the mean stress could never exceed the fatigue strength coefficient S_f', the fatigue strength at one reversal. Morrow's statement is expressed in the following form of constant life diagram:

$$S_{ar} = \frac{S_a}{1 - \frac{S_m}{S_f'}} \tag{4.5.5}$$

or

$$S_a = (S_f' - S_m) \times (2N_f)^b \tag{4.5.6}$$

The difference between the Goodman and the Morrow equations can be observed from the Haigh plot shown in Figure 4.37, in which the negative value of the slope of the line is termed as the *mean stress sensitivity factor, M*. If the M factor is given, the mean stress correction line in the Haigh diagram is

$$S_{ar} = S_a + M \times S_m \tag{4.5.7}$$

Another popular method is the Smith, Watson, and Topper (SWT) equation (Smith et al., 1970), in which the equivalent fully reversed stress amplitude is expressed as follows:

$$S_{ar} = \sqrt{S_{max} \times S_a} = S_f' \times (2N_f)^b \quad S_{max} > 0 \tag{4.5.8}$$

Mathematically, this equation predicts infinite life if $S_{max} \leq 0$ and the fatigue crack can be considered not to initiate under these conditions. This SWT formula is commonly used in assessing the mean stress effects for component life data (Dowling, 1998).

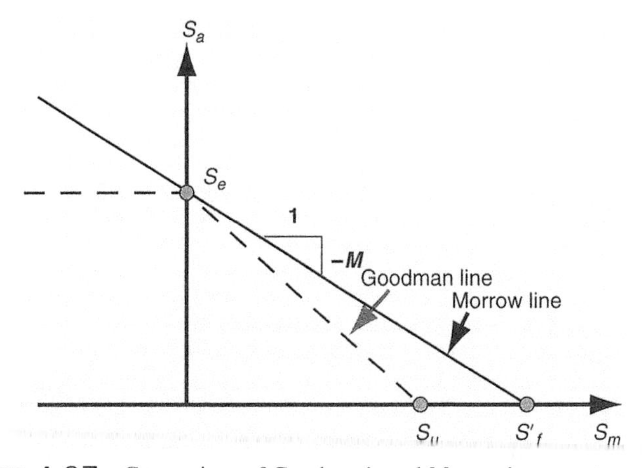

FIGURE 4.37 Comparison of Goodman's and Morrow's mean stress models.

For relatively small mean stress loading, the Morrow and the SWT approaches are considered better than the Goodman method. The Goodman mean stress correction formula should only be used if none of the fatigue properties are available. In general, the SWT model correlates to the experimental fatigue data for most of structural metals and appears to work particularly well for aluminum alloys.

For relatively large mean stress conditions, an empirical model based on the concept of the mean stress sensitivity factor was introduced. As illustrated in Figure 4.38, M factors were found to vary in different mean stress levels (Radaj and Sonsino, 1998). For example, the mean stress sensitivity factor for low mean stress loading ($-1 \le R < 0$) denotes M defined as follows:

$$M = \frac{S_{e,\ R=-1} - S_{e,\ R=0}}{S_{e,\ R=0}} \tag{4.5.9}$$

The mean stress sensitivity factor for loading with low, compressive mean stress levels ($-\infty \le R < -1$) denotes M_2 and varies from 0 to M. The mean stress sensitivity factor for higher mean stress levels ($0 \le R \le 1$ or $S_m > S_a$) denoting M_3 is usually lower than M by a factor of 3 ($M_3 \approx M/3$). This is based on the empirical observation that loading with high mean and small amplitude shows higher damaging effects than that predicted by M. It is also found by Schütz (1968) that the M factor for a material increases with a higher ultimate strength, as illustrated in Figure 4.39.

In the case of weldments, local notch plasticity may take place at very low nominal stresses due to the presence of high-tensile residual stress levels that are about two thirds of yield stress after welding. Thus, it is the common practice in the welding industry to predict fatigue life of weldments based on the nominal stress amplitude without mean stress corrections. According to Radaj and Sonsino (1998), this corresponds to the value of $M_3 = 0$ in the Haigh plot.

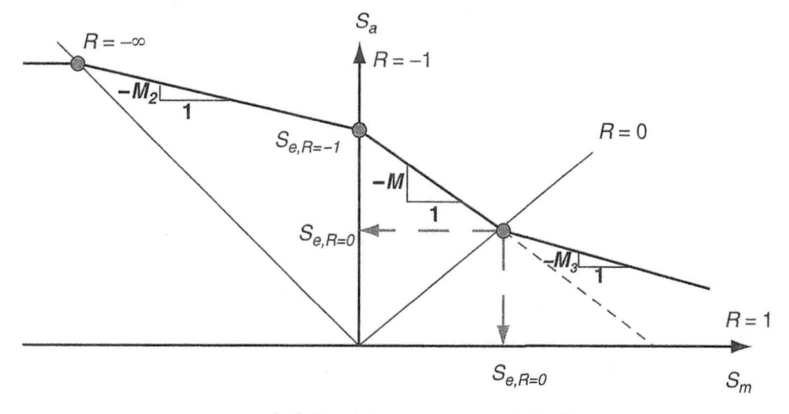

FIGURE 4.38 Mean stress sensitivity factors.

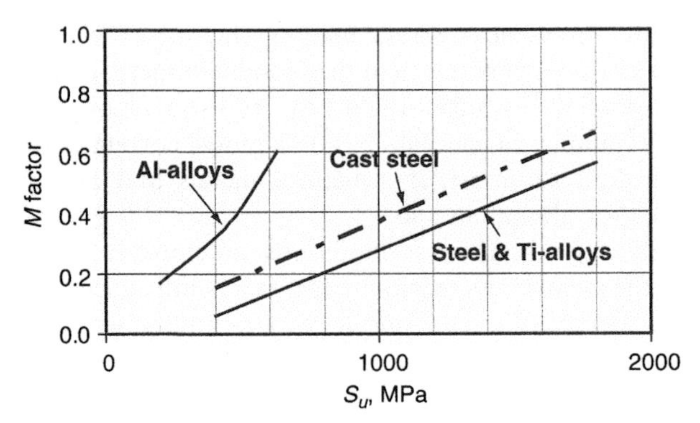

FIGURE 4.39 Effect of the ultimate strength of a material on the M factor. Adapted from Schütz (1968) and used with permission of FAT-Schriftenreihe.

If the baseline S–N curve was generated by specimens under $R = 0$ loading, it is required to convert any positive mean offset loading to an equivalent $R = 0$ loading. For a given mean stress sensitivity factor, the following conversion formula is used:

$$S_{ar,\,R=0} = \frac{S_a + M \times S_m}{M + 1} \tag{4.5.10}$$

This equation is popular in spot welded fatigue life prediction because single spot weld laboratory specimens cannot resist any compression that leads to local buckling of the metal sheet. Thus, these specimens are often subjected to $R = 0$ loading for the generation of a baseline S–N curve.

Any shear mean stress can be considered positive because signs of shear are arbitrarily chosen. Experimental fatigue data indicate that shear mean stress has little effect on fatigue strength of unnotched members under torsion. Where significant stress raisers are present in a component subjected to torsional loading, the state of stress at high-stress-concentration areas deviates from pure shear. Thus, experimental results under these conditions show a shear mean stress detrimental to the fatigue strength approximately as significant as that observed for bending stresses in other load cases. It is recommended to use the Goodman equation in $\tau_a - \tau_m$ for notched torsion members for which the ultimate shear strength S_{us} is given.

Example 4.10. A hot-rolled notched component made of SAE 1005 ($S_u = 321$ MPa) consists of a plate with a center hole of 4.0 mm. As shown in Figure 4.40, the plate has a width of 16 mm and a thickness of 5 mm. The elastic stress concentration factor K_t of 2.4 due to tension was calculated based on the geometric ratio. Fatigue tests with cyclic axial loading on the hot-rolled unnotched plates of the same material were conducted to generate a baseline S–N curve that has the following fatigue properties: $S_f' = 886$ MPa and $b = -0.14$.

(a) Based on Morrow's mean stress correction, determine the fatigue life of this notched plate subjected to a cyclic axial loading varying from +8000 N to −6000 N.
(b) Based on Morrow's mean stress correction, determine the fatigue life of this notched plate subjected to a variable-amplitude loading history as shown in Figure 4.41.

Solution. (a) From the given fatigue properties, the S–N equation for the unnotched component can be expressed:

$$S_a = S_f'(2N_f)^b = 886(2N_f)^{-0.14}$$

Calculating the fatigue limit at 10^6 cycles (smooth plate),

$$S_a = 886(2 \times 10^6)^{-0.14} = 116\,\text{MPa}$$

The plate to be analyzed has a hole, which acts as a stress raiser. The previous S–N equation must be adjusted. The component is made of steel.

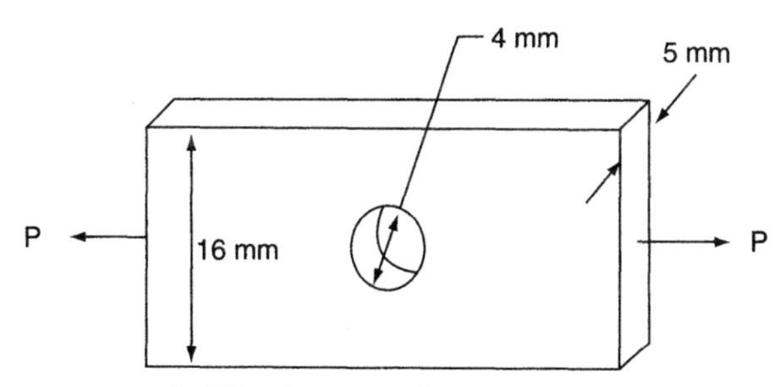

FIGURE 4.40 Geometry of a hot-rolled notched plate.

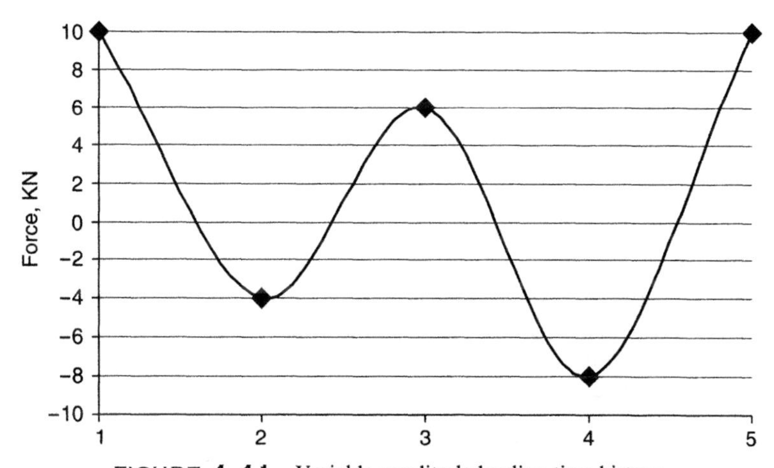

FIGURE 4.41 Variable amplitude loading-time history.

Therefore, Peterson's equation should be used to determine the fatigue strength reduction factor K_f.

$$K_f = 1 + (K_t - 1)q$$

Because the ultimate strength of the material is less than 560 MPa, the emperical notch sensitivity curve for steel should be used to determine q. For a notch radius of 2 mm and the ultimate strength of 321 MPa, q is determined to be 0.68 from the graph for axial loading. Thus,

$$K_f = 1 + (2.4 - 1)0.68 = 1.952$$

Collin's approach is used to adjust the S–N curve for a notch effect. The fatigue limit of the notched plate is calculated as follows:

$$S_{e,\text{notched}} = \frac{S_e}{K_f} = \frac{116}{1.952} = 59.5\,\text{MPa}$$

The adjusted S–N equation is obtained by determining the new slope between the fatigue strength values at 1 and 2×10^6 reversals:

$$b' = \frac{\log(59.5) - \log(886)}{\log(2 \times 10^6) - \log(1)} = \frac{-1.17}{6.30} = -0.186$$

$$S_{a,\text{notched}} = 886\left(2N_f\right)^{-0.186}$$

After determining the S–N curve for the notched plate, the loading of the plate is considered. First, the following loading must be converted into nominal stress.

$$P_{\max} = 8000\,\text{N} \qquad P_{\min} = -6000\,\text{N}$$

Because the plate is axially loaded, the nominal stress in the plate can be determined by the following elastic equation based on the net cross-sectional area:

$$S = \frac{P}{A_{\text{net}}}$$

For $P_{\max}$ and $P_{\min}$,

$$S_{\max} = \frac{P_{\max}}{A_{\text{net}}} = \frac{8000}{(16 - 4)(5)} = \frac{8000}{60} = 133\,\text{MPa}$$

$$S_{\min} = \frac{P_{\min}}{A_{\text{net}}} = \frac{-6000}{(16 - 4)(5)} = \frac{-6000}{60} = -100\,\text{MPa}$$

From the maximum and minumum stress, the nominal stress amplitude and mean stress is determined:

$$S_a = 117\,\text{MPa} \qquad S_m = 16.7\,\text{MPa}$$

Because this loading results in a mean stress, the Morrow equation is used to determine the fatigue life of the notched plate:

$$117 = (886 - 16.7)(2N_f)^{-0.186} \quad \text{and} \quad N_f = 24,100 \text{ cycles}$$

(b) Since the approach for determining the notched S–N equation is the same as Part (a), the previous results are used. Performing a rainflow count on the loading history and using the same S–N formula for the notched plate from Part (a), the life can be estimated in the following:

n_i (cycles)	P_{max} (N)	P_{min} (N)	S_{max} (MPa)	S_{min} (MPa)	S_a (MPa)	S_m (MPa)	N_f (cycles)	d
1	6000	−4000	100	−66.7	83.3	16.7	1.50E+5	6.67E−6
1	10000	−8000	167	−133	150	16.7	6.33E+3	1.58E−4
							$\sum_d =$	1.65E−4

Because the total damage due to one block of the load time history is $\sum d = 1.65 \times 10^{-4}$, the notched plate subjected to this variable loading will live for 6060 blocks of loading ($= 1.0/1.65 \times 10^{-4}$).

Example 4.11 (Application). A hollow shaft made of nodular cast iron is used to transmit torque within an automatic transmission. The shaft has a machined surface finish and an ultimate tensile strength of 552 MPa. The geometric dimensions of the notched shaft are shown in Figure 4.42. The elastic stress concentration factor K_t of 4.04 due to torsion was calculated based on the notch radius of 0.13 mm. By using the Goodman mean stress correction formula, determine the number of cycles to failure for the shaft subjected to 0 to 678 N·m torque loading.

Solution. Based on the net sectional properties, the maximum nominal shear stress due to the maximum applied torque on the shaft is calculated:

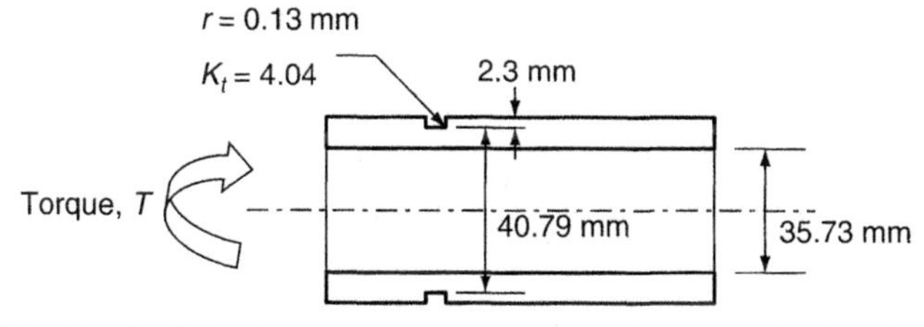

FIGURE 4.42 Geometry of an automatic transmission hollow shaft.

$$S_{max} = \frac{16T_{max}d_o}{\pi(d_o^4 - d_i^4)} = \frac{16 \times 678 \times 40.79}{\pi(40.79^4 - 35.73^4)} = 123.7\,\text{MPa}$$

The nominal shear stress amplitude and mean stress are then determined:

$$S_a = S_m = 61.85\,\text{MPa}$$

The ultimate shear strength S_{us} for the nodular cast iron is estimated:

$$S_{us} = 1.3 \times S_u = 1.3 \times 552 = 717.6\,\text{MPa}$$

Because of the stress raiser, Goodman's mean stress correction formula is used to produce the fully reversed nominal shear stress:

$$S_{ar} = \frac{S_a}{1 - \frac{S_m}{S_{us}}} = \frac{61.85}{1 - \frac{61.85}{717.6}} = 67.7\,\text{MPa}$$

To determine the number of cycles to failure, one needs to generate the S–N curve for the notched shaft under torsion. The fatigue limit for cast iron at 5×10^7 cycles is estimated to be

$$S_e = \frac{S_{be}C_L C_S C_D C_R}{K_f}$$

where

$$S_{be} = 0.4S_u \text{ for cast iron} = 0.4(552) = 221\,\text{MPa}$$

The load factor for torsion C_L is 0.8. Based on the graph of surface finish factors and using the curve for machined surfaces, the surface finish factor C_s is 0.76. The size factor C_D is obtained on the net outer diameter of the shaft

$$C_D = 1.189 \times d^{-0.097} = 1.189 \times (40.79)^{-0.097} = 0.82$$

The reliability factor C_R is 1.0 for the assumed reliability of 0.5.

An estimate of K_f for cast iron with spheroid graphite relies on Heywood's empirical equation:

$$K_f = \frac{K_t}{1 + \frac{2\sqrt{a'}}{\sqrt{r}}\left(\frac{K_t - 1}{K_t}\right)}$$

where

$$\sqrt{a'} = \frac{173.6}{S_u} = \frac{173.6}{552} = 0.314\,\text{mm}^{1/2}$$

Thus, $K_f = \dfrac{4.04}{1 + 2\dfrac{0.314}{\sqrt{0.13}}\left(\dfrac{4.04 - 1}{4.04}\right)} = 1.75$

The fatigue limit of the notched hollow shaft at 5×10^7 cycles is determined to be

$$S_e = \frac{S_{be} C_L C_S C_D C_R}{K_f} = \frac{221 \times 0.8 \times 0.76 \times 0.82 \times 1}{1.75} = 63.0\,\text{MPa}$$

The next step is to find out the fatigue strength at 1000 cycles, which is

$$S_{1000} = \frac{0.9 \times S_{us} \times C_R}{K_f'}$$

From the empirical graph of steels for K_f and K_f', one has the following relation:

$$\frac{K_f' - 1}{K_f - 1} = 0.1$$

The value of K_f' is then determined to be 1.05.

Thus, the fatigue strength at 1000 cycles is

$$S_{1000} = \frac{0.9 \times S_{us} \times C_R}{K_f'} = \frac{0.9 \times 717.6 \times 1}{1.05} = 615\,\text{MPa}$$

The fatigue life at 67.7 MPa is then determined as

$$\frac{\log(10^3) - \log(5 \times 10^7)}{\log(615) - \log(63.0)} = \frac{\log(10^3) - \log(N_f)}{\log(615) - \log(67.7)}$$

which leads to $N_f = 3.55 \times 10^7$ cycles.

Example 4.12 (Application). A semielliptic leaf spring in Figure 4.43 is subjected to a cyclic center load, $2F$, varying from 1500 to 6000 N. It is given that the central bolt used to hold the five leaves together causes a fatigue strength reduction factor K_f of 1.3. Each leaf spring has a length L of 1200 mm, a width w of 80 mm, and a thickness t of 7 mm.

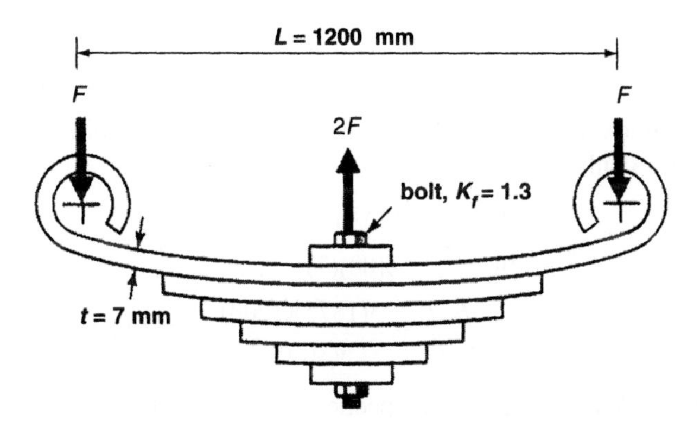

FIGURE 4.43 A semielliptic leaf spring with loads.

The leaf spring made of a steel alloy was heat-treated and then shot-peened prior to installation. The steel alloy has an ultimate strength of 700 MPa. The shot peening process on the leaf spring introduces a compressive residual stress of −350 MPa on the surface, increases the hardness to 250 BHN, and increases the surface roughness (AA) to 1.3 μm. Does this leaf spring subject to the cyclic loads have an infinite life?

Solution. Owing to the symmetry of the semielliptical leaf spring, it can be sectioned through the center of the bolt-hole and only one half of the overall system is considered. Each half of the leaf spring acts as a cantilever carrying half of the total load. The induced bending stress is the most dominant cause of failure. The shear stress due to the force, F, acting in the vertical direction can be neglected. The clamp load of the bolt is neglected in this example. Also, it is assumed that there is no friction between the leaves, so each leaf acts individually to resist exactly one fifth of the bending load at the bolt hole region. Therefore, one leaf can be analyzed, and the result applies to all five leaves.

The bending load applied to the leaf spring center varies from a maximum of 6000 N to a minimum of 1500 N. Because the problem has been simplified as stated previously, these loads are reduced by half at each end.

$$F_{max} = 3000\,\mathrm{N} \qquad F_{min} = 750\,\mathrm{N}$$

From the maximum and minumum loads, the mean load and load amplitude at both ends can be determined.

$$F_m = 1875\,\mathrm{N} \quad F_a = 1125\,\mathrm{N}$$

To determine the stress per leaf spring caused by the bending loads, an equation for elastic bending stress must be derived from the geometrical information provided. For bending loads, the nominal stress is $S = \frac{Mc}{I}$. From the geometry, it is known that $M = \frac{F(L/2)}{5}$, $c = \frac{t}{2}$, and $I = \frac{1}{12}\,wt^3$. Simplifying the equation provides the bending nominal stress equation for a leaf spring:

$$S = \frac{3FL}{5wt^2}$$

Using the derived nominal stress equation, the stress amplitude and mean stress can be calculated.

$$S_a = \frac{3F_aL}{5wt^2} = \frac{3(1125)(1200)}{5(80)(7)^2} = 207\,\mathrm{MPa}$$

$$S_{m,\,\mathrm{bending}} = \frac{3F_mL}{5wt^2} = \frac{3(1875)(1200)}{5(80)(7)^2} = 344\,\mathrm{MPa}$$

Because the leaf spring is shot peened, the residual stress of −350 MPa must be considered in the mean stress calculation.

$$S_m = S_{m,\ \mathrm{bending}} + S_{m,\ \mathrm{shotpeen}} = 344 + (-\ 350) = -5.61\ \mathrm{MPa}$$

Because the loading results in a compressive mean stress, it is conservative to ignore this stress and the equivelent fully reversed stress amplitude S_{ar} becomes 207 MPa.

Now, to determine whether the life is infinite or not, the value of the applied stress must be compared with the fatigue limit that can be estimated in the following:

$$S_e = \frac{S_{be} C_L C_D C_S C_R}{K_f}$$

where for steel $S_u = 3.45\ \mathrm{BHN} = 3.45(250) = 863 < 1400\ \mathrm{MPa}$,

$$S_{be} = 0.5 S_u = 0.5(863) = 431\ \mathrm{MPa}$$

The load factor for bending is $C_L = 1.000$.

Because the spring leaf is not circular, the equivalent diameter must be determined using

$$d_e = \sqrt{0.65wt} = \sqrt{0.65(80)(7)} = 19.1\ \mathrm{mm}$$

The size factor is then determined:

$$\text{For } d_e > 8\ \mathrm{mm},\ C_D = 1.189(d)^{-0.097} = 1.189(19.1)^{-0.097} = 0.89$$

The surface factor is found using the surface finish factor graph with $S_u = 863$ MPa and a surface finish of 1.3 μm:

$$C_s = 0.92$$

Because no reliability requirement is given, it is assumed that a reliability of 50% is adequate for this analysis. Thus,

$$C_R = 1.000$$

The fatigue strength reduction factor is $K_f = 1.3$ and is the information provided within the problem statement. Thus, the corrected fatigue limit of the leaf spring is

$$S_e = \frac{(431)(1.000)(0.826)(0.92)(1.000)}{1.3} = 272\ \mathrm{MPa}$$

Because the applied bending stress ($S_{ar} = 207$ MPa) is below the fatigue limit ($S_e = 272$ MPa), it is concluded that the leaf spring under the applied cyclic loads will have infinite life.

Example 4.13 (Application). According to a Proving Grounds (PG) endurance schedule, the manufactured rear axle shafts installed in a rear wheel drive (RWD) vehicle experienced two types of fatigue failure: cracking at the rotor pilot radius and at the spline end, as shown in Figures 4.44 and 4.45,

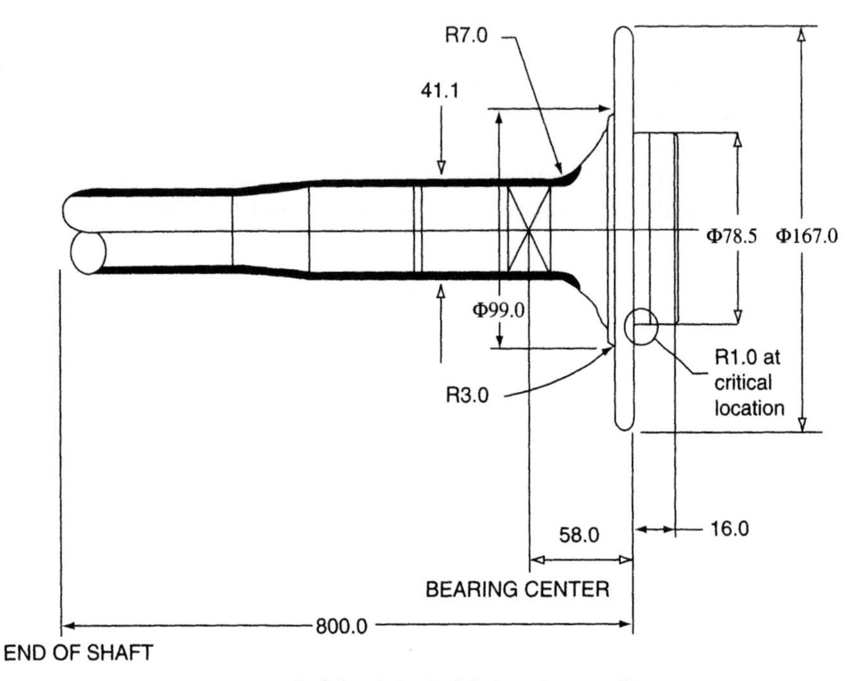

FIGURE 4.44 Axle shaft hub end rotor pilot.

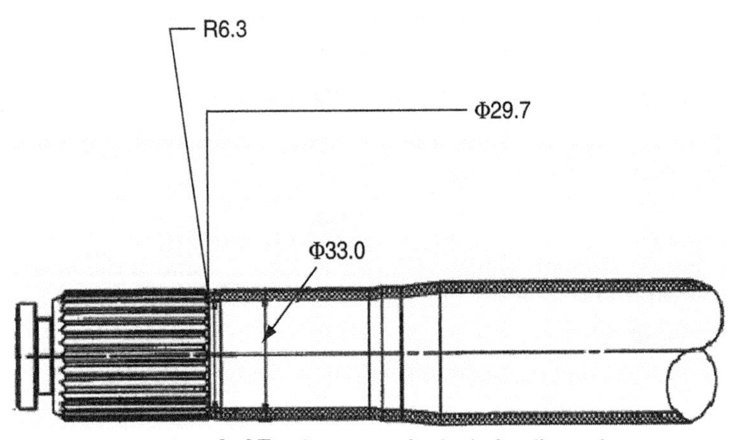

FIGURE 4.45 Geometry of axle shaft spline end.

respectively. Field investigation suggested that the cornering force at the tire patch is the primary loading that causes cracks at the rotor pilot radius and that the alternating torque levels on the axle shafts are responsible for the shear failure mechanism found at the spline end.

A new design with the detailed drawing dimensions shown in Figures 4.44 and 4.45 was proposed for improvement. The newly designed axle shaft made of SAE 1050 steel has the ultimate strength of 690 MPa. The entire shaft was heat treated to the Brinell hardness of 300 BHN except for the shaft hub end. The pilot radius and the spline end were machined to meet the dimensional specification. The K_t values at the rotor pilot radius in bending and at the spline end in torsion are 4.01 and 1.16, respectively.

According to an axle manufacturer's design manual, the axle shaft should have the capability of resisting the centrifugal force of 0.9 G due to vehicle cornering for 1,000,000 stress cycles, and the shafts should be designed for alternating the maximum slip torque levels for 8000 cycles.

The loads acting on the rear axle shaft can be calculated as follows:

(a) Effect of vehicle cornering: The cornering force at the tire patch induces bending moment and axial force on the flange end of the axle shaft. The bending moment and axial force are calculated as follows:

$$M = 0.9 \times (\mathrm{GAWR}/2) \times \mathrm{SLR}$$

$$P = 0.9 \times (\mathrm{GAWR}/2)$$

where GAWR is the gross axle weight rating (15 KN) and SLR is the static loaded radius of the tire (358 mm).

(b) Effect of maximum slip torque: The maximum slip torque for each axle shaft is given as follows:

$$T_{r,\,\max} = \frac{\mu \times \mathrm{GAWR}}{1 - \dfrac{\mu H}{L}} \times \mathrm{SLR} \times 0.5$$

where L is the wheel base (3330 mm), H is the vehicle C.G. height (625 mm), and μ is the coefficient of friction for a dry pavement (0.9).

Determine whether the newly designed axle shaft meets the two design guidelines by estimating the fatigue life of the axle shafts due to the vehicle cornering force and maximum alternating slip torque.

Solution. (a) To validate the design of the axle shaft, fatigue analyses for both durability criteria—cornering and torsion—should be performed. First, consider cornering loading on the rotor pilot. For the bending moment caused by the vehicle cornering,

$$M = 0.9(\mathrm{SLR})\left(\frac{\mathrm{GAWR}}{2}\right) = 0.9(0.358)\left(\frac{15000}{2}\right) = 2417\ \mathrm{N{\cdot}m}$$

As the shaft rotates one revolution, the stresses due to bending on the rotor pilot will alternate between tension and compression. Therefore, in one revolution of the shaft, there is one cycle of fully reversed applied moments ($M_a = 2417\ \mathrm{N{\cdot}m}$) acting on the rotor pilot area.

The axial load caused by vehicle cornering induces constant compression to the rotor pilot and zero load amplitude. Thus,

$$P_m = -0.9\left(\frac{\text{GAWR}}{2}\right) = -0.9\left(\frac{15000}{2}\right) = -6750 \text{ N}$$

For bending loading, the nominal bending stress amplitude in the shaft hub can be determined:

$$S_a = \frac{32M_a}{\pi d^3} = \frac{32 \times 2417 \times 1000}{\pi \times 78.5^3} = 50.9 \text{ MPa}$$

For the axial loading, the nominal mean stress in the shaft is determined:

$$S_m = \frac{4P_m}{\pi(d^2)} = \frac{4(-6750)}{\pi(78.5^2)} = -1.39 \text{ MPa}$$

Because of the compressive mean stress on the pilot radius, it is conservative to consider the fully reversed nominal bending stress (±50.9MPa) acting alone in this fatigue analysis.

To determine whether the shaft subjected to applied bending stresses of ±50.9MPa can survive 1,000,000 cycles, the fatigue limit must be estimated. This value can be obtained as follows:

$$S_e = \frac{S_{be} C_L C_D C_S C_R}{K_f}$$

where

$S_{be} = 0.5S_u = 0.5(690) = 345$ MPa, for steel where $S_u = 690 < 1400$ MPa

The load factor for bending $C_L = 1.0$.

The size factor is

$$C_D = 1.189(d)^{-0.097} = 1.189(78.5)^{-0.097} = 0.779, \text{ for } d = 78.5\,mm > 8\,mm$$

Because $S_u = 690$MPa, comparing this value to the machined curve on the graph of surface finish factor leads to $C_s = 0.75$. Because no reliability requirement is given, it is assumed that a median value (a reliability of 50%) is adequate, i.e., $C_R = 1.000$.

Because the stress concentration factor at the rotor pilot radius due to bending is known as $K_t = 4.01$, Peterson's equation is used to determine K_f for the steel shaft. As $S_u = 690 > 560$ MPa, the equation for the emperical notch sensitivity can be used to determine q. For bending,

$$a = 0.0254 \times \left(\frac{2079}{S_u}\right)^{1.8} = 0.0254 \times \left(\frac{2079}{690}\right)^{1.8} = 0.185 \text{ mm}$$

$$q = \frac{1}{1 + \frac{a}{r}} = \frac{1}{1 + \frac{0.185}{1.00}} = 0.844$$

Estimates of K_f can be found by

$$K_f = 1 + (K_t - 1)q$$

Thus,

$$K_f = 1 + (4.01 - 1)0.844 = 3.54$$

As a result, the fatigue limit is

$$S_e = \frac{S_{be} C_L C_D C_S C_R}{K_f} = \frac{345(1.0)(0.779)(0.75)(1.0)}{3.54} = 56.9\text{ MPa}$$

Because the applied bending nominal stress amplitude (50.9 MPa) is less than the fatigue limit (56.9 MPa), the newly designed axle shaft under vehicle cornering should survive for infinite life and will have no problem meeting the design guideline (>1,000,000 cycles).

(b) The torsional loaded spline end is checked here. First, the maximum slip torque must be determined and then converted into nominal shear stress. The maximum slip torque is determined as follows:

$$T = 0.5(\text{SLR})\left(\frac{\mu(\text{GAWR})}{1 - \dfrac{\mu(H)}{L}}\right) = 0.5(0.358)\left(\frac{0.9(15000)}{1 - \dfrac{0.9(0.625)}{3.33}}\right) = 2908\text{ N}\cdot\text{m}$$

The design criterion for the shaft in torsion requires that this load is applied cyclically, i.e., $T_a = 2908\,\text{N}\cdot\text{m}$. The nominal shear stress amplitude is obtained:

$$S_a = \frac{16T_a}{\pi(d)^3} = \frac{16(2908)(1000)}{\pi(29.7)^3} = 565\text{ MPa}$$

To determine the fatigue life of the shaft subjected to fully reversed nominal shear stress amplitude $S_a = 565\text{MPa}$, first estimate the S–N curve for this notched shaft. The fatigue limit is estimated to be

$$S_e = \frac{S_{be} C_L C_D C_S C_R}{K_f}$$

Because the shaft is heat treated in this area, the ultimate strength of the material with the heat treatment must be used.

$$S_u = 3.45(\text{BHN}) = 3.45(300) = 1035\text{ MPa for BHN} < 500\text{ BHN}$$

$$S_{be} = 0.5S_u = 0.5(1035) = 518\text{ MPa for steel where } S_u < 1400\text{ MPa}$$

The load factor for torsion is $C_L = 0.58$. The size factor for $d = 29.7 > 8$ mm is

$$C_D = 1.189(d)^{-0.097} = 1.189(29.7)^{-0.097} = 0.856$$

Comparing $S_u = 1034$ MPa to the machined curve on the graph of the surface finish leads to the surface factor $C_s = 0.69$. Because no reliability requirement is given, it is assumed that the median is adequate and the reliability factor for a reliability of 50% is $C_R = 1.000$.

Because the stress concentration factor at the spline end under torsion is given as $K_t = 1.16$, Peterson's equation can be used to determine K_f. The ultimate strength of the material is greater than 560 MPa, and therefore, the emperical notch sensitivity formula should be used to determine q.

For torsional loading,

$$a = 0.01524 \times \left(\frac{2079}{S_u}\right)^{1.8} = 0.01524 \times \left(\frac{2079}{1034}\right)^{1.8} = 0.0536\,\text{mm}$$

$$q = \frac{1}{1 + \frac{a}{r}} = \frac{1}{1 + \frac{0.0536}{6.3}} = 0.992$$

Estimating K_f using Peterson's equation,

$$K_f = 1 + (K_t - 1)_q = 1 + (1.16 - 1)0.992 = 1.16$$

Thus, the fatigue limit of the notched shaft is

$$S_e = \frac{S_{be}C_L C_D C_S C_R}{K_f} = \frac{(518)(0.58)(0.856)(0.69)(1.000)}{1.16} = 153.0\text{ MPa}$$

The fatigue strength at 1000 cycles is estimated to be

$$S_{1000} = \frac{0.9 S_{us} C_R}{K_f'}$$

where the ultimate shear strength S_{us} for steels is approximated to be $0.8S_u =$ 827 MPa, and $C_R = 1.0$ K_f' can be found from the emperical relationship between K_f and K_f'. For steel with $S_u = 1034$ MPa,

$$S_{1000}' = \frac{K_f' - 1}{K_f - 1} = 0.384$$

$$K_f' = 1 + (1.16 - 1)0.384 = 1.06$$

Therefore, the fatigue strength at 1000 cycles of the notched component is

$$S_{1000} = \frac{0.9 S_{us} C_R}{K_f'} = \frac{0.9(827)(1.0)}{1.06} = 702\text{ MPa}$$

After both S_{1000} and S_e have been determined, they can be used to determine the number of cycles for the shaft under $S_a = 565$ MPa. Determining the life at the applied stress,

$$\frac{\log(10^3) - \log(10^6)}{\log(702) - \log(153)} = \frac{\log(10^3) - \log(N_f)}{\log(702) - \log(565)}, \text{ solving for } N_f = 2680 \text{ cycles.}$$

Because the cycles to failure (2680 cycles) are less than 8000 cycles at the alternating maximum slip torque levels, the spline area will require redesigning.

4.6 COMBINED PROPORTIONAL LOADS

The previous sections deal with the cases where a structural component is subject to cyclic uniaxial loading. However, actual components are often subjected simultaneously to multiple loads in several directions, and the components may be under a multiaxial state of stresses. In many cases, these loads are applied in such a way that the principal stress orientation in the member does not vary with time (proportional loading). Under multiaxial proportional loads, a slightly different stress-life approach should be taken. This approach is described as follows:

1. The baseline *S–N* curve is established for bending ($C_L = 1.0$) and without correcting the fatigue strength reduction factors (K_f' and K_f).

2. For ductile materials, the multiaxial stresses can be combined into an equivalent uniaxial bending stress amplitude. This effective stress can be found by using the von Mises criterion with the stresses calculated from the loads and the corresponding fatigue strength reduction factors. For brittle materials, the maximum principal stress theory is recommended.

   For a ductile material in a biaxial stress state (i.e., plane stress state), the local stress amplitudes are

$$\sigma_{x,a} = S_{x,a} \times K_{f,N,\text{ axial/bending}} \tag{4.6.1}$$

$$\sigma_{y,a} = S_{y,a} \times K_{f,N,\text{ axial/bending}} \tag{4.6.2}$$

$$\tau_{xy,a} = S_{xy,a} \times K_{f,N,\text{ torsion}} \tag{4.6.3}$$

   and the local mean stresses are

$$\sigma_{x,m} = S_{x,m} \times K_{f,N,\text{ axial/bending}} \tag{4.6.4}$$

$$\sigma_{y,m} = S_{y,m} \times K_{f,N,\text{ axial/bending}} \tag{4.6.5}$$

$$\tau_{xy,m} = S_{xy,m} \times K_{f,N,\text{ torsion}} \tag{4.6.6}$$

   The equivalent bending stress amplitude, according to the von Mises theory, is expressed as follows:

$$\sigma_{eq,a} = \sqrt{\sigma_{x,a}^2 + \sigma_{y,a}^2 - \sigma_{x,a} \times \sigma_{y,a} + 3\tau_{xy,a}^2} \tag{4.6.7}$$

Note that the sign conventions are important in the equivalent stress calculation. When two normal stresses peak at the same time, both amplitudes ($\sigma_{x,a}$ and $\sigma_{y,a}$) should have the same algebraic sign. When one is at the valley while the other is at a peak, the two should have the opposite algebraic sign.

3. The multiaxial mean stresses can be converted into an equivalent uniaxial mean stress. There are two approaches for the equivalent mean stress calculation. One derived from the von Mises theory is as follows:

$$\sigma_{eq,m} = \sqrt{\sigma_{x,m}^2 + \sigma_{y,m}^2 - \sigma_{x,m}\sigma_{y,m} + 3\tau_{xy,m}^2} \tag{4.6.8}$$

The other is derived from the Sines experimental observations (Sines, 1955). Sines found that a torsional mean stress did not affect the fatigue life of a component subjected to alternating torsion or bending stresses. Thus, ignoring the effect of torsional mean stress on a fatigue life, the equivalent mean stress is calculated as follows:

$$\sigma_{eq,m} = \sigma_{x,m} + \sigma_{y,m} \tag{4.6.9}$$

The latter approach is preferable and has been used by Socie et al. (1997) for estimating fatigue life under proportional multiaxial loading.

4. Finally, the Goodman mean stress correction equation is used to determine the equivalent stress amplitude with zero mean stress. By using the Goodman mean stress correction formula, the equivalent fully reversed bending stress is

$$S_{ar} = \frac{\sigma_{eq,a}}{1 - \dfrac{\sigma_{eq,m}}{\sigma_u}} \tag{4.6.10}$$

5. An iterative process to determine the fatigue life of a ductile component is required because the fatigue strength reduction factor is dependent on fatigue life. It is important to make an initial guess of a fatigue life for the appropriate fatigue strength reduction factors. After the fatigue strength reduction factors are determined, the equivalent mean and bending stress amplitude can be calculated by Equations 4.6.7 and 4.6.9, and the equivalent fully reversed bending stress can be computed by Equation 4.6.10. The stress S_{ar} is then used in the S–N curve for the fatigue life of the ductile component. The iterative procedure will continue until the guessed fatigue life for the fatigue strength reduction factor is identical to the life found from the S–N curve.

Example 4.14. An automatic transmission shaft shown in Figure 4.46 is subjected to the following proportional loads:

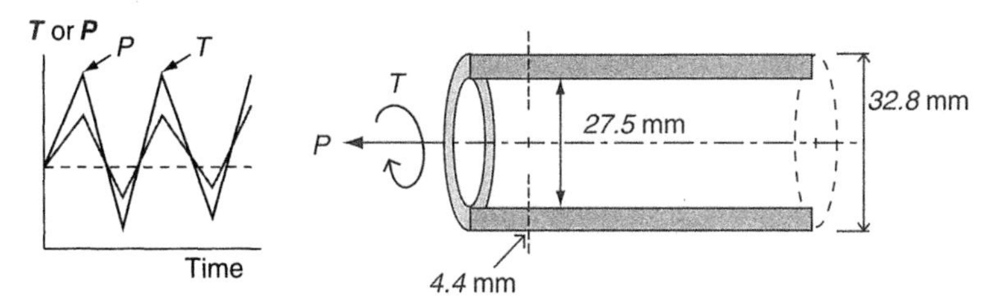

FIGURE 4.46 A notched tube subjected to combined loading.

(1) A torsional load (T) alternating from 860 to 340 N·m
(2) An axial load (P) varying from 21000 to 5900 N

The tubular shaft made of ductile steel with an ultimate tensile strength of 1000 MPa is machined. The elastic stress concentration factors due to torsional and axial loading are $K_{t,\text{torsion}} = 3.75$ and $K_{t,\text{axial}} = 3.23$, respectively. Based on Heywood's S–N approach, determine the number of cycles to failure with R99.9 for the shaft subjected to combined loading.

Solution. To determine the fatigue life of the tubular shaft under combined loading, first estimate the S–N curve for this component due to bending without the notch. The fatigue limit is estimated to be

$$S_e = S_{be} C_L C_D C_S C_R$$

where the bending fatigue limit for steels where $S_u = 1000\text{MPa} < 1400\text{MPa}$ is

$$S_{be} = 0.5 S_u = 0.5(1000) = 500\text{MPa}$$

The load factor for bending is $C_L = 1.0$. The size factor for $d = 32.8\text{mm} > 8.0\,\text{mm}$ is

$$C_D = 1.189(d)^{-0.097} = 0.869(1.29)^{-0.097} = 0.85$$

Entering the machined curve on the surface finish factor graph with $S_u = 1000$ MPa, the surface factor is $C_s = 0.68$.

The reliability factor for a reliability of 99.9% is $C_R = 0.753$. Thus, the bending fatigue limit is

$$S_e = (500)(1.0)(0.85)(0.68)(0.753) = 218\text{MPa}$$

The fatigue strength at 1000 cycles is estimated to be

$$S_{1000} = 0.9 S_u C_R = 0.9(1000)0.753 = 678\text{MPa}$$

After both S_{1000} and S_e have been determined, they are used to graph the S–N curve for the component with no notch effect.

The loading is now considered. First, the applied loads must be converted into local stresses and then adjusted for any mean stress.

Torsional loading: $T_{max} = 860\,\text{N}\cdot\text{m}$; $T_{min} = 340\,\text{N}\cdot\text{m}$
Axial loading: $P_{max} = 21000$ N; $P_{min} = 5900$ N
From the maximum and minumum loads, the mean load and load amplitude are determined:

$$P_m = 13450\,\text{N};\ T_m = 600\text{N}\cdot\text{m}$$

$$P_a = 7550\,\text{N};\ T_a = 260\text{N}\cdot\text{m}$$

For axial loading, the nominal mean stress and stress amplitude in the shaft are determined:

$$S_{x,m} = \frac{4P_m}{\pi\left(d_o^2 - d_i^2\right)} = \frac{4(13450)}{\pi(0.0328^2 - 0.0275^2)} = 53.6 \text{ MPa}$$

$$S_{x,a} = S_{x,m}\left(\frac{P_a}{P_m}\right) = 53.8\left(\frac{7550}{13500}\right) = 30.1 \text{ MPa}$$

For torsional loading, the nominal stress mean and amplitude in the shaft can be determined:

$$S_{xy,m} = \frac{16T_m d}{\pi\left(d_o^4 - d_i^4\right)} = \frac{16(600 \times 1000)32.8}{\pi(32.8^4 - 27.5^4)} = 171 \text{ MPa}$$

$$S_{xy,a} = S_{xy,m}\left(\frac{T_a}{T_m}\right) = 171\left(\frac{260}{600}\right) = 74.1 \text{ MPa}$$

The steel shaft to be analyzed has a hole that acts as a stress riser. Therefore, the fatigue strength reduction factor, which is dependent on the fatigue life, must be adjusted by interpreting K_f and K_f' at 10^6 and 10^3 cycles. Because the component is made of steel, Peterson's equation should be used to determine K_f. Also, because of $S_u = 1000$ MPa > 560MPa, the emperical notch sensitivity formula could be used to determine q.
For axial loading,

$$a = 0.0254 \times \left(\frac{2079}{S_u}\right)^{1.8} = 0.0254\left(\frac{2079}{1000}\right)^{1.8} = 0.0948\,\text{mm}$$

$$q_{\text{axial}} = \frac{1}{1 + \frac{a}{r}} = \frac{1}{1 + \frac{0.0948}{4.4/2}} = 0.96$$

For torsional loading,

$$a = 0.01524 \times \left(\frac{2079}{S_u}\right)^{1.8} = 0.01524 \times \left(\frac{2079}{1000}\right)^{1.8} = 0.0569\,\text{mm}$$

$$q = \frac{1}{1 + \frac{a}{r}} = \frac{1}{1 + \frac{0.0569}{4.4/2}} = 0.97$$

K_f is estimated by using Peterson's equation,

$$K_f = 1 + (K_t - 1)q$$

For axial loading at 10^6 cycles,

$$K_{f,\text{ axial}} = 1 + \left(K_{t,\text{ axial}} - 1\right)q_{\text{axial}} = 1 + (3.23 - 1)0.96 = 3.14$$

For torsional loading at 10^6 cycles:

$$K_{f,\text{ torsion}} = 1 + (K_{t,\text{ torsion}} - 1)q_{\text{torsion}} = 1 + (3.75 - 1)0.98 = 3.67$$

K_f' can be obtained from the emperical relationship between K_f and K_f':

$$\frac{K_f' - 1}{K_f - 1} = f'\ 0.36$$

$$K_f' = 1 + \left(K_f - 1\right)0.36$$

For axial loading at 10^3 cycles,

$$K'_{f,\text{ axial}} = 1 + (K_{f,\text{ axial}} - 1)q'_{1000} = 1 + (3.14 - 1)0.36 = 1.77$$

For torsional loading at 10^3 cycles,

$$K'_{f,\text{ torsion}} = 1 + (K_{f,\text{ torsion}} - 1)q'_{1000} = 1 + (3.67 - 1)0.36 = 1.96$$

Making an initial guess of the number of cycles to failure and estimating the corresponding fatigue strength factor by the following linear relationship give

$$\frac{\log(10^3) - \log(10^6)}{\log\left(K_f'\right) - \log(K_f)} = \frac{\log(10^3) - \log\left(N_{f,\text{ guess}}\right)}{\log\left(K_f'\right) - \log\left(K_{f,N,\text{ guess}}\right)}$$

The initial guess of the fatigue life is $N_{f,\text{ guess}} = 10{,}000$ cycles.
For axial loading,

$$\frac{\log(10^3) - \log(10^6)}{\log(1.77) - \log(3.14)} = \frac{\log(10^3) - \log(10^4)}{\log(1.77) - \log\left(K_{f,\text{ axial},\ 10^4}\right)},$$

solving for $K_{f,\text{ axial},\ 10^4} = 2.14$

For torsional loading,

$$\frac{\log(10^3) - \log(10^6)}{\log(1.96) - \log(3.67)} = \frac{\log(10^3) - \log(10^4)}{\log(1.96) - \log\left(K_{f,\text{ torsion},\ 10^4}\right)},$$

solving for $K_{f,\text{ torsion},\ 10^4} = 2.42$

After the K_f factors for tension and torsion have been determined based on the initial guess, the local stresses are determined.

For axial loading,

$$\sigma_{x,a} = S_{x,a}K_{f,N,\text{axial}} = (30.1)(2.14) = 64.4\,\text{MPa}$$
$$\sigma_{x,m} = S_{x,m}K_{f,N,\text{axial}} = (53.6)(2.14) = 115\,\text{MPa}$$

For torsional loading,

$$\tau_{xy,a} = S_{xy,a}K_{f,N,\,Torsion} = (74.2)(2.42) = 180\,\text{MPa}$$
$$\tau_{xy,m} = S_{xy,m}K_{f,N,\,Torsion} = (171)(2.42) = 414\,\text{MPa}$$

Calculating the equivalent bending stress amplitude and mean,

$$\sigma_{eq,a} = \sqrt{\sigma_{x,a}^2 + \sigma_{y,a}^2 - \sigma_{x,a}\sigma_{y,a} + 3\tau_{xy,a}^2}$$
$$\sigma_{eq,a} = \sqrt{64.4^2 + 0^2 - 64.4(0) + 3(180)^2} = 318\,\text{MPa}$$
$$\sigma_{eq,m} = \sigma_{x,m} + \sigma_{y,m} = 115 + 0 = 115\,\text{MPa}$$

Because the local equivalent stresses do not result in a nonzero mean stress, the Goodman equation is used to determine the equivelent stress amplitude with zero mean stress:

$$S_{ar} = \frac{\sigma_{eq,a}}{\left(1 - \frac{\sigma_{eq,m}}{S_u}\right)} = \frac{318}{\left(1 - \frac{115}{1000}\right)} = 359\,\text{MPa}$$

Determining the life at this stress value:

$$\frac{\log(10^3) - \log(10^6)}{\log(678) - \log(218)} = \frac{\log(10^3) - \log(N_f)}{\log(678) - \log(359)}, \text{ solving for } N_f$$

$$N_f = 48,000 \text{ cycles}$$

Because the cycles to failure do not match the initial guess, this process of iterations continues by using the average of the calculated and the guessed lives from the previous iteration as the next guessed value. Table 4.12 illustrates the important calculated parameters for the iteration process. From the data shown in the table, $N_f = 26{,}000$ cycles.

TABLE 4.12 Iteration Process for Estimating the Fatigue Life of a Component with Combined Stresses, Using the Modified Sines Method

N_{guess}	$K_{f,\text{axial}}$	$K_{f,\text{torsional}}$	$\sigma_{x,a}$	$\sigma_{x,m}$	$\sigma_{xy,m}$	$\sigma_{xy,m}$	$\sigma_{eq,a}$	$\sigma_{eq,m}$	S_{ar}	N_f
10,000	2.14	2.42	64.9	115	180	414	318	115	359	48,000
29,000	2.34	2.66	70.9	126	197	455	349	126	399	25,200
27,100	2.32	2.65	70.3	125	197	453	348	125	398	25,600
26,400	2.32	2.64	70.3	125	196	451	347	125	397	26,000
26,200	2.32	2.64	70.3	125	196	451	347	125	397	26,000

REFERENCES

Adib, H. and Pluvinage, G., Theoretical and numerical aspects of the volumetric approach for fatigue life prediction in notched components, *International Journal of Fatigue*, Vol. 25, 2003, pp. 67–76.

American Society for Testing and Materials (ASTM), Standard practice for statistical analysis of linear or linearized stress-life (S–N) and strain-life (e-N) fatigue data, ASTM Standard, E739-91, West Conshohocken, PA, 1998, pp. 631–637.

Bannantine, J. A., Comer, J. J., and Handrock, J. L., *Fundamentals of Metal Fatigue Analysis*, Prentice Hall, New York, 1990.

British Standard Institution, Methods of fatigue testing. Part 5: Guide to the application of statistics, British Standard 3518, 1996.

Collins, J. A. *Failure of Materials in Mechanical Design—Analysis, Prediction, and Prevention*, 2nd ed., Wiley, New York, 1993.

Dixon, W. J. and Mood, A. M., A method for obtaining and analyzing sensitivity data, *Journal of the American Statistical Association*, Vol. 43, 1948, pp. 109–126.

Dowling, N. E., *Mechanical Behavior of Materials: Engineering Methods for Deformation, Fracture, and Fatigue*, 2nd ed., Prentice Hall, New York, 1998.

Gerber, W. Z., Calculation of the allowable stresses in iron structures, *Z. Bayer Archit Ing Ver*, Vol. 6, No. 6, 1874, pp. 101–110.

Goodman, J., *Journal of Mechanics Applied to Engineering*, 1st ed., Longmans, Green, New York, 1899.

Haibach, E., Modifizierte Lineare Schadensakkumulations-hypothese zur Berücksichtigung des Dauerfestigkeitsabfalls mit Fortschreitender Schädigung, LBF TM No 50/70 (Lab fur Betriebsfestigkeit Darmstadt, FRG, 1970).

Haigh, B. P., Experiments on the fatigue of brasses, *Journal of the Institute of Metals*, Vol. 18, 1917, pp. 55–86.

Heywood, R. B., *Designing Against Failure*, Chapman & Hall, London, 1962.

Horger, O. J., *ASME Handbook: Metals Engineering – Design*, 2nd ed., McGraw-Hill, New York, 1965.

Japan Society of Mechanical Engineers, Standard method of statistical fatigue testing, JSME S 002-1981, 1981.

Johnson, R. C., Specifying a surface finish that won't fail in fatigue, *Machine Design*, Vol. 45, No. 11, 1973, p.108.

Juvinall, R. C., *Engineering Considerations of Stress, Strain, and Strength*, McGraw-Hill, New York, 1967.

Juvinall, R. C. and Marshek, K.M., *Fundamentals of Machine Component Design*, 3rd ed., John Wiley & Sons, New York, 2000, 314 pp.

Kececioglu, D. B., *Robust Engineering Design-by-Reliability with Emphasis on Mechanical Components & Structural Reliability*, Vol. 1, DEStech Publications, 2003, pp. 185–218.

Krause, D. E., Cast iron—a unique engineering material, ASTM Special Publication 455, American Society for Testing and Materials, West Conshohocken, PA , 1969, pp. 3–29.

Kuguel, R. A., Relation between theoretical stress concentration factor and fatigue notch factor deduced from the concept of highly stressed volume, *American Society for Testing and Materials Proceedings*, Vol. 61, 1961, pp. 732–748.

Kuhn, P., and Figge, I. E., "Unified Notch-Strength Analysis for Wrought Aluminum Alloys," NASA TN D-125 9, 1962.

Lassociation Francaise de Normalisation, "Essais de Fatigue" Tour Europe, Paris, la Dedense, Normalisation Francaise A 03-405, 1991.

Lee, Y., Lu, M., Segar, R., Welch, C., and Rudy, R., A reliability-based cumulative fatigue damage assessment in crack initiation, *International Journal of Materials and Product Technology*, Vol. 14, No. 1, 1999, pp. 1–16.

Lieberman, G. J., Tables for one-sided statistical tolerance limits, *Industrial Quality Control*, Vol. XIV, No.10, 1958, pp. 7–9.

Lin, S., Lee, Y., and Lu, M., Evaluation of the staircase and the accelerated test methods for fatigue limit distribution, *International Journal of Fatigue*, Vol. 23, 2001, pp. 75–83.

McGreevy, T. E. and Socie, D. F., Competing roles of microstructure and flaw size, *Journal of Fatigue & Fracture of Engineering Materials & Structures*, Vol. 22, 1999, pp. 495–508.

Metal Power Industries Federation, *Material Standards for PM Structural Parts,* 2000 ed., MPIF Standard 35, 2000.

Morrow, J., *Fatigue Design Handbook, Advances in Engineering*, Vol. 4, SAE, Warrendale, PA, 1968, pp. 21–29.

Murakami Y. and Nagata, J., Effects of small defects on fatigue properties of materials, mechanism of ultralong life fatigue and applications to fatigue design of car components, SAE Paper 2002-01-0575, 2002.

Nakazawa H. and Kodama, S. Statistical S–N Testing Method with 14 Specimens: JSME standard method for determination of S–N curve, *Statistical Research on Fatigue and Fracture* (Tanaka, T., Nishijima,S., and Ichikawa, M., Eds.), *Current Japanese Materials Research,* Vol. 2, 1987, pp. 59–69.

Neuber, H., *Theory of Notch Stress*, J.W. Edwards, Ann Arbor, MI, 1946.

Peterson, R. E., Analytical approach to stress concentration effects in aircraft materials, Technical Report 59-507, U. S. Air Force – WADC Symposium on Fatigue Metals, Dayton, Ohio, 1959.

Peterson, R. E., *Stress Concentration Factors*, Wiley, New York, 1974.

Pilkey, W. D., *Peterson's Stress Concentration Factors*, 2nd ed., Wiley, New York, 1997.

Qylafku, G., Azari, Z., Kadi, N., Gjonaj, M., and Pluvinage, G., Application of a new model proposal for fatigue life prediction on notches and key-seats, *International Journal of Fatigue*, Vol. 21, 1999, pp. 753–760.

Radaj, D. and Sonsino, C. M., *Fatigue Assessment of Welded Joints by Local Approaches*, Abington Publishing, Cambridge, 1998.

Richards, C. W., *Engineering Materials Science*, Wadsworth, San Francisco, 1961.

Schütz, W., View points of material selection for fatigue loaded structures (in German), Laboratorium für Betriebsfestigkeit LBF, Darmstadt, Bericht Nr. TB-80, 1968.

Shen, C. The statistical analysis of fatigue data, Ph.D. Thesis, Department of Aerospace and Mechanical Engineering, University of Arizona, 1994.

Shen, C. L., Wirshing, P. H., and Cashman, G. T., Design curve to characterize fatigue strength, *Journal of Engineering Materials and Technology*, Vol. 118, 1996, pp. 535–541.

Siebel, E. and Stieler, M., Significance of dissimilar stress distributions for cycling loading, *VDI-Z.*, Bd 97, No. 5, 1955, pp. 121-126 (in German).

Sines, G., Failure of materials under combined repeated stress superimposed with static stresses, Tech. Note 3495, National Advisory Council for Aeronautics, Washington, D.C., 1955.

Smith, K. N., Watson, P., and Topper, T. H., A stress-strain function for the fatigue of metals, *Journals of Mate*rials, Vol. 5, No. 4, 1970, pp. 767–778.

Socie, D., Reemsnyder, H., Downing, S., Tipton, S., Leis, B., and Nelson, D., Fatigue life prediction. In *SAE Fatigue Design Handbook*, 3rd ed., Society of Automotive Engineers, Warrendale, PA, 1997, pp. 320–323.

Soderberg, C. R., Fatigue of safety and working stress, *Transactions of the American Society of Mechanical Engineers* Vol. 52 (Part APM-52-2), 1930, pp.13–28.

Stanzl, S. E., Tschegg, E. K., and Mayer, H., Lifetime measurements for random loading in the very high cycle fatigue range, *International Journal of Fatigue*, Vol. 8, No. 4, 1986, pp. 195–200.

Stephens, R. I., Fatemi, A., Stephens, R. R., and Fuchs, H. O., *Metal Fatigue in Engineering*, 2nd ed., Wiley, New York, 2001.

Tryon, R. G. and Dey, A., Reliability-based model for fatigue notch effect, SAE Paper 2003-01-0462, 2003.
Waisman, J.L. and Sines, G., *Metal Fatigue*, McGraw-Hill, New York,1959.
Williams, C. R., Lee, Y., and Rilly, J. T., A practical method for statistical analysis of strain-life fatigue data, *International Journal of Fatigue*, Vol. 25/5, 2003, pp. 427–436.
Wilson, J. S. and Haigh, B. P., Stresses in bridges, *Engineering (London)*, 1923, pp. 446–448.
Wirshing, P. H., Statistical summaries of fatigue data for design purposes, NASA Contract Report 3697, N83-29731, 1983.
Yao, W., Stress field intensity approach for predicting fatigue life, *International Journal of Fatigue*, Vol. 15, No. 3, 1993, pp. 243–245.
Zhang, J. and Kececioglu, D. B., New approaches to determine the endurance strength distribution. In *Proceedings of the 4th ISSAT International Conference on Reliability & Quality in Design*, Seattle, WA, 1998, pp. 297–301.

APPENDIX 4-A

DERIVATION OF RELIABILITY FACTORS, C_{S_e}

The effects of scatter on the required reliability of a structural component under fatigue loading should be incorporated into the design procedure by introducing a reliability correction factor C_R that shifts the median fatigue limit μ_{S_e} downward to a fatigue limit to a specific reliability level $S_{e,R}$. The reliability correction factor is defined as

$$S_{e,R} = C_R \times \mu_{S_e}$$

Assuming that the fatigue limits are normally distributed and have the coefficient of variations $C_{S_e} = 0.08$, the reliability correction factors to the reliability levels of 0.90, 0.95, and 0.99 are determined as follows:

A random variable for a fatigue limit is denoted by S_e and the standard deviation of the fatigue limit (σ_{S_e}) is calculated as

$$\sigma_{S_e} = C_{S_e} \times \mu_{S_e}$$

Denote a reliability level as R, and the probability of failure follows:

$$P[S_{e,R} < S_e] = 1 - R$$

Introducing a new variable,

$$Z = \frac{S_{e.r} - \mu_{S_e}}{\sigma_{S_e}}$$

then,

$$P[S_{e,R} < S_e] = -\varphi(Z) = 1 - R$$

where $\varphi(z)$ is the standard normal density function and

$$Z = -\varphi^{-1}(1 - R)$$

The fatigue limit with a reliability R can be written as follows:

$$S_{e,R} = \mu_{S_e} - \phi^{-1}(1 - R) \times \sigma_{S_e} = \mu_{S_e} - \phi^{-1}(1 - R) \times C_{S_e} \times \mu_{S_e}$$

and

$$C_R = \frac{S_{e,R}}{\mu_{S_e}} = 1 - \phi^{-1}(1 - R) \times C_{S_e}$$

Based on the coefficient of variations ($C_{S_e} = 0.08$), the reliability factors for $R = 0.9, 0.95$, and 0.99 are summarized as follows:

R	$1 - R$	$\phi^{-1}(1 - R)$	C_R
0.90	0.1	−1.282	0.897
0.95	0.05	−1.645	0.868
0.99	0.001	−2.326	0.814

5

STRAIN-BASED FATIGUE ANALYSIS AND DESIGN

YUNG-LI LEE
DAIMLERCHRYSLER
DARRYL TAYLOR
DAIMLERCHRYSLER

5.1 INTRODUCTION

In the previous chapters, the analysis and design of components using stress-based fatigue life prediction techniques were presented. This approach to the fatigue analysis of components works well for situations in which only elastic stresses and strains are present. However, most components may appear to have nominally cyclic elastic stresses, but notches, welds, or other stress concentrations present in the component may result in local cyclic plastic deformation. Under these conditions, another approach that uses the local strains as the governing fatigue parameter (the local strain–life method) was developed in the late 1950s and has been shown to be more effective in predicting the fatigue life of a component.

The local strain–life method is based on the assumption that the life spent on crack nucleation and small crack growth of a notched component can be approximated by a smooth laboratory specimen under the same cyclic deformation at the crack initiation site. This is illustrated in Figure 5.1. By using this concept it is possible to determine the fatigue life at a point in a cyclically loaded component if the relationship between the localized strain in the specimen and fatigue life is known. This strain–life relationship is typically represented as a curve of strain versus fatigue life and is generated by conducting strain-controlled axial fatigue tests on smooth, polished specimens of the material. Strain-controlled axial fatigue testing is recommended because the material at stress concentrations and notches in a component may be under cyclic plastic deformation even when the bulk of the component behaves elastically during cyclic loading.

The local strain–life method can be used proactively for a component during early design stages. Fatigue life estimates may be made for various potential design geometries and manufacturing processes prior to the existence of any actual components provided the material properties are available. This will result in a reduction in the number of design iterations by identifying and rejecting unsatisfactory designs early in the design process. This reduces the design cycle and gets product to market quickly. The local strain–life approach is preferred if the load history is irregular or random and where the mean stress and the load sequence effects are thought to be of importance. This method also provides a rational approach to differentiate the high-cycle fatigue and the low-cycle fatigue regimes and to include the local notch plasticity and mean stress effect on fatigue life.

Because the material data used are only related to the laboratory specimen, the fabrication effects in the actual component, such as surface roughness/finish, residual stress, and material properties alteration due to cold forming, and welding, may not appropriately be taken into account in the local strain–life approach. The effect of surface finish may be included by testing laboratory specimens with the same surface condition, but the extra cost and time involved may decrease the benefit of using the local strain–life approach.

This chapter presents the strain–life fatigue methodology in the following sequence: experimental test programs, determination and estimates of cyclic stress–strain and fatigue properties, mean stress correction models, and notch analysis.

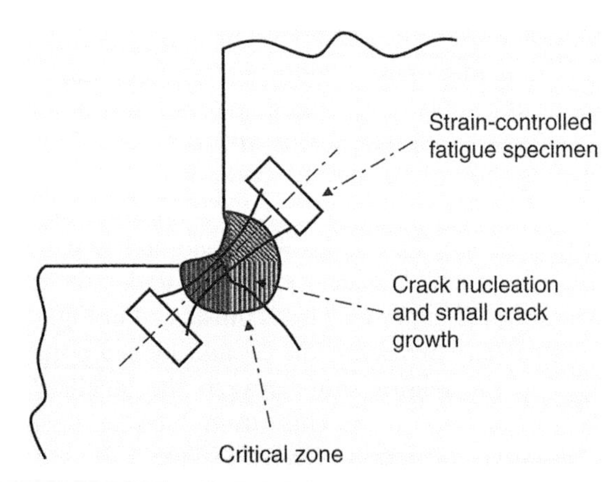

FIGURE 5.1 Concept of the local strain life approach.

5.2 EXPERIMENTAL TEST PROGRAM

5.2.1 SPECIMEN FABRICATION

According to the relevant standards, cylindrical specimens of approximately 12 mm and 6 mm diameter are normally tested for monotonic tensile and strain–life fatigue testing, respectively, though flat coupons may also be used. The recommended specimens have uniform or hourglass test sections. The specimen surface has to be longitudinally polished and checked carefully under magnification to ensure complete removal of machine marks within the test section. ASTM Standards E8M (ASTM, 2001) and E606 (ASTM, 1998a) list various types of sample configurations and dimensions for use in monotonic and constant-amplitude axial fatigue tests. Figure 5.2 shows configurations and dimensions of two typical fatigue test samples.

5.2.2 TEST EQUIPMENT

To perform strain–life fatigue analyses, several monotonic and cyclic material properties are needed. In this section, the test procedures, test equipment, and test samples used to determine these properties are described.

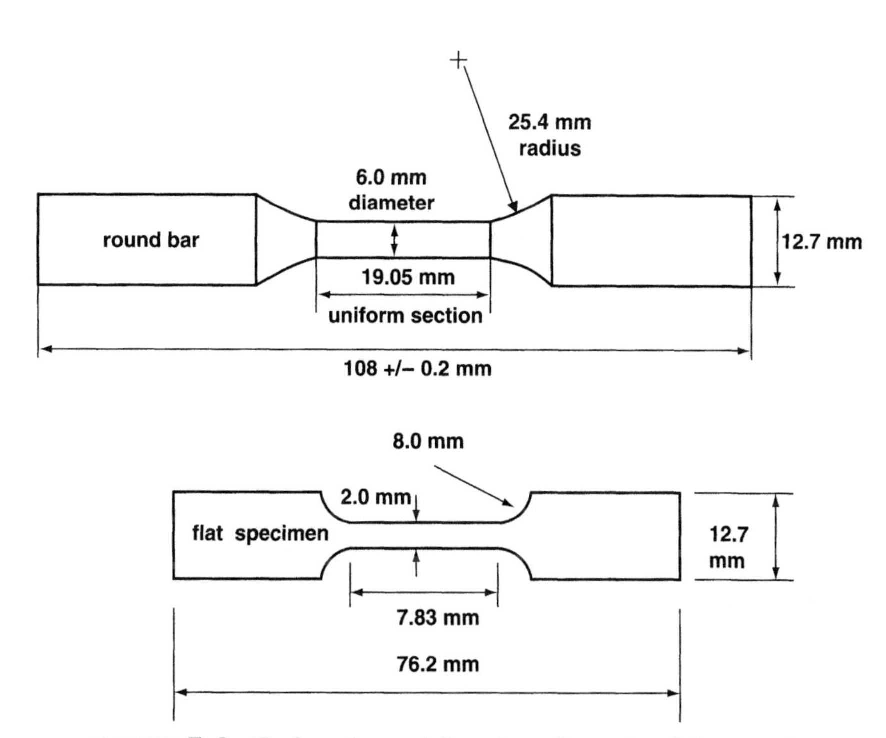

FIGURE 5.2 Configurations and dimensions of exemplary fatigue samples.

5.2.2.1 Apparatus

A commercial closed-loop servohydraulic axial load frame is used to conduct the tests. An extensometer or a high-quality clip gauge is used to measure the strain. The test machine uses the strain output to provide feedback to the servocontrollers. Figure 5.3 illustrates the schematic diagram of a strain-controlled fatigue testing system layout. To protect the specimen surface from the knife edge of the extensometer, ASTM E606 recommends the use of a transparent tape or epoxy to cushion the attachment. Most of the tests are conducted at room temperature. To minimize temperature effects on the extensometer and load cell calibration, ASTM E606 requires that temperature fluctuations be maintained within $\pm 2°C$ ($\pm 3.6°F$).

5.2.2.2 Alignment

Alignment of the load path components is essential for the accurate measurement of strain–life material constants. Misalignment of the load path components (such as load cell, grips, specimens, and actuator) can result from both tilt and offset between the center line of the load path system. According to ASTM E606, the maximum bending strain should not exceed 5% of the minimum axial strain range imposed during any test program. For example, if the minimum axial strain range is 6000 microstrains, the maximum allowable bending strain should be equal to or less than 300

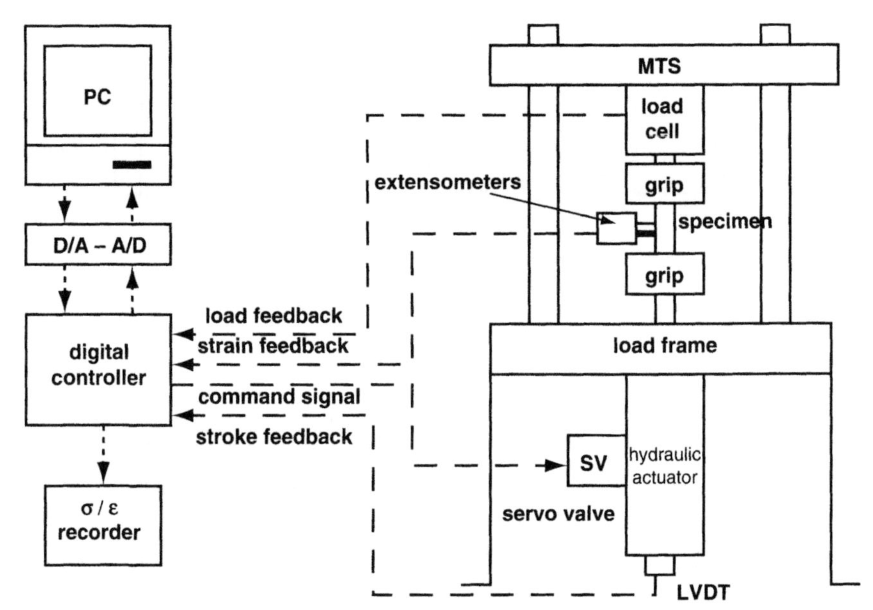

FIGURE 5.3 Schematic diagram of the strain-controlled fatigue testing system layout. Reproduced from Yang (1994).

microstrains. ASTM Standard E1012 (ASTM, 1999) documents methods to verify specimen alignment.

5.2.3 TEST METHODS AND PROCEDURES

5.2.3.1 Monotonic Tension Tests

ASTM Standard E8M (ASTM, 2001) defines the test method and requirements for the determination of the stress–strain relationship for a material from monotonic tension tests. Two specimens are used to obtain the monotonic properties. The specimens are tested to fracture under strain or displacement control. For the elastic and initial region (0 to 0.5% mm/mm), a strain rate about three quarters of the maximum allowable rate is chosen. After yielding (0.5 to 15% mm/mm), the strain rate is increased by a factor of 3. After the tension tests are concluded, the broken specimens are carefully reassembled. The final gage lengths of the fractured specimens, the final diameter, and the necking radius are then measured to determine the final strain state of the specimens.

5.2.3.2 Constant-Amplitude Axial Fatigue Tests

Constant-amplitude axial fatigue tests provide information about the cyclic and fatigue behavior of materials. All constant-amplitude axial fatigue tests are performed according to ASTM E606 to develop a strain–life curve over a range of approximately 100 to 5,000,000 cycles. In general, for each material of interest, at least 18 specimens are tested at various strain amplitudes (either with three duplicate tests per strain level or with one test for each strain amplitude). Failure of the specimens is defined when the maximum load decreases by 50% because of a crack or cracks being present. Each specimen is cycled by using fully reversed triangular waveform loads at an applied frequency between 0.20 and 3.0 Hz.

At long lives, the total strain becomes quite small, and the control of these quantities requires accurate instrumentation and extreme precision in the test procedure. A temporary solution of this problem is to use stress control at long lives. Tests with anticipated lives exceeding 1 million cycles are changed to stress control mode if the shape of the hysteresis loop and the stress amplitude are stabilized. Higher load frequencies in the range of 20 to 35 Hz can be used to shorten the overall test time and cost. As per ASTM E606, the frequency or strain rate should not increase the test sample temperature by more than the room temperature plus 2°C (3.6°F).

Incremental step-stress testing is an alternative approach that can be used to determine the cyclic stress–strain behavior. As illustrated in Figure 5.4, these tests are performed under fully reversed strain control using a constant ramp time. The specimen is cycled at ±0.1% increments from ±0.1% to

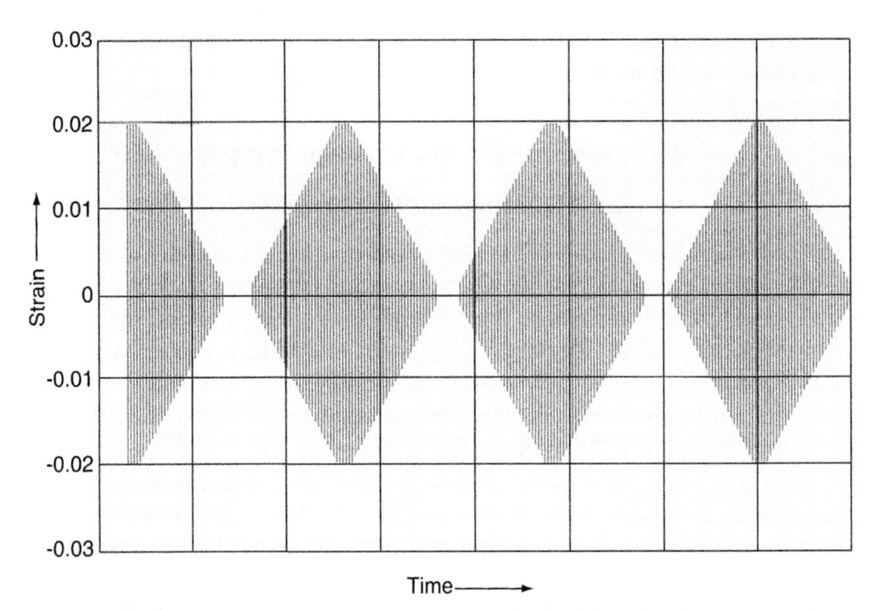

FIGURE 5.4 Strain–time and stress–time plots obtained by using the incremental step stress test for a material.

a maximum strain amplitude. The specimen is then unloaded starting with the maximum strain and continues in a similar progression to ±0.1%. The entire process is repeated until the cyclic stress–strain curve is stabilized, which means that there is no change in the stress–strain values between the last two loading blocks.

5.3 ANALYSIS OF MONOTONIC AND CYCLIC STRESS–STRAIN BEHAVIOR OF MATERIALS

Once monotonic and cyclic stress–strain data are generated by using the tests described in Section 5.2, these data are analyzed to determine the material constants needed for strain–life analyses.

5.3.1 MONOTONIC MECHANICAL PROPERTIES

The following parameters are determined from monotonic tensile tests:

S_{ys} = 0.2% Offset yield strength
S_u = Ultimate tensile strength
%RA = Percent reduction area
%EI = Percent elongation
E = Modulus of elasticity

K = Monotonic strength coefficient
n = Monotonic strain hardening exponent
σ_f = True fracture strength
ε_f = True fracture ductility

The engineering stress (S) and strain (e) relationship is determined by monotonic tensile tests performed on smooth, cylindrical specimens. Engineering stresses and strains are defined by using the original cross-sectional area (A_o) and original length (l_o) of the test specimen. For example, the engineering stress is

$$S = \frac{P}{A_o} \tag{5.3.1}$$

where P is the applied load. The engineering strain is

$$e = \frac{l - l_o}{l_o} = \frac{\Delta l}{l_o} \tag{5.3.2}$$

where l is the instantaneous length.

Four parameters can be measured directly from the engineering stress–strain relationship: yield strength (S_y), ultimate tensile strength (S_u), percentage elongation (%El), and reduction in area (%RA). A typical engineering stress and strain curve is plotted in Figure 5.5. The yield strength represents the limit of elastic behavior and is commonly defined as the stress associated with 0.2% plastic strain. The ultimate tensile strength is the maximum load carrying capability of a specimen and is defined as

$$S_u = \frac{P_{max}}{A_o} \tag{5.3.3}$$

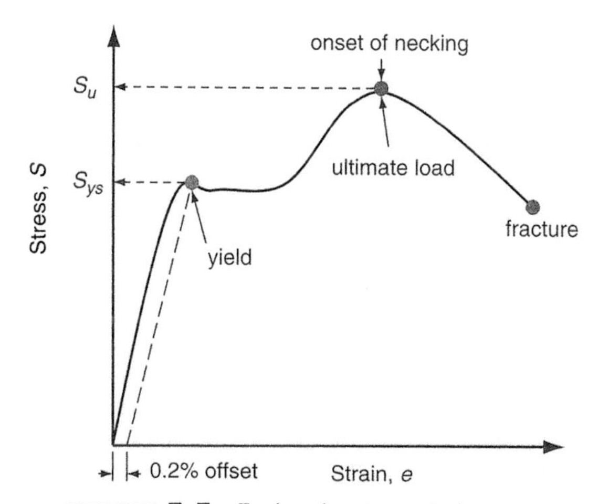

FIGURE 5.5 Engineering stress–strain curve.

Both percent elongation and reduction in area are measured by assembling a fractured tensile specimen together and measuring the final length (l_f) and final cross-sectional area (A_f). These two parameters are the measure of metal ductility prior to fracture and are defined as follows:

$$\%\text{EI} = 100 \times \frac{l_f - l_o}{l_o} \tag{5.3.4}$$

$$\%\text{RA} = 100 \times \frac{A_f - A_o}{A_o} \tag{5.3.5}$$

Instead of the original dimensions of the specimen, the true stress (σ) and the true strain (ε) are defined by using the instantaneous cross-sectional area (A) and length (l). They are expressed as follows:

$$\sigma = \frac{P}{A} \tag{5.3.6}$$

$$\varepsilon = \int_{l_o}^{l} \frac{dl}{l} = \ln \frac{l}{l_o} \tag{5.3.7}$$

Based on the assumption of constant volume during tensile testing, true stress and true strain are calculated from engineering stress and engineering strain, according to the following relationships:

$$\sigma = S(1 + e) \tag{5.3.8}$$

$$\varepsilon = \ln(1 + e) \tag{5.3.9}$$

These relationships are only valid up to the onset of necking in the specimen.

Assuming that the total strain can be decomposed into elastic (ε^e) and plastic strain (ε^p) components, the true stress versus true strain plot shown in Figure 5.6 is often described by the following equation (Ramberg and Osgood, 1943):

$$\varepsilon = \varepsilon^e + \varepsilon^p = \frac{\sigma}{E} + \left(\frac{\sigma}{K}\right)^{1/n} \tag{5.3.10}$$

where E is the modulus of elasticity. The material constants K and n describe the monotonic stress-plastic strain relationship and can be evaluated by the least-squares fit to the data plotted on a log–log graph as illustrated in Figure 5.7:

$$\sigma = K(\varepsilon^p)^n \tag{5.3.11}$$

In accordance with ASTM Standard E739 (ASTM, 1998b), when performing the least-squares fit, the logarithms of the stresses are assumed to be statistically dependent and the logarithms of the plastic strain statistically independent. It is suggested that any plastic strain less than a threshold value of 0.0005 mm/mm (Williams et al., 2003) be ignored to exclude the measurement errors.

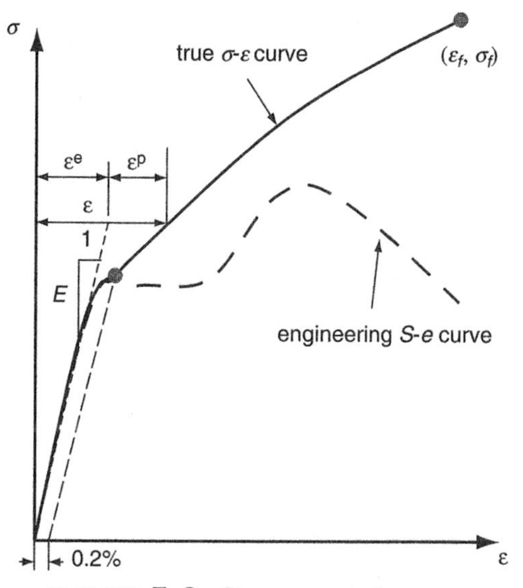

FIGURE 5.6 True stress–strain curve.

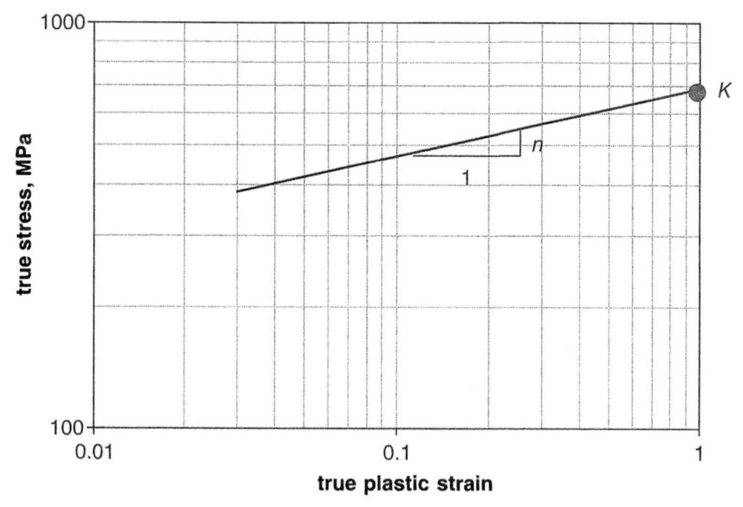

FIGURE 5.7 Monotonic true stress versus true plastic strain curve.

The true fracture strength is the true stress at the point of fracture during a monotonic tensile test and is corrected for necking by applying the Bridgman correction factor (Bridgman, 1944):

$$\sigma_f = \frac{\dfrac{P_f}{A_f}}{\left(1 + \dfrac{4R}{D_f}\right) \ln\left(1 + \dfrac{D_f}{4R}\right)} \tag{5.3.12}$$

where P_f is the load at fracture, R is the neck radius, and D_f is the diameter at fracture. The true fracture ductility ε_f is the true strain at final fracture and is calculated as follows:

$$\varepsilon_f = \ln\left(\frac{A_o}{A_f}\right) = \ln\left(\frac{100}{100 - \%RA}\right) \tag{5.3.13}$$

5.3.2 CYCLIC MATERIAL PROPERTIES

5.3.2.1 Transient Cyclic Response

The transient cyclic response of a material describes the process of change in the resistance of a material to deformation due to cyclic loading. If a material is repeatedly cycled under fully reversed strain-controlled loading, the material may respond in one of the following ways: cyclic hardening, cyclic softening, remaining stable, or some combination of these responses.

Figures 5.8 and 5.9 demonstrate transient cyclic hardening and transient cyclic softening, respectively, under strain-controlled loading. In transient cyclic hardening, the stress developed in each successive strain reversal increases as the number of cycles increases. In transient cyclic softening, the stress decreases as the number of cycles increases. In both cases, the rate of change of the applied stress will gradually reduce and the stress magnitude will reach a stable level (a steady-state condition) and remain stable for the rest of the fatigue life until the detection of the first fatigue crack. The

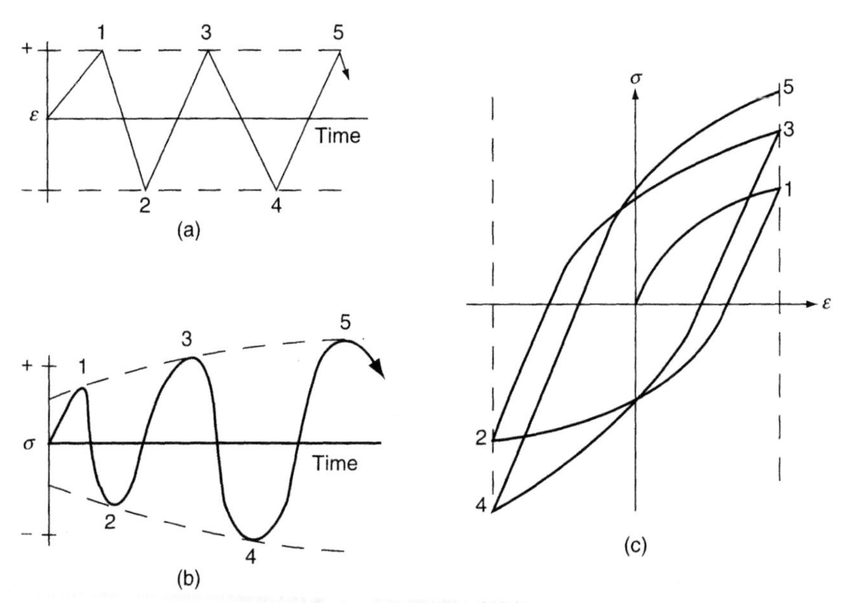

FIGURE 5.8 Transient behavior—cyclic hardening.

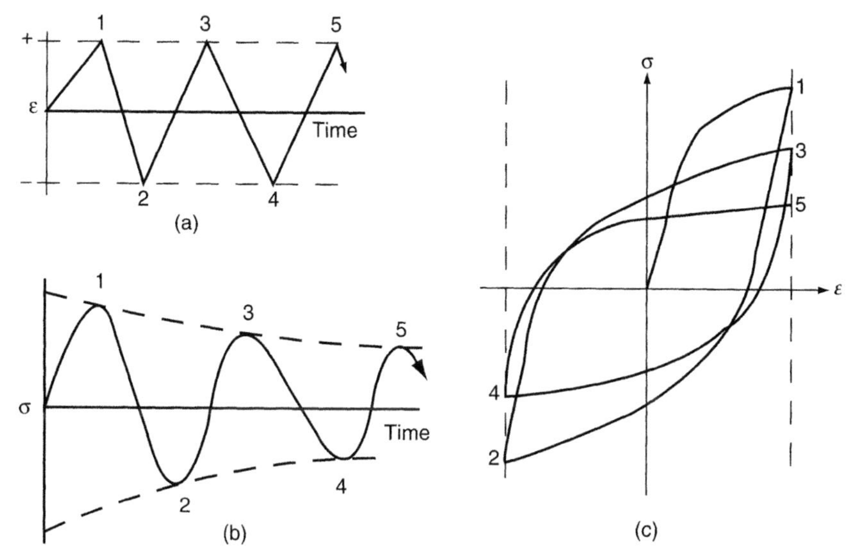

FIGURE 5.9 Transient behavior—cyclic softening.

transient cyclic behavior is handled by using the cyclically stable behavior in the local strain–life approach since the percentage of the number of cycles spent on this transient response is relatively small compared to the total fatigue life.

The transient phenomenon is believed to be associated with stability of the dislocation substructure within the metal crystal lattice of a material. In general, soft materials such as aluminum alloys with low dislocation densities tend to harden and hard materials such as steels tend to soften.

5.3.2.2 Steady-State Cyclic Stress–Strain Behavior

The properties determined from the steady-state cyclic stress–strain response are the following:

σ'_y = 0.2% Offset cyclic yield strength
K' = Cyclic strength coefficient
n' = Cyclic strain hardening exponent

Fatigue life can be characterized by the steady-state behavior because for constant strain–amplitude controlled testing, the stress–strain relationship becomes stable after rapid hardening or softening in the initial cycles corresponding to the first several percent of the total fatigue life. The cyclic stable stress–strain response is the hysteresis loop and is identified in Figure 5.10. The hysteresis loop defined by the total strain range ($\Delta\varepsilon$) and the total stress range ($\Delta\sigma$) represents the elastic plus plastic work on a material undergoing loading and unloading. Usually, the stabilized hysteresis loop is taken at half of the fatigue life.

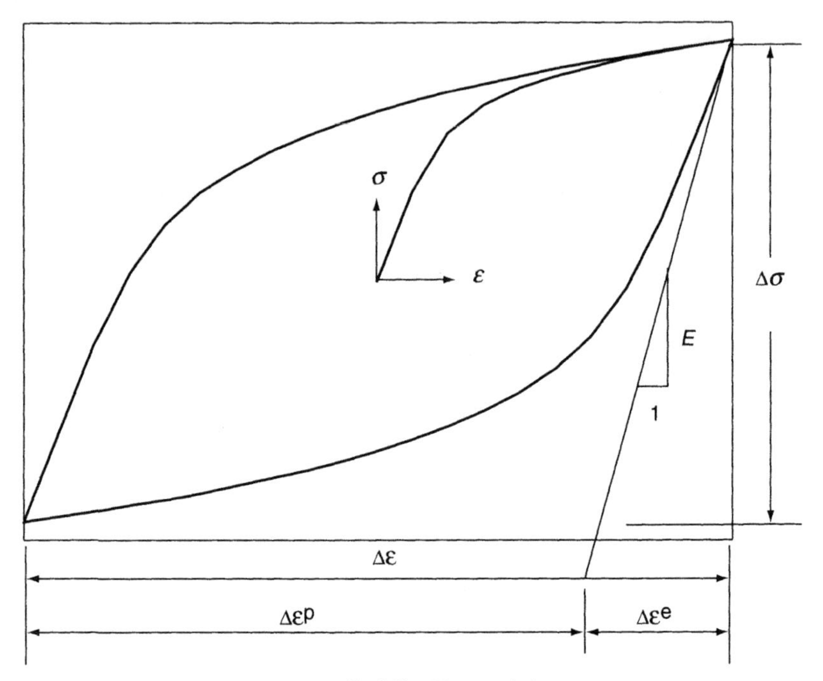

FIGURE 5.10 Hysteresis loop.

When a family of stabilized hysteresis loops with various strain amplitude levels is plotted on the same axes as shown in Figure 5.11, a cyclic stress–strain curve is defined by the locus of the loop tips and has the following form similar to the monotonic stress–strain response:

$$\varepsilon = \varepsilon^e + \varepsilon^p = \frac{\sigma}{E} + \left(\frac{\sigma}{K'}\right)^{1/n'} \tag{5.3.14}$$

where $'$ represents the parameters associated with cyclic behavior to differentiate them from those associated with monotonic behavior.

The cyclic yield stress (σ_y') is the stress at 0.2% plastic strain on a cyclic stress–strain curve. This value is determined by constructing a line parallel to the modulus of elasticity through the 0.2% strain offset at the zero stress point. The stress at which the line intersects the cyclic stress–strain curve is taken as the cyclic yield stress.

Masing's assumption (Masing, 1926) states that the stress amplitude versus strain amplitude curve can be described by the cyclic stress–strain curve, meaning

$$\varepsilon_a = \varepsilon_a^e + \varepsilon_a^p = \frac{\sigma_a}{E} + \left(\frac{\sigma_a}{K'}\right)^{1/n'} \tag{5.3.15}$$

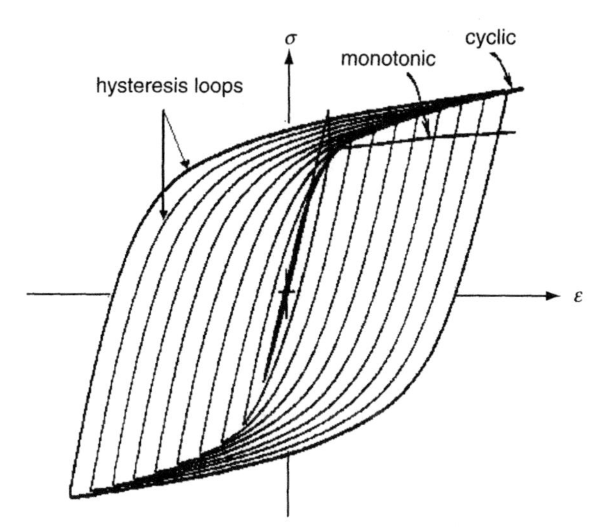

FIGURE 5.11 Cyclic stress-strain curve.

It also states that the material should have symmetric behavior in tension and compression. This is true for many homogeneous metals. Equation 5.3.15 can be rewritten in terms of strain range and stress range:

$$\frac{\Delta\varepsilon}{2} = \frac{\Delta\varepsilon^e}{2} + \frac{\Delta\varepsilon^p}{2} = \frac{\Delta\sigma}{2E} + \left(\frac{\Delta\sigma}{2K'}\right)^{1/n'} \tag{5.3.16}$$

which can be reduced to the following stabilized hysteresis loop equation:

$$\Delta\varepsilon = \frac{\Delta\sigma}{E} + 2\left(\frac{\Delta\sigma}{2K'}\right)^{1/n'} \tag{5.3.17}$$

Equation 5.3.17 has been widely used for describing and tracking the stress–strain behavior under variable amplitude loading conditions.

The cyclic strength coefficient K' and the cyclic strain hardening exponent n' are the intercept and slope of the best fit line, respectively, to the true stress amplitude $\frac{\Delta\sigma}{2}$ versus true plastic strain amplitude $\frac{\Delta\varepsilon^P}{2}$ data plotted on a log–log scale:

$$\frac{\Delta\sigma}{2} = K'\left(\frac{\Delta\varepsilon^p}{2}\right)^{n'} \tag{5.3.18}$$

where the true plastic strain amplitude is calculated by the following equation:

$$\frac{\Delta\varepsilon^p}{2} = \frac{\Delta\varepsilon}{2} - \frac{\Delta\sigma}{2E} \tag{5.3.19}$$

When performing the least-squares fit to determine K' and n', the logarithms of the true plastic strain amplitude are the statistically independent variable, and the logarithms of the stress amplitude are the statistically dependent variable. It is recommended (Williams et al., 2003) that plastic strain amplitudes less than 0.0005 mm/mm be neglected in the regression analysis

The cyclic stress–strain curve reflects the resistance of a material to cyclic deformation and can be different from the monotonic stress–strain curve. Superimposed monotonic and cyclic curves are shown in Figure 5.12. Typically, metals with a high monotonic strain–hardening exponent ($n > 0.2$) will harden whereas those with a low monotonic strain hardening exponent ($n < 0.1$) will cyclically soften. A rule of thumb (Bannantine et al., 1990) is that the material will harden if $S_u/S_y > 1.4$ and the material will soften if $S_u/S_y < 1.2$. For ratios of S_u/S_y between 1.2 and 1.4, the material can exhibit hardening, softening, or both. These figures indicate that use of monotonic material properties for fatigue life predictions can sometimes lead to inaccurate results.

Example 5.1. A notched component made of SAE 1137 carbon steel has the following material properties: $E = 209000$ MPa, $K' = 1230$ MPa, $n' = 0.161$. When the component is subjected to cyclic loading, a strain gage placed at the notch root records a fully reversed constant strain amplitude of 10,000 microstrains ($\mu\varepsilon$). Model the hysteresis stress–strain response at the notch root.

Solution. To determine the stress–strain response and hysteresis loop, we must first determine the initial loading case and then the initial reversal and all of the following. We first consider the initial loading.

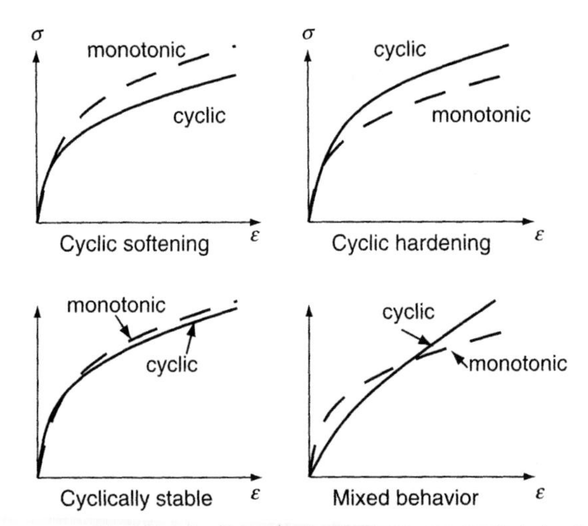

FIGURE 5.12 Cyclic and monotonic stress–strain behavior.

The initial loading starts from 0.0 $\mu\varepsilon$ and extends to the first 10,000 $\mu\varepsilon$ (= 0.010). To define the curve generated by the stress–strain relationship during initial loading, it is important to analyze multiple points along the strain path. For the initial loading, steps from 0.002 to 0.01 strains are used. To determine the stress at each intermediate step of the strain, the cyclic stress–strain equation is used:

$$\varepsilon_1 = \frac{\sigma_1}{E} + \left(\frac{\sigma_1}{K'}\right)^{\frac{1}{n'}}$$

For example, in the case of strain changing from 0 to 0.01, the stress σ_1 at the strain of 0.01 is determined:

$$0.01 = \frac{\sigma_1}{209000} + \left(\frac{\sigma_1}{1230}\right)^{\frac{1}{0.161}}$$

Solving for σ_1,

$$\sigma_1 = 557.46\,\text{MPa}$$

Table 5.1 shows the values for σ and ε at each increment in the initial loading condition. Having defined the initial loading condition, we can now move on to the first loading reversal. Again, it is important to define the curve so that multiple points along the strain are considered. To determine the change in stress, the following Masing model is used:

$$\Delta\varepsilon = \frac{\Delta\sigma}{E} + 2\left(\frac{\Delta\sigma}{2K'}\right)^{\frac{1}{n'}}$$

For example, in the case of the strain changing from 0.01 to −0.01, the stress increment $\Delta\sigma$ corresponding to the strain increment $\Delta\varepsilon = 0.02$ is obtained by

$$0.02 = \frac{\Delta\sigma}{209000} + 2\left(\frac{\Delta\sigma}{2 \times 1230}\right)^{\frac{1}{0.161}}$$

Solving for $\Delta\sigma$ results

TABLE 5.1 Initial Loading: True Stress and Strain Data Points

ε	σ (MPa)
0.002	342.98
0.004	447.03
0.006	497.56
0.008	531.56
0.010	557.46

$$\Delta\sigma = 1114.92\,\text{MPa}$$

Because Masing's model calculates the change in stress, to determine the stress at this point ($\varepsilon = -0.01$), the change must be compared to any stress value in this reversal with respect to the reference stress $\sigma = 557.46\,\text{MPa}$, the ending point of this reversal.

$$\sigma_2 = \sigma_1 - \Delta\sigma = 557.46 - 1114.92 = -557.46\,\text{MPa}$$

Table 5.2 shows the values for σ and ε at each increment in the first loading reversal. All the stress and the strain increments in each step are calculated with respect to the reference point.

For the second loading reversal, the same procedure is used as that for the first. For example, in the case of strain changing from -0.010 to 0.010, $\Delta\sigma$ corresponding to $\Delta\varepsilon = 0.020$ is determined by

$$0.02 = \frac{\Delta\sigma}{209000} + 2\left(\frac{\Delta\sigma}{2 \times 1230}\right)^{\frac{1}{0.161}}$$

Solving for $\Delta\sigma$,

$$\Delta\sigma = 1114.92\,\text{MPa}$$

Again, when comparing to the reference stress ($\sigma_2 = -557.46\,\text{MPa}$), the stress at this point is calculated as

$$\sigma_3 = \sigma_2 + \Delta\sigma = -557.46 + 1114.92 = 557.46\,\text{MPa}$$

Table 5.3 shows the tabulated values for σ and ε at each increment in the second reversal. From the points calculated for the initial, first, and second reversals, the hysteresis loop can be graphed in Figure 5.13, in which the local $\sigma - \varepsilon$ coordinate is located at each reference point.

TABLE 5.2 First Loading Reversal: True Stress and Strain Data Points

ε	$\Delta\varepsilon$	$\Delta\sigma$ (MPa)	σ (MPa)
0.010	Reference	0.0	557.46
0.008	0.002	411.70	145.76
0.006	0.004	685.96	−128.5
0.004	0.006	815.30	−257.84
0.002	0.008	894.06	−336.6
0.000	0.010	950.70	−393.24
−0.002	0.012	995.12	−437.66
−0.004	0.014	1031.79	−474.33
−0.006	0.016	1063.12	−505.66
−0.008	0.018	1090.52	−533.06
−0.010	0.020	1114.92	−557.46

TABLE 5.3 Second Loading Reversal: True Stress and Strain Data Points

ε	$\Delta\varepsilon$	$\Delta\sigma$ (MPa)	σ (MPa)
−0.010	Reference	0.0	−557.46
−0.008	0.002	411.70	−145.76
−0.006	0.004	685.96	128.5
−0.004	0.006	815.30	257.84
−0.002	0.008	894.06	336.6
0.000	0.010	950.70	393.24
0.002	0.012	995.12	437.66
0.004	0.014	1031.79	474.33
0.006	0.016	1063.12	505.66
0.008	0.018	1090.52	533.06
0.010	0.020	1114.92	557.46

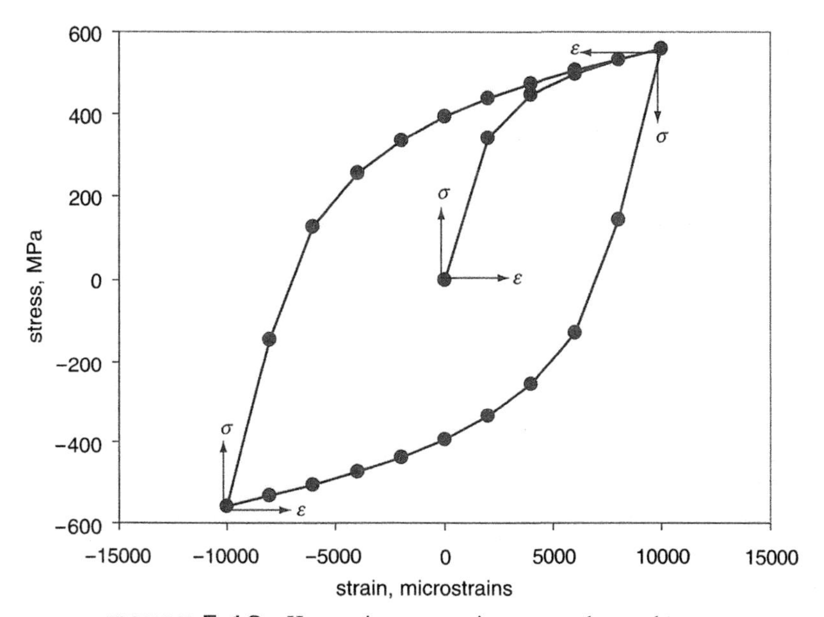

FIGURE 5.13 Hysteresis stress-strain curve at the notch root.

5.3.2.3 Constant-Amplitude Fatigue Behavior

The properties that are determined from stabilized hysteresis loops and strain–life data are the following:

σ_f' = Fatigue strength coefficient

b = Fatigue strength exponent, usually varying between −0.04 and −0.15 for metals

ε_f' = Fatigue ductility coefficient

c = Fatigue ductility exponent, usually varying between –0.3 and –1.0 for metals

$2N_T$ = Transition fatigue life in reversals

Based on the proposal by Morrow (1965), the relation of the total strain amplitude (ε_a) and the fatigue life in reversals to failure ($2N_f$) can be expressed in the following form:

$$\varepsilon_a = \varepsilon_a^e + \varepsilon_a^p = \frac{\sigma_f'}{E}(2N_f)^b + \varepsilon_f'(2N_f)^c \tag{5.3.20}$$

Equation 5.3.20, called the strain–life equation, is the foundation for the strain-based approach for fatigue. This equation is the summation of two separate curves for elastic strain amplitude–life ($\varepsilon_a^e - 2N_f$) and for plastic strain amplitude-life ($\varepsilon_a^p - 2N_f$). Dividing the Basquin (1910) equation by the modulus of elasticity gives the equation for the elastic strain amplitude-life curve:

$$\varepsilon_a^e = \frac{\Delta\varepsilon^e}{2} = \frac{\sigma_a}{E} = \frac{\sigma_f'}{E}(2N_f)^b \tag{5.3.21}$$

Both Manson (1953) and Coffin (1954) simultaneously proposed the equation for the plastic strain amplitude-life curve:

$$\varepsilon_a^p = \frac{\Delta\varepsilon^p}{2} = \varepsilon_f'(2N_f)^c \tag{5.3.22}$$

When plotted on log–log scales, both curves become straight lines as shown in Figure 5.14. Besides the modulus of elasticity E, the baseline strain

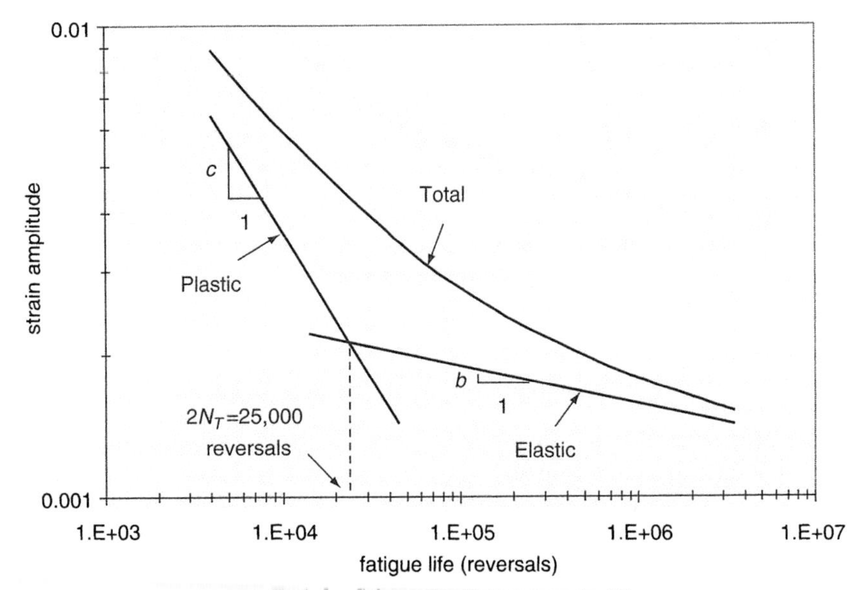

FIGURE 5.14 Schematic of a total strain–life curve.

life expression is defined by four regression parameters: fatigue strength coefficient (σ_f'), fatigue strength exponent (b), fatigue ductility coefficient (ε_f'), and fatigue ductility exponent (c).

When determined by using a least-squares analysis, these parameters are valid in the range of the experimental data. This emphasizes the necessity for a broad range in experimental data and a sufficient number of data points required in each of the elastic strain dominant and the plastic strain dominant regions. Instead of using the nonlinear least-squares method for the strain–life model (Langlais and Vogel, 1995), the approach preferred here is to use the linear model for fitting each linear range of the data because it gives less weight to misleading low-cycle and high-cycle data and provides better results (Yan et al., 1992; Williams et al., 2003). The choice of the linear range of data depends on the transition fatigue life.

The transition fatigue life point is defined as the fatigue life when the magnitude of plastic strain amplitude is equal to that of elastic strain amplitude. As shown in Figure 5.14, the transition fatigue life ($2N_T$) is the intersection of the elastic and plastic strain-life curves. In this case, a fatigue life of 25,000 reversals is just an example of where the transition life may occur. The region to the left of this point where fatigue life is less than the transition fatigue life is considered the plastic strain dominant region and is referred to as the low-cycle fatigue (LCF) region. The region to the right where fatigue life is higher than the transition fatigue life is the elastic strain dominant region and is referred to as the high-cycle fatigue (HCF) region. Figure 5.15 shows the

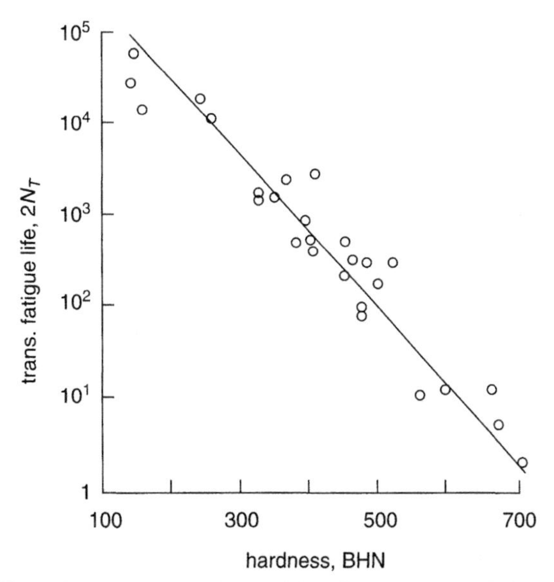

FIGURE 5.15 Relationship between transition life and hardness for steels. Adapted with permission from Landgraf (1970), ASTM STP467, Copyright ASTM International.

order-of-magnitude variation in transition life for steels with various hardness values where steels with high hardness and ultimate strength have lower transition fatigue life.

The following two examples demonstrate how strain–life fatigue parameters are determined by the least-squares analysis. Different from the ASTM standard E739 (ASTM 1998b), an approach for the statistical analysis of strain–life fatigue data (Williams et al., 2003) is introduced here and has the following unique features:

(1) A linear regression model is restricted to the linear range of the data, e.g., either the low-cycle fatigue data or the high-cycle fatigue data.
(2) A threshold of plastic strain amplitude of 0.0005 is suggested. Below this value, the data points can be neglected to avoid measurement errors.
(3) A method based on a modification of the Owen tolerance interval (Shen et al., 1996) is used to quantify the statistical variation of fatigue data.

In these examples, the introduction of the transition fatigue life is important to accurately determine the fatigue strength parameters (b and σ_f') and the fatigue ductility parameters (c and ε_f'). The elastic strain dominant and the plastic strain dominant regions need to be emphasized, respectively. By using the transition fatigue life as the boundary between these two regions, the data points in each region can be accurately represented and appropriately weighted in this analysis.

Example 5.2. A test company conducted strain-controlled constant-amplitude axial fatigue tests on 16 samples with one sample for each strain amplitude. A summary of the test results is shown in Table 5.4. The samples were made of SAE 1137 mild carbon steel. Throughout the tests, failures were defined when load drop was 50% of the initial value. BKE denotes failure detected between knife edges and NOKE denotes failure detected outside the knife edges, with no other cracks in between. Based on the available experimental data, determine the median and the R90C90 cyclic and fatigue properties.

Solution. First, the fracture location must be considered for each sample. If the sample did not fail, the data point should be excluded from this analysis, because this analysis technique is valid only for failure occurrences. In this case, the data point with Sample 13 is not considered in this analysis.

If the sample displayed a failure that occurs outside of the knife edges (test section) with no other cracks evident within the test section, a NOKE failure, that point is circumspect. To determine the validity of this point, the duration before failure must be checked. If the failure was obviously premature, occurring after a small fraction of the anticipated number of cycles with respect to the life trend, the point should be excluded and a note with an

TABLE 5.4 Constant-Amplitude Axial Fatigue Test Data for SAE 1137 Carbon Steel

Sample Number	Modulus (MPa)	Total Strain Amplitude (mm/mm)	Max. Stress at Half-Life (MPa)	Min. Stress at Half-Life (MPa)	Total Fatigue Life (cycles)	Failure Location
16	208229	0.00900	545	−561	2117	BKE
15	207815	0.00800	532	−550	2351	BKE
12	206850	0.00700	515	−529	3699	BKE
10	211676	0.00600	500	−513	5502	BKE
4	208919	0.00500	461	−467	7384	BKE
9	215124	0.00450	456	−466	10035	BKE
1	206850	0.00400	437	−444	15264	BKE
3	207540	0.00350	420	−427	20884	BKE
2	210297	0.00300	401	−409	38552	BKE
5	208229	0.00250	374	−379	66838	BKE
8	209608	0.00225	354	−366	92467	BKE
6	210298	0.00200	342	−358	218749	BKE
11	209963	0.00188	330	−336	336546	NOKE
7	206161	0.00175	316	−322	1663979	BKE
13	207367	0.00150	288	−295	5000000	NONE

explanation made. The data point for Sample 11 is valid from the aspect of the life trend. Other measurements can be checked to verify the decision to exclude a point. For example, one can check the minimum and maximum stresses observed during an individual test. The absolute value of the minimum and maximum should be approximately equal. If not, a mean or residual stress has likely developed because of improper testing procedure. Such effects should not occur in testing of this nature.

Next, the elastic and plastic strain amplitudes are calculated. The measured data allow one to determine the location of the transition fatigue life. The plastic strain and elastic strain amplitudes are required to be calculated for each sample in the test. A new data set for fatigue data analysis is generated in Table 5.5. Based on the same magnitude of the calculated elastic and plastic strain amplitudes in Table 5.5, the transition life is roughly estimated as 22,000 reversals.

Once the transition life is estimated, the data to be analyzed are chosen such that an intersection of the plastic and elastic strain curve is ensured. The choice of the data points in each region is very subjective. Extra data points beyond the elastic-strain or the plastic-strain dominant region are selected because the transition fatigue life has a stochastic nature. In this example, all the data points with viable lives less than and the two data points immediately greater than the transition life are selected for obtaining fatigue ductility parameters. Fatigue strength parameters are developed by using all the data points with viable lives greater than and the two points immediately less than the transition life.

TABLE 5.5 Summary of Useful Fatigue Test Data for SAE 1137 Mild Carbon Steel

Sample Number	Total Strain Amplitude (mm/mm)	Average Stress Amplitude at Half-Life (MPa)	Modulus (MPa)	Total Fatigue Life (reversals)	Calculated Elastic Strain Amplitude (mm/mm)	Calculated Plastic Strain Amplitude (mm/mm)
16	0.00900	553	208229	4234	0.002656	0.006344
15	0.00800	540	207815	4702	0.002603	0.005397
12	0.00700	522	206850	7398	0.002523	0.004477
10	0.00600	507	211676	11004	0.002394	0.003606
4	0.00500	464	208919	14768	0.002221	0.002779
9	0.00450	461	215124	20070	0.002144	0.002356
1	0.00400	441	206850	30528	0.00213	0.001870
3	0.00350	424	207540	41774	0.002041	0.001459
2	0.00300	405	210297	77104	0.001925	0.001075
5	0.00250	376	208229	133676	0.001806	0.000695
8	0.00225	360	209608	184934	0.001718	0.000532
6	0.00200	350	210298	437498	0.00163	0.000337
11	0.00188	333	209963	673092	0.001587	0.000293
7	0.00175	319	206161	3327958	0.001548	0.000202

Note: Underlined values indicate which data points were used for the low-cycle and high-cycle fatigue data analysis.

Low-Cycle Fatigue Data Analysis

To begin the fatigue ductility parameter analysis, the following data are required: total life and calculated plastic strain amplitude for the plastic region data points. Because the plastic strain amplitude is the control parameter, the statistical independent variable X for the analysis will be the logarithms of the plastic strain amplitude, $\log(\varepsilon_a^p)$, and the statistical dependent variable Y will be the logarithms of the fatigue life, $\log(2N_f)$. A linear regression analysis by using the least-squares method on the data selected is performed. The R^2 regression statistic should be more than 0.90. A lower R^2 value suggests that substantial variations due to factors other than the linear relationship between the two variables exist. Given the intercept (A), slope (B), and variables X and Y, the equation for the regression line takes the linear form:

$$Y = \hat{A} + \hat{B}X$$

where

$$\hat{A} = (-1/c)\log(\varepsilon_f')$$
$$\hat{B} = 1/c$$

and

$$c = 1/\hat{B} \tag{5.3.23}$$

$$\varepsilon_f' = 10^{(-c \times \hat{A})} \tag{5.3.24}$$

The linear regression analysis results in $\hat{A} = 0.06933$, $\hat{B} = -1.611$, $R^2 = 0.9942$, and $s = 0.03011$. By using Equations 5.3.23 and 5.3.24, the fatigue ductility parameters are found to be $c = -0.6207$ and $\varepsilon_f' = 1.104$. Because the number of data points in the analysis exceeds five, a reliability and confidence analysis can be performed. For this example, an R90C90 analysis is illustrated. Following the statistical analysis method described in Section 4.2.2, the R90C90 regression line Y_{R90C90} can be determined by shifting the median regression line toward the left with the following equation:

$$Y_{R90C90} = Y - K_{\mathrm{OWEN}} \times s \tag{5.3.25}$$

where s is the standard error of Y on X. In this case for the sample size $n = 8$, the Owen tolerance factor, K_{OWEN}, is 2.441, and the standard error (s) from the linear regression analysis is 0.03011. Following the procedure in Section 4.2.2 and the assumption of a constant fatigue ductility exponent ($c = -0.6207$), the R90C90 fatigue ductility coefficient ($\varepsilon'_{f,R90C90}$) is determined to be 0.9938.

High-Cycle Fatigue Data Analysis

The analysis for the cyclic properties relevant to the elastic-strain dominant region (b and σ_f') is performed using the same method and criteria previously detailed, but the half-life stress amplitude is used instead of the plastic strain amplitude. The elastic strain amplitude could be used, but the coefficient produced would have to be multiplied by the average value of the modulus of elasticity to produce the desired strength coefficient. To generate the values of σ_f' and b when performing the least squares on the high-cycle fatigue data, the logarithms of the stress amplitude, $\log(\sigma_a)$, are the statistical independent variable, and the logarithms of the reversals to failure, $\log(2N_f)$, are the statistical dependent variable. Again, the regression equation assumes the following linear form:

$$Y = \hat{A} + \hat{B}X$$

where

$$\hat{A} = (-1/b)\log(\sigma_f')$$

$$\hat{B} = 1/b$$

and

$$b = 1/\hat{B} \tag{5.3.26}$$

$$\sigma_f' = 10^{(-b \times \hat{A})} \tag{5.3.27}$$

The linear regression analysis results in $\hat{A} = 37.11$, $\hat{B} = -12.36$, $R^2 = 0.9517$, and $s = 0.1742$. By using Equations 5.3.26 and 5.3.27, the fatigue strength parameters are determined to be $b = -0.0809$ and $\sigma_f' = 1006\,\text{MPa}$.

Because the number of data points in the analysis exceeds five, a reliability and confidence analysis can be performed. In this case, for a sample size $n = 10$, the Owen tolerance factor, K_{OWEN}, is 2.253, and the standard error, s, from the linear regression analysis is 0.1742. Following the procedure in Section 4.2.2 and using the assumption of constant fatigue strength exponent ($b = -0.0809$), the R90C90 fatigue strength coefficient ($\sigma'_{f,R90C90}$) is determined to be 936 MPa.

Once the four fatigue parameters have been derived, the total strain–life curve can be assembled. The graph in Figure 5.16 consists of the total strain–life curve, plastic strain–life curve, and elastic strain–life curve. These three curves are plotted on a log–log scale graph with the experimental data superimposed on the graph. This aids in demonstrating the quality of the curve fit. If the R90C90 curves were generated, these curves can be superimposed with the median curves for design, as shown in Figure 5.16.

Analyses of the Cyclic Stress–Strain Behavior

The cyclic strength coefficient K' and the cyclic strain hardening exponent n' are related by Equation 5.3.18. Instead of using the nonlinear least-squares

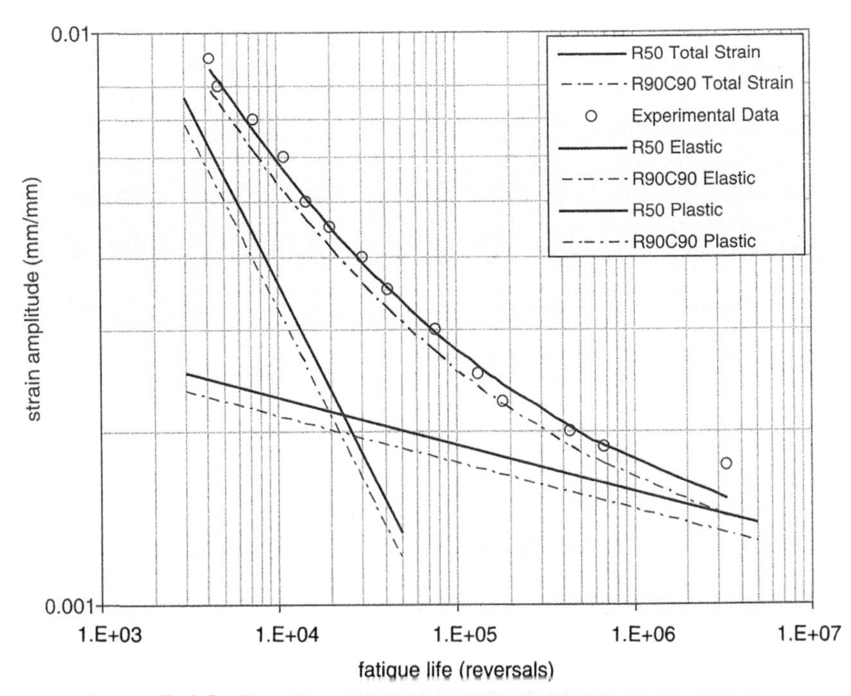

FIGURE 5.16 Total strain–life curves, R50 and R90C90, for SAE 1137 steel.

method (Langlais and Vogel, 1995) for the Masing equation, Equation 5.3.15, the fitting procedure illustrated here is to use the linear least-squares analysis to fit the entire range of the logarithms of stress and plastic strain data. When performing the least-squares fit, the logarithms of the true plastic strain amplitude, $\log(\varepsilon_a^p)$, are the statistical independent variable X and the logarithms of the stress amplitude, $\log(\sigma_a)$, are the statistical dependent variable Y. Plastic strain amplitudes of less than 0.0005 are neglected to avoid measurement errors. Given the intercept A, slope B, and variables X and Y, the equation for the regression line takes the linear form:

$$Y = \hat{A} + \hat{B}X$$

where

$$\hat{A} = \log(K')$$

$$\hat{B} = n'$$

and

$$n' = \hat{B} \tag{5.3.28}$$

$$K' = 10^{\hat{A}} \tag{5.3.29}$$

In an analysis of the data presented in Table 5.5, all the sample data in which plastic strain amplitude exceeds 0.0005 are included. The linear regression analysis results in $\hat{A} = 3.090$, $\hat{B} = 0.1608$, $R^2 = 0.9905$, and $s = 0.008126$. Thus, $n' = 0.1608$ and $K' = 1230\,\text{MPa}$.

Reliability and confidence analysis for the cyclic stress–strain parameters can be performed. In this case, for the sample size $n = 11$, the Owen tolerance factor, K_{OWEN}, is 2.162, and the standard error, s, from the linear regression analysis is 0.008126. With the assumption of constant cyclic strain hardening exponent ($n' = 0.1608$), the R90C90 cyclic strength coefficient (K'_{R90C90}) is determined to be 1181 MPa. The R50 and R90C90 cyclic stress–strain curves for SAE 1137 carbon steel are illustrated in Figure 5.17.

If these simple power laws (Masing equation, Basquin equation, and Coffin and Manson equation) are perfect fits to the test data, only four of the six constants are independent. Solving Equations 5.3.21 and 5.3.22 to eliminate fatigue life ($2N_f$) and then comparing the results with Equation 5.3.18 gives the following relation between these six fatigue properties:

$$K' = \frac{\sigma'_f}{\left(\varepsilon'_f\right)^{n'}} \tag{5.3.30}$$

$$n' = \frac{b}{c} \tag{5.3.31}$$

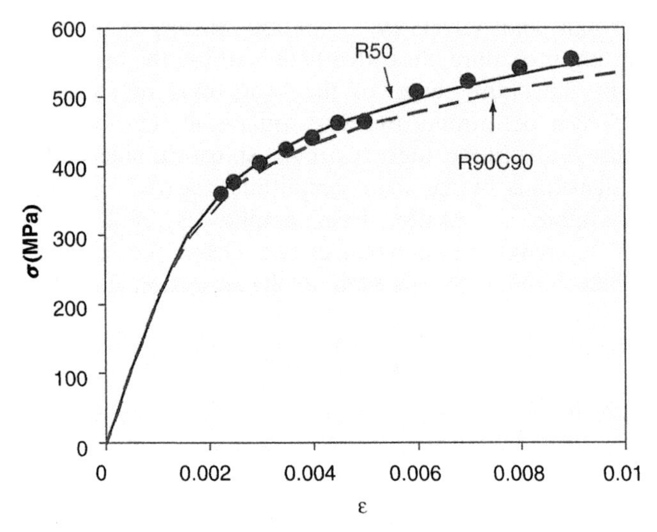

FIGURE 5.17 Cyclic stress–strain curves, R50 and R90C90, for SAE 1137 steel.

However, because of the different choice of the fitting data range, these equations are no longer valid. A discrepancy arises because the Basquin equation and the Coffin and Manson equation apply only locally and the equations are incorrect for data outside the range as a result of the local fitting procedure. Similarly, plastic strain components are in error outside the range in which the Massing equation (Equation 5.3.18) was fit. Consequently, elastic and plastic strains calculated from the strain life curve disagree with the elastic and plastic strains computed from the cyclic stress–strain curve, or

$$\frac{\sigma_f'}{E}(2N_f)^b \neq \frac{\sigma_a}{E} \tag{5.3.32}$$

and

$$\varepsilon_f'(2N_f)^c \neq \left(\frac{\sigma_a}{K'}\right)^{1/n'} \tag{5.3.33}$$

Determination of the Transition Fatigue Life

The transition fatigue life in reversals ($2N_T$) can be calculated by equating the elastic and plastic strain components from the strain–life equation:

$$\frac{\sigma_f'}{E}(2N_T)^b = \varepsilon_f'(2N_T)^c \tag{5.3.34}$$

and

$$2N_T = \left(\frac{\varepsilon_f' E}{\sigma_f'}\right)^{1/b-c} \tag{5.3.35}$$

TABLE 5.6 Data Analysis Summary for SAE 1137 Carbon Steel

		R50	R90C90
Average Young's modulus (MPa)	E	209000	209000
Fatigue strength coefficient (MPa)	σ_f'	1006	936
Fatigue strength exponent	b	−0.0809	−0.0809
Fatigue ductility coefficient	ε_f'	1.104	0.9938
Fatigue ductility exponent	c	−0.6207	−0.6207
Cyclic strength coefficient (MPa)	K'	1230	1181
Cyclic strain hardening exponent	n'	0.1608	0.1608
Transition fatigue life (reversals)	$2N_T$	23600	22200

Therefore, the transition fatigue lives for R50 and R90C90 are obtained as 23,700 reversals and 22,300 reversals, respectively. A typical summary of this analysis is presented in Table 5.6.

Example 5.3. Repeat the same fatigue data analysis demonstrated in Example 5.2 for the SAE D4512, a low-ductility, high-strength alloy used in the industrial case components. The strain-controlled axial fatigue test results are summarized in Table 5.7. Determine the median and the R90C90 strain–life curves.

Solution. A brief material analysis is provided for SAE D4512, which is a steel used in industrial cast components. This is a low-ductility, high-strength alloy that does not exhibit the transition fatigue life point within a commonly tested total strain range. The transition life point for this material occurs beyond the experimental data range at approximately 800 reversals. In such a case, the criterion for data selection is plastic strain being greater than 0.0005 mm/mm rather than the data's relationship to the position of the transition life point.

The data presented in Table 5.7 are analyzed by the same method discussed in the example for SAE 1137 carbon steel. The only difference is how the data are selected. The data are reviewed using the same failure location criteria and the same minimum plastic strain requirements. The selection of data to use for the plastic strain analysis is the primary difference. In the situation in which the transition life point is not present in the data, all data with a calculated plastic strain greater than 0.0005 mm/mm are used to develop the fatigue ductility coefficient and exponent. To determine the fatigue strength coefficient and exponent, all data points that were not excluded because of an improper failure are used. Figure 5.18 shows the median and R90C90 total strain–life curves, the median and R90C90 plastic and elastic regression curves, and the experimental data. Figure 5.19 depicts the R50 and R90C90 cyclic stress–strain curves. A typical summary of this analysis is presented in Table 5.8.

TABLE 5.7 Summary of Useful Fatigue Test Data for SAE 4512

Sample Number	Total Strain Amplitude (mm/mm)	Average Stress Amplitude at Half-Life (MPa)	Modulus (MPa)	Total Fatigue Life (reversals)	Calculated Elastic Strain Amplitude (mm/mm)	Calculated Plastic Strain Amplitude (mm/mm)	Failure Location
14	0.00500	477	175695	1066	0.002715	0.002285	BKE
6	0.00450	469	177762	3780	0.002638	0.001862	BKE
9	0.00400	451	173628	5914	0.002598	0.001402	BKE
11	0.00350	433	172250	14842	0.002514	0.000986	BKE
16	0.00315	422	172939	21126	0.002440	0.000710	BKE
4	0.00300	411	171561	33064	0.002396	0.000604	BKE
15	0.00275	403	172939	22138	0.002330	0.000420	BKE
19	0.00265	395	173284	31888	0.002280	0.000370	BKE
1	0.00250	376	172939	72100	0.002174	0.000326	BKE
8	0.00235	373	175695	114936	0.002123	0.000227	BKE
10	0.00215	354	175695	125840	0.002015	0.000135	BKE
2	0.00200	329	174317	314982	0.001887	0.000113	OKE
13	0.00183	303	172250	1337430	0.001759	0.000071	BKE
3	0.00175	289	174317	438390	0.001658	0.000092	BKE
12	0.00165	278	172939	930654	0.001608	0.000043	BKE
5	0.00150	257	174317	11025548	0.001474	0.000026	NONE
7	0.00100	171	174317	305602	0.000981	0.000019	NONE

Note: Underlined values indicate which data points were used for the elastic analysis and the plastic analysis. The last two rows of data points were excluded from the analysis for no cracks.

5.4 MEAN STRESS CORRECTION METHODS

In designing for durability, the presence of a nonzero mean normal stress can influence fatigue behavior of materials because a tensile or a compressive normal mean stress has been shown to be responsible for accelerating or decelerating crack initiation and growth. Experimental data support that compressive mean normal stresses are beneficial and tensile mean normal stresses are detrimental to fatigue life. This has been observed under conditions when the mean stress levels are relatively low compared to the cyclic yield stress and the fatigue behavior falls in the long-life regime where elastic strain is dominant.

In conjunction with the local strain life approach, many models have been proposed to quantify the effect of mean stresses on fatigue behavior. The commonly used models in the ground vehicle industry are those by Morrow (1968) and by Smith, Watson, and Topper (Smith et al., 1970). These equations are empirically based and should be compared with test data to determine which model is the most appropriate for the material and test conditions of interest. The two models are described in the following sections.

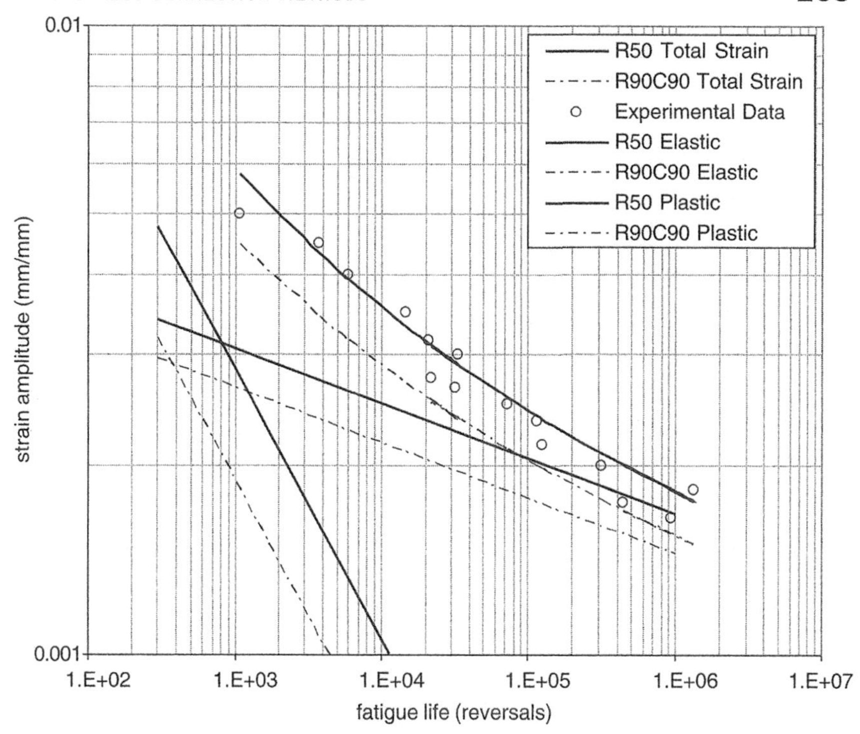

FIGURE 5.18 Total strain–life curves, R50 and R90C90, for SAE D4512 steel.

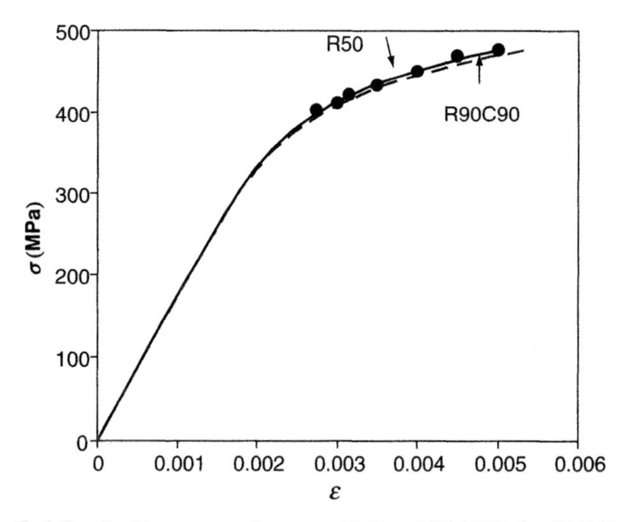

FIGURE 5.19 Cyclic stress–strain curves, R50 and R90C90, for SAE D4512 steel.

TABLE 5.8 Data Analysis Summary for SAE D4512 Steel

		R50	R90C90
Average Young's modulus (MPa)	E	174000	174000
Fatigue strength coefficient (MPa)	σ_f'	978	884
Fatigue strength exponent	b	−0.0876	−0.0876
Fatigue ductility coefficient	ε_f'	0.0556	0.0371
Fatigue ductility exponent	c	−0.4305	−0.4305
Cyclic strength coefficient (MPa)	K'	939	926
Cyclic strain hardening exponent	n'	0.111	0.111
Transition fatigue life (reversals)	$2N_T$	799	329

5.4.1 MORROW'S MEAN STRESS CORRECTION METHOD

Morrow has proposed the following relationship when a mean stress is present:

$$\varepsilon_a = \frac{\sigma_f' - \sigma_m}{E}\left(2N_f\right)^b + \varepsilon_f'\left(2N_f\right)^c \tag{5.4.1}$$

This equation implies that mean normal stress can be taken into account by modifying the elastic part of the strain–life curve by the mean stress (σ_m). As illustrated in Figure 5.20, the model indicates that a tensile mean stress would reduce the fatigue strength coefficient σ_f' whereas a compressive mean stress would increase the fatigue strength coefficient. This equation has been extensively cited for steels and used with considerable success in the long-life regime when plastic strain amplitude is of little significance.

Walcher, Gray, and Manson (Walcher et al., 1979) noted that for other materials such as Ti-6Al-4V, σ_f' is too high a value for mean stress correction

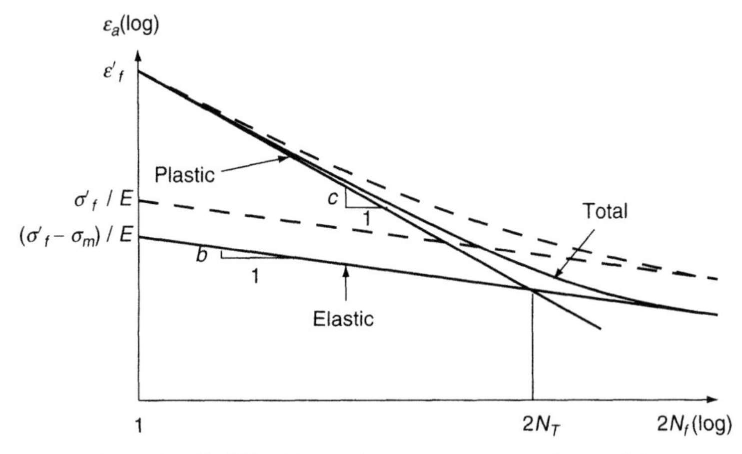

FIGURE 5.20 Morrow's mean stress correction model.

and an intermediate value of $k_m\sigma_f'$ can be determined by experiment. Thus, a generic formula was proposed:

$$\varepsilon_a = \frac{k_m\sigma_f' - \sigma_m}{E}(2N_f)^b + \varepsilon_f'(2N_f)^c \tag{5.4.2}$$

This equation requires additional test data to determine $k_m\sigma_f'$.

5.4.2 SMITH–WATSON–TOPPER (SWT) MODEL

Smith, Watson, and Topper (Smith et al., 1970) proposed a method that assumes that the amount of fatigue damage in a cycle is determined by $\sigma_{max}\varepsilon_a$, where σ_{max} is the maximum tensile stress and ε_a the strain amplitude. Also, the SWT parameter is simply a statement that "$\sigma_a\varepsilon_a$ for a fully reversed test is equal to $\sigma_{max}\varepsilon_a$ for a mean stress test." Thus, this concept can be generalized and expressed in the following mathematical form (Langlais and Vogel, 1995):

$$\sigma_{max}\varepsilon_a = \sigma_{a,\,rev}\ \varepsilon_{a,\,rev} \qquad \sigma_{max} > 0 \tag{5.4.3}$$

where $\sigma_{a,\,rev}$ and $\varepsilon_{a,\,rev}$ are the fully reversed stress and strain amplitudes, respectively, that produce an equivalent fatigue damage due to the SWT parameter. The value of $\varepsilon_{a,\,rev}$ should be obtained from the strain–life curve (Equation 5.3.20) and the value of $\sigma_{a,\,rev}$ from the cyclic stress–strain curve (Equation 5.3.15). The SWT parameter predicts no fatigue damage if the maximum tensile stress becomes zero and negative. The solutions to Equation 5.4.3 can be obtained by using the Newton–Raphson iterative procedure.

For a special case of Equation 5.4.3, in which materials behave ideally and satisfy the compatibility condition (i.e., $n' = b/c$ and $K' = \sigma_f'/\left(\varepsilon_f'\right)^{n'}$), the maximum tensile stress for fully reversed loading is given by

$$\sigma_{max} = \sigma_a = \sigma_f'(2N_f)^b \tag{5.4.4}$$

and by multiplying the strain–life equation, the SWT mean stress correction formula is expressed as follows:

$$\sigma_{max}\varepsilon_a = \frac{\left(\sigma_f'\right)^2}{E}(2N_f)^{2b} + \sigma_f'\varepsilon_f'(2N_f)^{b+c} \qquad \sigma_{max} > 0 \tag{5.4.5}$$

The SWT formula has been successfully applied to grey cast iron (Fash and Socie, 1982), hardened carbon steels (Koh and Stephens, 1991; Wehner and Fatemi, 1991), and microalloyed steels (Forsetti and Blasarin, 1988). The SWT formula is regarded as more promising for use. As illustrated in Figure 5.21, a set of fatigue data with several levels of mean stress values coalesces into a single curve. This indicates that the data points can be represented by the SWT parameter.

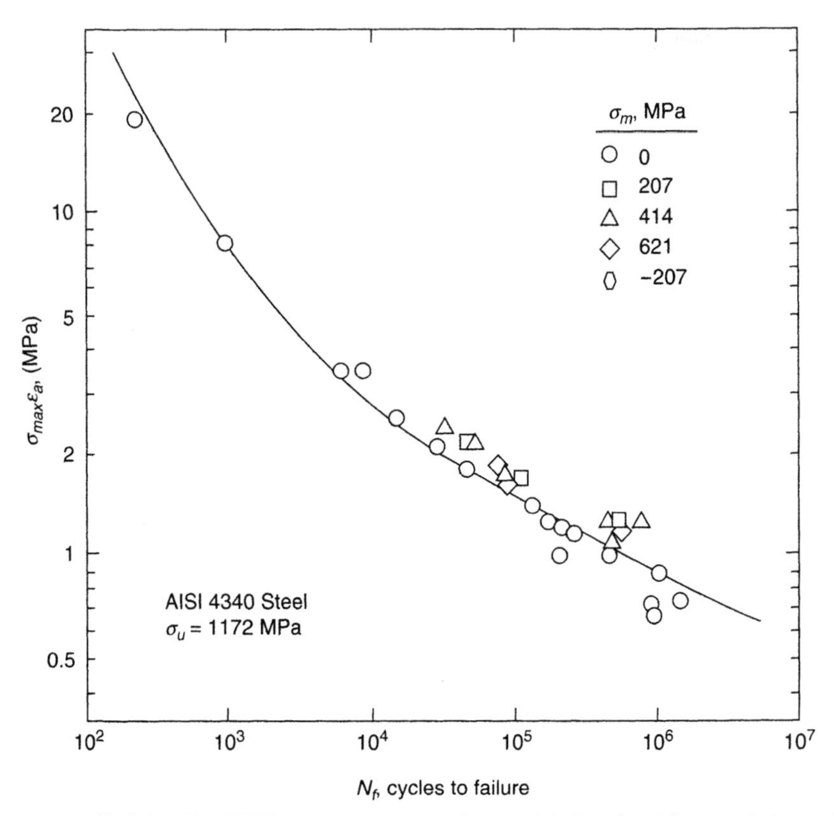

FIGURE 5.21 The SWT mean stress correction model. Reprinted by permission of Pearson Education, Inc., Upper Saddle River, NJ, from "Mechanical Behavior of Materials," 2nd edition by Dowling (1998).

Example 5.4. A component is made of steel with the following cyclic properties: $E = 210000\,\text{MPa}$, $\sigma_f' = 1,100\,\text{MPa}$, $b = -0.1$, $\varepsilon_f' = 0.6$, and $c = -0.5$. If the component were cycled under bending loads that induce a fully reversed strain amplitude of 0.004, would you expect to shot peen the component for fatigue life increase?

Solution. An appropriate shot-peening process will produce compressive residual stress on the surface. The beneficial residual stress will enhance the fatigue strength of the part in the high-cycle fatigue region, but the residual stress will gradually dissipate in the low-cycle fatigue region because of the phenomenon of mean stress relaxation. The transition fatigue life will serve as a criterion for our decision making. From Equation 5.3.35, the transition fatigue life can be calculated as follows:

$$2N_T = \left(\frac{\varepsilon_f' E}{\sigma_f'}\right)^{1/b-c} = 140425 \text{ reversals}$$

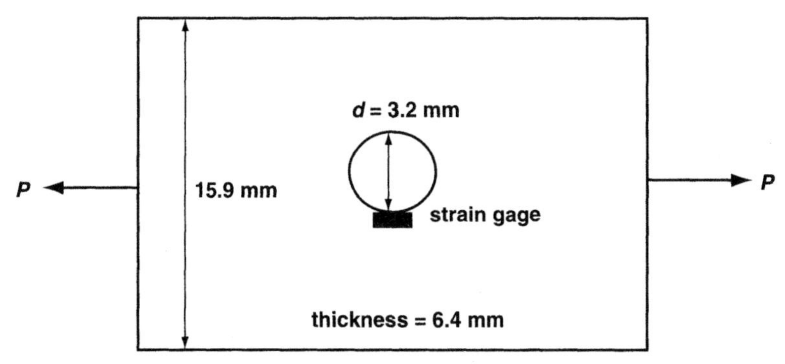

FIGURE 5.22 Configuration and dimension of a plate with a center hole.

The total strain amplitude at the fatigue life transition can be determined by calculating either the elastic strain amplitude or the plastic strain amplitude at that life and multiplying the result by a factor of 2:

$$\varepsilon_a^p = \frac{\Delta\varepsilon^p}{2} = \varepsilon_f^{'}(2N_f)^c = 0.6 \times 140425^{-0.5} = 0.0016$$

The total strain amplitude is $0.0032(= 2 \times 0.0016)$. The applied strain amplitude of 0.0040 is higher than the strain amplitude at the transition life. This indicates that the loading can be considered to fall within the low-cycle fatigue region. Under this circumstance, shot peening should not be considered to improve the fatigue strength of the component. This can also be done by comparing the calculated fatigue life of the component to the transition life.

Example 5.5. A single active strain gage was placed at the notch root of a notched plate as shown in Figure 5.22. A notched component made of SAE 1137 carbon steel has the following material properties: $E = 209000$ MPa, $K' = 1230$ MPa, $n' = 0.161$, $\sigma_f' = 1006$ MPa, $b = -0.0809$, $\varepsilon_f' = 1.104$, and $c = -0.6207$. The recorded strain time history because of the applied load is repetitive and is shown in Figure 5.23.

1. Plot the cyclic stress–strain response (hysteresis loop).
2. Estimate the fatigue life of the notched plate with the SWT formula.
3. Estimate the fatigue life of the notched plate with the Morrow equation.

Solution. To determine the hysteresis loop, we must first establish the initial loading case and then the first loading reversal and all the subsequent cycles. First, the initial loading is considered. The initial loading starts from zero strain and extends to 5000 $\mu\varepsilon$. To define the curve generated by the stress–strain relationship during the loading, it is important to analyze multiple points along the strain path. For the initial loading, steps of 0.0005 to a value

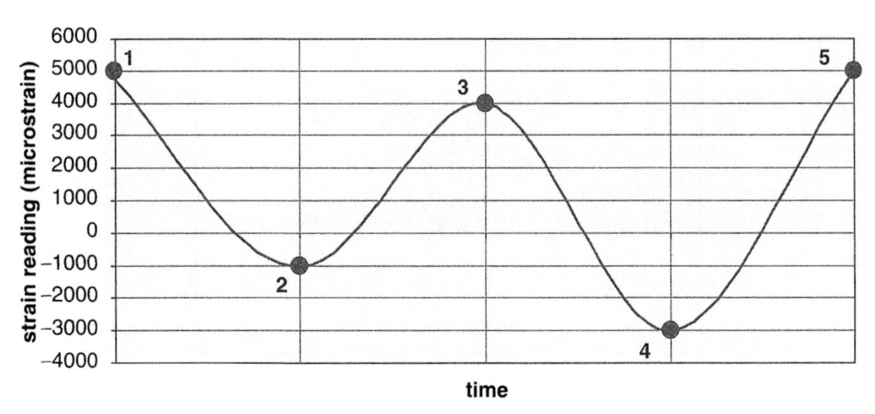

FIGURE 5.23 Recorded strain reading in microstrains at the notch root.

of 0.005 are used. To determine the stress at each intermediate step of the strain, the cyclic stress–strain equation is used.

$$\varepsilon_1 = \frac{\sigma_1}{E} + \left(\frac{\sigma_1}{K'}\right)^{\frac{1}{n'}}$$

For example, σ_1 at $\varepsilon_1 = 0.005$ is obtained by

$$0.005 = \frac{\sigma_1}{209000} + \left(\frac{\sigma_1}{1230}\right)^{\frac{1}{0.161}}$$

Solving for σ_1

$$\sigma_1 = 475.35\,\text{MPa}$$

Table 5.9 shows the tabulated values for σ and ε at each increment in the initial loading.

TABLE 5.9 Initial Loading to Point 1: True Stress and Strain Data Points

ε_1	σ_1 (MPa)
0.0005	104.45
0.0010	205.85
0.0015	288.10
0.0020	342.98
0.0025	380.21
0.0030	407.65
0.0035	429.24
0.0040	447.03
0.0045	462.17
0.0050	475.35

Having defined the initial loading condition, the first loading reversal can now be examined ($\varepsilon_1 = 0.0050 \rightarrow \varepsilon_2 = -0.0010$). Again, it is important to define the curve so that multiple points along the strain path are considered with respect to the reference point (Point 1). To determine the change in stress and strain, Masing's model is used:

$$\Delta\varepsilon = \frac{\Delta\sigma}{E} + 2\left(\frac{\Delta\sigma}{2K'}\right)^{\frac{1}{n'}}$$

For example, in the case of strain change from 0.0050 to −0.0010 ($\Delta\varepsilon = 0.0060$), the stress change, $\Delta\sigma$, is obtained:

$$0.0060 = \frac{\Delta\sigma}{209000\ \text{MPa}} + 2\left(\frac{\Delta\sigma}{2 \times 1230}\right)^{\frac{1}{0.161}}$$

Solving for $\Delta\sigma$

$$\Delta\sigma = 815.30\ \text{MPa}$$

Because Masing's model calculates the change in stress, the change must be compared with the reference stress ($\sigma_1 = 475.35$ MPa) to determine the stress at Point 2.

$$\sigma_2 = \sigma_1 - \Delta\sigma = 475.35 - 815.30 = -339.95\ \text{MPa}$$

Table 5.10 shows the tabulated values for σ and ε at each increment in the first loading reversal. For the second loading reversal ($\varepsilon_2 = -0.0010 \rightarrow \varepsilon_3 = 0.0040$), the same procedure is used as for the first, except that Point 2 serves as the new reference point. For example, in the case of $\Delta\varepsilon = 0.0050$, $\Delta\sigma$, is determined:

$$0.0050 = \frac{\Delta\sigma}{209000} + 2\left(\frac{\Delta\sigma}{2 \times 1230}\right)^{\frac{1}{0.161}}$$

Solving for $\Delta\sigma$,

$$\Delta\sigma = 760.42\ \text{MPa}$$

Again, because Masing's model calculates the change in stress, the change must be compared with the reference stress ($\sigma_2 = -339.95$ MPa) to determine the stress at Point 3.

$$\sigma_3 = \sigma_2 + \Delta\sigma = -339.95 + 760.42 = 420.47\ \text{MPa}$$

For the third and fourth loading reversals, the same procedure is used. When a material response returns to its previously experienced deformation, it will remember the past path to reach such a state and will follow the path with additional increase in deformation. This is the so-called memory effect observed in materials undergoing complex loading histories. In the third

TABLE 5.10 First Loading Reversal (Points 1 to 2): True Stress and Strain Data Points

ε	Δε	Δσ (MPa)	σ (MPa)
0.0050	Reference 1	0.00	475.35
0.0045	0.0005	104.50	370.85
0.0040	0.0010	208.91	266.44
0.0035	0.0015	312.37	162.98
0.0030	0.0020	411.70	63.65
0.0025	0.0025	501.15	−25.8
0.0020	0.0030	576.20	−100.85
0.0015	0.0035	636.88	−161.53
0.0010	0.0040	685.96	−210.61
0.0005	0.0045	726.38	−251.03
0.0000	0.0050	760.42	−285.07
−0.0005	0.0055	789.69	−314.34
−0.0010	0.0060	815.30	−339.95

TABLE 5.11 Second Loading Reversal (Points 2 to 3): True Stress and Strain Data Points

ε	Δε	Δσ (MPa)	σ (MPa)
−0.0010	Reference 2	0.00	−339.95
−0.0005	0.0005	104.50	−235.45
0.0000	0.0010	208.91	−131.04
0.0005	0.0015	312.37	−27.58
0.0010	0.0020	411.70	71.70
0.0015	0.0025	501.15	161.20
0.0020	0.0030	576.20	236.25
0.0025	0.0035	636.88	296.93
0.0030	0.0040	685.96	346.01
0.0035	0.0045	726.38	386.43
0.0040	0.0050	760.42	420.47

reversal, after the smaller reversal from Points 3 to 2 returns to the previously experienced maximum deformation, Point 2, the rest of the loading reversal will follow the path from Points 1 to 4. Tables 5.11–5.13 show the tabulated values for σ and ε at each increment in the second, third, and fourth loading reversals. From the points calculated for the initial and reversals, the hysteresis loops are graphed in Figure 5.24.

After determining the hysteresis loops, an estimate of the fatigue life using the SWT method is needed. Ignoring the initial loading, a rainflow cycle counting analysis is performed on the strain results. By using SWT Equation 5.4.3, the damage calculation results are determined in Table 5.14. The total damage and fatigue life prediction are obtained as follows:

$$\Sigma d_i = 1.78 \times 10^{-5} + 8.09 \times 10^{5} = 9.87 \times 10^{-5}$$

TABLE 5.12 Third Loading Reversal (Points 3 to 4): True Stress and Strain Data Points

ε	Δε	Δσ (MPa)	σ (MPa)
0.0040	Reference 3	0.00	420.47
0.0035	0.0005	104.50	315.97
0.0030	0.0010	208.91	211.56
0.0025	0.0015	312.37	108.10
0.0020	0.0020	411.70	8.77
0.0015	0.0025	501.15	−80.68
0.0010	0.0030	576.20	−155.73
0.0005	0.0035	636.88	−216.41
0.0000	0.0040	685.96	−265.49
−0.0005	0.0045	726.38	−305.91
−0.0010	0.0050	760.42	−339.95
0.0050	Reference 1	0.00	475.35
−0.0015	0.0065	838.04	−362.69
−0.0020	0.0070	858.48	−383.13
−0.0025	0.0075	877.05	−401.70
−0.0030	0.0080	894.06	−418.71

TABLE 5.13 Fourth Loading Reversal (Points 4 to 5): True Stress and Strain Data Points

ε	Δε	Δσ (MPa)	σ (MPa)
−0.0030	Reference 4	0.00	−418.71
−0.0025	0.0005	104.50	−314.21
−0.0020	0.0010	208.91	−209.80
−0.0015	0.0015	312.37	−106.34
−0.0010	0.0020	411.70	−7.01
−0.0005	0.0025	501.15	82.44
0.0000	0.0030	576.20	157.49
0.0005	0.0035	636.88	218.17
0.0010	0.0040	685.96	267.25
0.0015	0.0045	726.38	307.67
0.0020	0.0050	760.42	341.71
0.0025	0.0055	789.69	370.98
0.0030	0.0060	815.30	396.59
0.0035	0.0065	838.04	419.33
0.0040	0.0070	858.48	439.77
0.0045	0.0075	877.05	458.34
0.0050	0.0080	894.06	475.35

$$\text{Life} = \frac{1}{\Sigma d_i} = \frac{1}{9.87 \times 10^{-5}} = 10,100 \text{ blocks}$$

For comparison with the SWT method, an estimate of the fatigue life by using Morrow's method is generated. By using the rainflow counting analysis given previously and the Morrow equation (Equation 5.4.1), the damage calculation is presented in Table 5.15. The total damage and fatigue life prediction are given as follows:

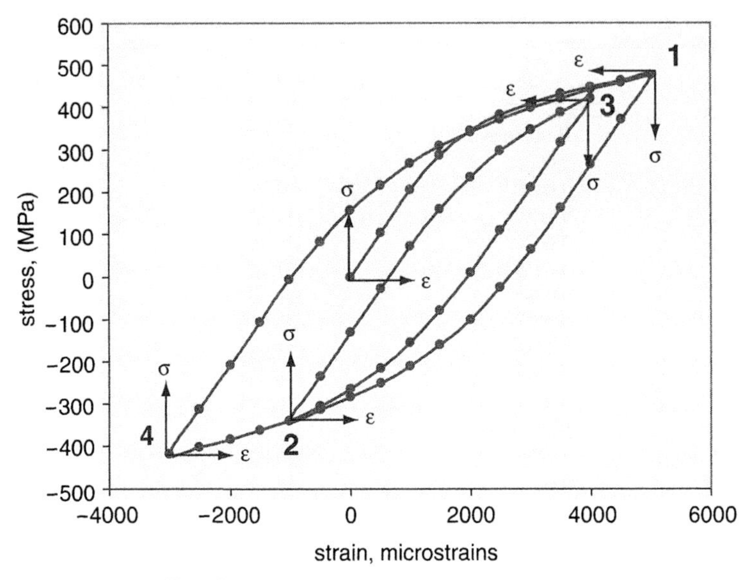

FIGURE 5.24 Hysteresis stress strain curves at the notch root.

TABLE 5.14 Summary of Rainflow Cycle Counting and Damage Calculation Results Based on the SWT Formula

n_i	From–To (με)	ε_{max} (με)	ε_{min} (με)	σ_{max} (MPa)	σ_{min} (MPa)	ε_a (με)	σ_{max} (MPa)	N_f (cycles)	d_i
1	−1000 to 4000	4000	−1000	420.47	−339.95	2500	420.47	56316	1.78E-05
1	5000 to −3000	5000	−3000	475.35	−418.71	4000	475.35	12354	8.09E-5

TABLE 5.15 Summary of Rainflow Cycle Counting and Damage Calculation Results Based on the Morrow Formula

n_i	From–To (με)	ε_{max} (με)	ε_{min} (με)	σ_{max} (MPa)	σ_{min} (MPa)	ε_a (με)	σ_m (MPa)	N_f (cycles)	d_i
1	−1000 to 4000	4000	−1000	420.47	−339.95	2500	40.26	67384	1.48E-05
1	5000 to −3000	5000	−3000	475.35	−418.71	4000	28.32	13596	7.35E-05

$$\Sigma d_i = 1.48 \times 10^{-5} + 7.35 \times 10^{-5} = 8.83 \times 10^{-5}$$

$$\text{Life} = \frac{1}{\Sigma d_i} = \frac{1}{8.83 \times 10^{-5}} = 11300 \text{ blocks}$$

5.5 ESTIMATION OF CYCLIC AND FATIGUE PROPERTIES

If no experimental data are available, an estimate of the cyclic and fatigue behavior of a material can be helpful in the design stage. However, the parameter estimation does not eliminate the need for real data. In Boardman (1982), the following parameters for most steels that harden below 500 BHN have been empirically developed:

a. Ultimate tensile strength, S_u: A commonly used approximation for the ultimate tensile strength based on Brinell hardness, BHN < 500, for low- and medium-strength carbon and alloy steels is

$$S_u(\text{MPa}) \approx 3.45 \times \text{BHN} \tag{5.5.1}$$

b. Fatigue strength coefficient, σ_f': An approximation relates the fatigue strength coefficient σ_f' to the true fracture strength σ_f with a range from $0.9\sigma_f$ to $1.2\sigma_f$. However, tests indicate the fatigue strength coefficient shows a linear relation with the ultimate strength, S_u, in the form

$$\sigma_f'(\text{MPa}) \approx \sigma_f \approx S_u + 345 \text{ MPa} \tag{5.5.2}$$

c. Fatigue strength exponent b:

$$b \approx -\frac{1}{6}\log\left(\frac{\sigma_f'}{0.5 \times S_u}\right) \tag{5.5.3}$$

d. Fatigue ductility coefficient ε_f': The fatigue ductility coefficient ε_f' is thought to be of the same order as the true fracture ductility, ε_f.

$$\varepsilon_f' \approx \varepsilon_f \approx \ln\left(\frac{100}{100 - \%\text{RA}}\right) \tag{5.5.4}$$

e. Fatigue ductility exponent c:
   1. Strong steels, for which $\varepsilon_f \approx 0.5$, $c = -0.5$
   2. Ductile steels, for which $\varepsilon_f \approx 1.0$, $c = -0.6$

f. Transition fatigue life $2N_T$ (reversals): For steels, it is found that the transition fatigue life has a dependence on Brinell hardness. Because hardness varies inversely with ductility, the transition life decreases as the hardness increases. The transition life is estimated as follows:

$$2N_T \approx \ln^{-1}(13.6 - 0.0185 \times \text{BHN}) \tag{5.5.5}$$

g. Cyclic strain hardening exponent n':

$$n' \approx \frac{b}{c} \tag{5.5.6}$$

h. Cyclic stress coefficient K':

$$K' \approx \frac{\sigma_f'}{\left(\varepsilon_f'\right)^{n'}} \tag{5.5.7}$$

For materials including steels, aluminum, and titanium alloys, Baumel and Seeger (1990) proposed a method of approximating the cyclic properties from monotonic stress strain relationships. The results of the so-called uniform material law are documented in Table 5.16, where $\psi = 1$ if $\frac{S_u}{E} \leq 3 \times 10^{-3}$ or $\psi = \left(1.375 - 125\frac{S_u}{E}\right) \leq 0$ if $\frac{S_u}{E} > 3 \times 10^{-3}$. Other estimating procedures and their uses are included elsewhere (Muralidharan and Manson, 1988; Roessle and Fatemi, 2000; Park and Song, 2003).

Example 5.6. A notched component is made of SAE 4340 steel and has a reported hardness of 242 BHN and modulus of elasticity of 205,000 MPa. Estimate the fatigue and cyclic material properties based on Boardman's approach.

Solution. The first step in this analysis is to determine the fatigue properties for the material of the component.

$$S_u = 3.45(\text{BHN}) = 3.45(242) = 835\,\text{MPa}$$

$$\sigma_f' = S_u + 345\text{MPa} = 835 + 345 = 1180\,\text{MPa}$$

TABLE 5.16 Estimated Cyclic Material Properties on Uniform Material Law

Cyclic Material Parameters	Steels, Unalloyed and Low-Alloy	Aluminum and Titanium Alloys
σ_f'	$1.67S_u$	$1.67\,S_u$
b	−0.087	−0.095
ε_f'	0.59ψ	0.35
C	−0.58	−0.69
K'	$1.65S_u$	$1.61S_u$
n'	0.15	0.11
S_e	$0.45S_u$	$0.42S_u$

Source: From Baumel and Seeger (1990).

$$b = -\frac{1}{6}\log\left(\frac{2\sigma_f'}{S_u}\right) = -\frac{1}{6}\log\left(\frac{2(1180)}{835}\right) = -0.075$$

$$2N_t = \ln^{-1}(13.6 - 0.0185\text{BHN}) = \ln^{-1}(13.6 - 0.0185(242)) = 9164 \text{ reversals}$$

Because the material is ductile steel, the fatigue ductility exponent c is estimated to be -0.6.

By using the following equation to determine the elastic strain amplitude at the transition point,

$$\frac{\Delta\varepsilon^e}{2} = \frac{\sigma_f'}{E}(2N_T)^b = \frac{1180}{205000}(9164)^{-0.075} = 0.0029 \text{ at } 2N_T \text{ reversals}$$

Because the elastic and plastic strain amplitudes are equal at the transition life point,

$$\frac{\Delta\varepsilon_p}{2} = \varepsilon_f'(2N_T)^c = 0.0029, \text{ solving for } \varepsilon_f'$$

$$\varepsilon_f' = \frac{0.0029}{(2N_T)^c} = \frac{0.0029}{(9164)^{-0.6}} = 0.691$$

Also, Equations (5.5.6) and (5.5.7) give

$$n' = \frac{b}{c} = \frac{-0.076}{-0.6} = 0.125$$

$$K' = \frac{\sigma_f'}{\left(\varepsilon_f'\right)^{n'}} = \frac{1180}{(0.691)^{0.125}} = 1236\,\text{MPa}$$

5.6 NOTCH ANALYSIS

Notch analysis is used to relate the nominal stress or strain changes in a component to the local stress and strain response at a notch. This can be used to predict the crack initiation life of notched components by using fatigue life data from smooth laboratory specimens.

Neuber (1961) first analyzed a grooved shaft subjected to monotonic torsional loading and derived a rule for nonlinear material behavior at the notch root. Neuber observed, as illustrated in Figure 5.25, that after local yielding occurs, the local true notch stress (σ) is less than the stress predicted by the theory of elasticity (σ^e) whereas the local true notch strain (ε) is greater than that estimated by the theory of elasticity. Normalizing the local true notch stress with respect to the nominal stress (S) and the true notch strain to the nominal strain (e) leads to the true stress concentration (K_σ) and the true strain concentration (K_ε) factors, respectively.

Figure 5.26 schematically illustrates the relation of three concentration factors (K_t, K_σ, K_ε) versus the true stress with regard to the yield stress

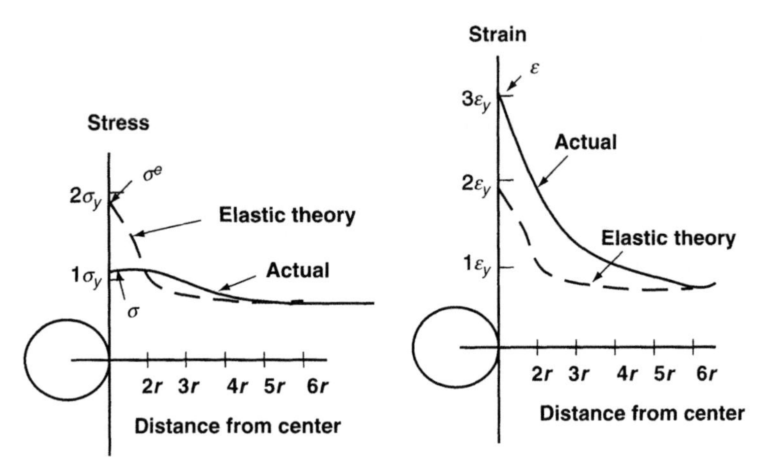

FIGURE 5.25 Stresses and strains at a notch.

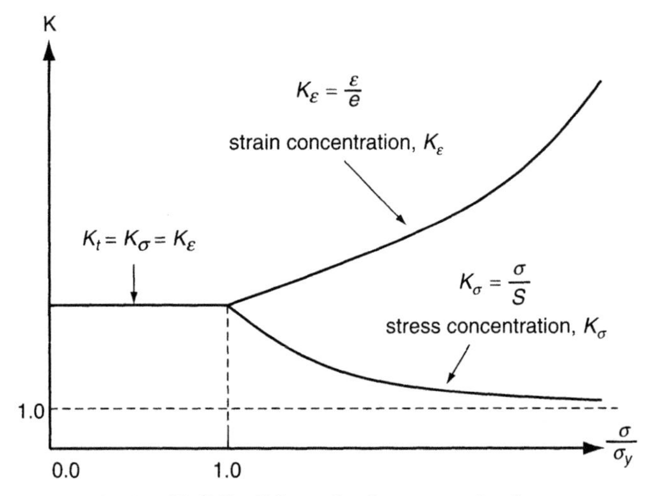

FIGURE 5.26 Schematic of concentration factors.

(σ_y). Neuber hypothesized that the elastic stress concentration factor (K_t) is the geometric mean of the true stress and strain concentration factors, i.e.,

$$K_t = \sqrt{K_\sigma K_\varepsilon} \tag{5.6.1}$$

Squaring both sides of Equation 5.6.1 leads to the following expression

$$\sigma\varepsilon = (SK_t)(eK_t) \tag{5.6.2}$$

When below the proportional limit, the product of $S\ K_t$ is the local notch stress calculated on an elastic material. The physical interpretation of the Neuber rule is shown in Figure 5.27.

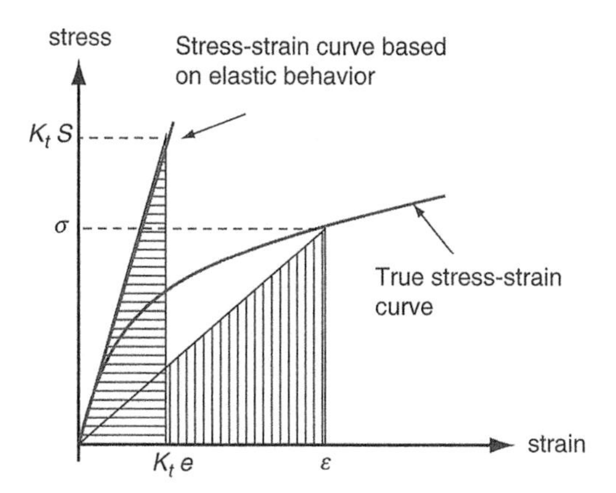

FIGURE 5.27 Interpretation of the Neuber model.

For cyclic stress and strain behavior at a notch root under a plane stress condition, the Neuber rule can be revised by replacing K_t with the fatigue strength reduction factor (K_f) (Wetzel, 1968; Topper et al., 1969) as follows:

$$K_f = \sqrt{K_\sigma K_\varepsilon} \tag{5.6.3}$$

and

$$K_\sigma = \frac{\sigma_a}{S_a} \tag{5.6.4}$$

$$K_\varepsilon = \frac{\varepsilon_a}{e_a} \tag{5.6.5}$$

where S_a and e_a are the nominal stress and strain amplitudes, respectively, and σ_a and ε_a are the local true stress and strain amplitudes, respectively. The K_f factor was introduced in the Neuber rule to improve fatigue life predictions and to include the stress gradient effect. However, this has been criticized for both accounting twice the notch root plasticity effects and incorporating the S–N empiricism into the more fundamentally satisfying theory. By knowing these conflicting results, we recommend that the modified Neuber rule (Equation 5.6.3) be used for local notch stress/strain estimates and fatigue life predictions. The fatigue strength reduction factor was covered in detail in Section 4.4.

During cyclic loading, it is assumed that the material follows the cyclic stress–strain curve for the initial loading and the hysteresis stress–strain behavior for the subsequent loading reversals. Therefore, in terms of the initial cyclic stress–strain curve the modified Neuber equation can be reduced to

$$\sigma_1 \varepsilon_1 = K_f^2 S_1 e_1 \tag{5.6.6}$$

and, in terms of the hysteresis stress–strain curve, to

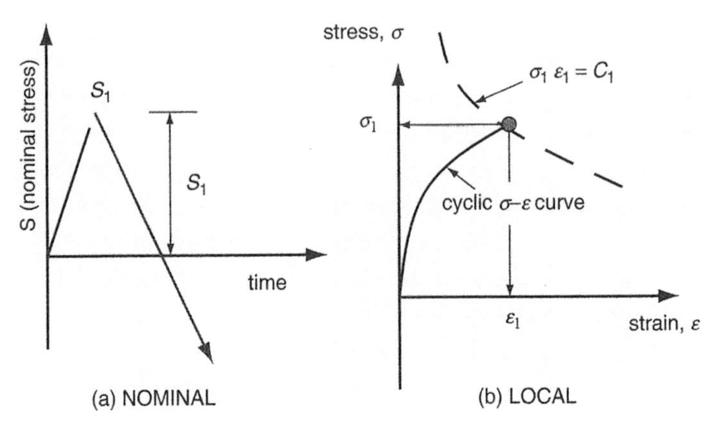

FIGURE 5.28 Initial nominal stress loading with the modified Neuber rule.

$$\Delta\sigma\Delta\varepsilon = K_f^2 \Delta S \Delta e \tag{5.6.7}$$

Equations 5.6.6 and 5.6.7 represent the equations of a hyperbola (i.e., XY = constant) for given K_f and nominal stress/strain data. This constant which is the right-handed side of Equation 5.6.7 is often referred to Neuber's constant. To solve for the two unknowns X and Y (i.e., σ_1 and ε_1 or $\Delta\sigma$ and $\Delta\varepsilon$), an additional equation for cyclic material behavior is required.

Figure 5.28 shows the application of the modified Neuber model to a notched plate subjected to an initial loading where the nominal strain e_1 and K_f are known based on the applied nominal stress S_1 and the notch geometry. The Neuber constant C_1 can be calculated and a hyperbola of the local stress and strain ($\sigma_1\varepsilon_1 = C_1$) is created. Then, a cyclic stress–strain curve for the material is needed to intersect the hyperbola for the solution of (σ_1,ε_1).

Figure 5.29 shows the application of the modified Neuber model to a notched plate subjected to a first loading reversal, where the nominal strain $\Delta\varepsilon$ and K_f are known based on the applied nominal stress ΔS and the notch geometry. The Neuber constant C_2 can be calculated and a hyperbola of the local stress and strain changes ($\Delta\sigma\Delta\varepsilon = C_2$) is generated. A hysteresis stress-strain curve for the material is needed to intersect the hyperbola for the solution of ($\Delta\sigma$,$\Delta\varepsilon$).

Depending on the nominal stress-strain behavior, the equations for the local notch stress and strain solutions are discussed in the following sections.

5.6.1 NOMINALLY ELASTIC BEHAVIOR

When the bulk of a notched component behaves elastically and plasticity takes place locally at the notch root (i.e., a nominally elastic condition), the following equations hold:

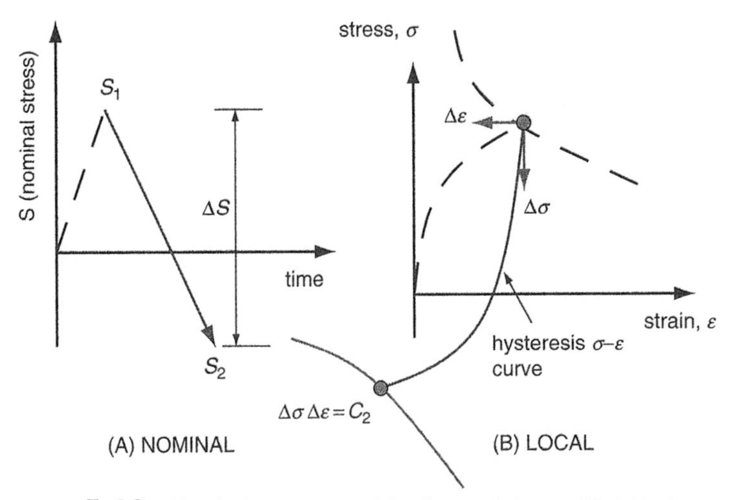

FIGURE 5.29 Nominal stress reversal loading with the modified Neuber rule.

- For the initial cyclic stress-strain curve

$$e_1 = \frac{S_1}{E} \tag{5.6.8}$$

$$\varepsilon_1 = \frac{\sigma_1}{E} + \left(\frac{\sigma_1}{K'}\right)^{1/n''} \tag{5.6.9}$$

- For the hysteresis stress-strain curve,

$$\Delta e = \frac{\Delta S}{E} \tag{5.6.10}$$

$$\Delta\varepsilon = \frac{\Delta\sigma}{E} + 2\left(\frac{\Delta\sigma}{2K'}\right)^{1/n'} \tag{5.6.11}$$

The assumption of nominally elastic material behavior works well when nominal stress is below 30% of the cyclic yield stress.

Substituting the elastic nominal stress–strain and the local cyclic stress–strain relations (Equations 5.6.8–5.6.11) into the modified Neuber equations (Equations 5.6.6 and 5.6.7) results in the following equations:

$$\frac{\sigma_1^2}{E} + \sigma_1\left(\frac{\sigma_1}{K'}\right)^{1/n'} = \frac{(K_f S_1)^2}{E} \tag{5.6.12}$$

and

$$\frac{(\Delta\sigma)^2}{E} + 2\Delta\sigma\left(\frac{\Delta\sigma}{2K'}\right)^{1/n'} = \frac{(K_f \Delta S)^2}{E} \tag{5.6.13}$$

Given K_f and the nominal stress data, these equations for the local stress can be solved by using the Newton–Raphson iteration technique. Once the local stress is determined, Equation 5.6.9 or 5.6.11 is used to obtain the corresponding local strain value.

If the residual stress (σ_r) and the residual strain (ε_r) at a notch root due to a welding process exist prior to any operating load reversals, Equation 5.6.12 for initial loading needs to be modified. The following modifications have been proposed:

- Lawrence et al. (1982):

$$\frac{\sigma_1^2}{E} + \sigma_1 \left(\frac{\sigma_1}{K'}\right)^{1/n'} = \frac{\left(K_f S_1 + \sigma_r\right)^2}{E} \tag{5.6.14}$$

- Reemsnyder (1981):

$$\left(1 - \frac{\sigma_r}{\sigma_1}\right)^2 \left[\frac{\sigma_1^2}{E} + \sigma_1 \left(\frac{\sigma_1}{K'}\right)^{1/n'}\right] = \frac{\left(K_f S_1\right)^2}{E} \tag{5.6.15}$$

- Baumel and Seeger (1989):

$$\sigma_1 (\varepsilon_1 - \varepsilon_r) = \frac{\left(K_f S_1\right)^2}{E} \tag{5.6.16}$$

Molsky and Glinka (1981) proposed another notch analysis method that assumes that the strain energy density at the notch root W_e is related to the energy density due to the nominal stress W_s by a factor of K_t^2. That means

$$W_e = K_t^2 W_s \tag{5.6.17}$$

The K_t factor, not K_f, should be used in the strain energy density method. Figure 5.30 illustrates the physical interpretation of the strain energy density

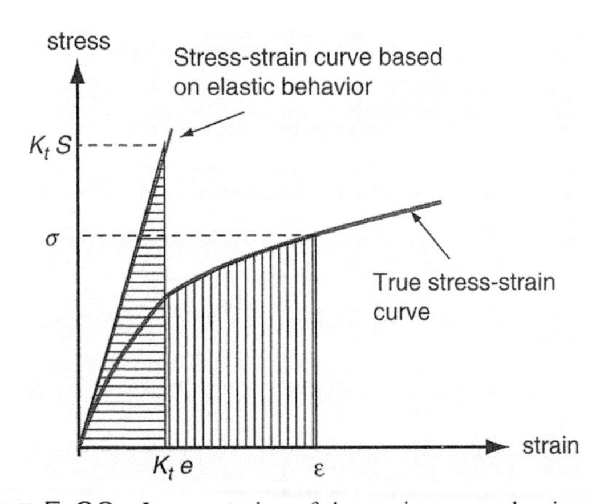

FIGURE 5.30 Interpretation of the strain energy density method.

method. If nominally elastic behavior of the notched specimen is assumed, the following strain energy equations can be obtained:

$$W_s = \frac{1}{2}\frac{S_a^2}{E} \tag{5.6.18}$$

and

$$W_e = \frac{\sigma_a^2}{2E} + \frac{\sigma_a}{1+n'}\left(\frac{\sigma_a}{K'}\right)^{1/n'} \tag{5.6.19}$$

Substituting Equations 5.6.18 and 5.6.19 into Equation 5.6.17 leads to the well-known energy density formula:

$$\frac{\sigma_a^2}{E} + \frac{2\sigma_a}{1+n'}\left(\frac{\sigma_a}{K'}\right)^{1/n'} = \frac{(K_t S_a)^2}{E} \tag{5.6.20}$$

For initial loading case, Equation 5.6.20 can be reduced to

$$\frac{\sigma_1^2}{E} + \frac{2\sigma_1}{1+n'}\left(\frac{\sigma_1}{K'}\right)^{1/n'} = \frac{(K_t S_1)^2}{E} = \frac{\left(\sigma_1^e\right)^2}{E} \tag{5.6.21}$$

For stabilized hysteresis behavior, Equation 5.6.20 can be reduced to

$$\frac{(\Delta\sigma)^2}{E} + \frac{4\Delta\sigma}{1+n'}\left(\frac{\Delta\sigma}{2K'}\right)^{1/n'} = \frac{(K_t \Delta S)^2}{E} = \frac{(\Delta\sigma^e)^2}{E} \tag{5.6.22}$$

where σ_1^e and $\Delta\sigma^e$ are the local stress and the stress change at a notch root based on an elastic finite element analysis, respectively.

Again it should be emphasized that the formulas presented in this section are only valid to local yielding, plane stress, and a uniaxial state of stress. Although these fail to account for gross yielding of a net section, the assumption of nominally elastic behavior of a component has been considered very useful in the majority of the automotive design applications. Notch analyses for the multiaxial stress and strain state is beyond the scope of our discussion and can be found elsewhere (Hoffman and Seeger, 1985; Barkey et al., 1994; Lee et al., 1995, Moftakhar et al., 1995; Koettgen et al., Gu and Lee, 1997).

5.6.2 NOMINALLY GROSS YIELDING OF A NET SECTION

When nonlinear net section behavior is considered, the nominal stress (S) and nominal strain (e) need to follow a nonlinear material relationship and Neuber's rule has to be modified. Seeger and Heuler (1980) proposed a modified version of nominal stress S^M to account for general yielding and is defined as follows:

$$S^M = S\left(\frac{\sigma_y}{S_p}\right) \tag{5.6.23}$$

where S_p is the nominal stress at the onset of general yielding of a net section area. This equation indicates that $S^M = \sigma_y$ when S is equal to S_p. The ratio of σ_y/S_p is a function of the notch geometry and the type of loading. For instance, a value of 0.667 is found for a beam with a rectangular section under bending.

Because the elastic notch stress (σ^e) calculated by the elastic finite element analysis is independent of the definition of nominal stress, a modified elastic stress concentration (K^M) associated with S^M is introduced in the following form:

$$\sigma^e = S^M K_t^M = SK_t \tag{5.6.24}$$

The Neuber rule can be rewritten:

$$\sigma\varepsilon = \left(K_t^M\right)^2 S^M e^M \tag{5.6.25}$$

The new form of Neuber's rule can be rearranged and extended to

$$\sigma\varepsilon = S^M K_t^M e^M K_t^M \left(\frac{S^M E}{S^M E}\right) \tag{5.6.26}$$

or

$$\sigma\varepsilon = \frac{(SK_t)^2}{E}\left(\frac{e^M E}{S^M}\right) \tag{5.6.27}$$

Equation 5.6.27 is the generalized Neuber's rule for nonlinear net section behavior where S^M and e^M lie on the cyclic stress–strain curve and have the following expression:

$$e^M = \frac{S^M}{E} + \left(\frac{S^M}{K'}\right)^{1/n'} \tag{5.6.28}$$

If the S^M–e^M curve remains in the elastic range, the factor $(e^M E/S^M)$ becomes unity and the equation reduces to the known Neuber equation.

There are two other expressions for the modified nominal stress. For example, S^M can be rewritten as follows:

$$S^M = L\left(\frac{\sigma_y}{L_p}\right) \tag{5.6.29}$$

where L is the applied load and L_p is the plastic limit load for elastic-perfectly plastic material. Another alternative is expressed in the following equation:

$$S^M = S\left(\frac{K_t}{K_p}\right) \tag{5.6.30}$$

where K_p, called the limited load factor, is defined as

$$K_p = \frac{\text{Load at onset of general yielding}}{\text{Load at first notch yielding}} = \frac{L_p}{L_y} \tag{5.6.31}$$

where L_y is the load producing first yielding of a net section. A finite element analysis with the elastic-perfectly plastic material can be utilized to determine L_p and L_y.

Example 5.7. A plate with a center hole has the dimensions shown in Figure 5.31. The plate is made of SAE 1005 steel that has the following properties: $E = 207000\,\text{MPa}$, $S_u = 320\,\text{MPa}$, $K' = 1240\,\text{MPa}$, $n' = 0.27, \sigma'_f = 886\,\text{MPa}$, $b = -0.14, \varepsilon'_f = 0.28$, and $c = -0.5$.

The notched plate is subjected to a variable nominal stress time history in Figure 5.32. The elastic stress concentration factor (K_t) for the notched plate under tension is calculated as 2.53. It is assumed that the notched plate follows nominally elastic behavior.

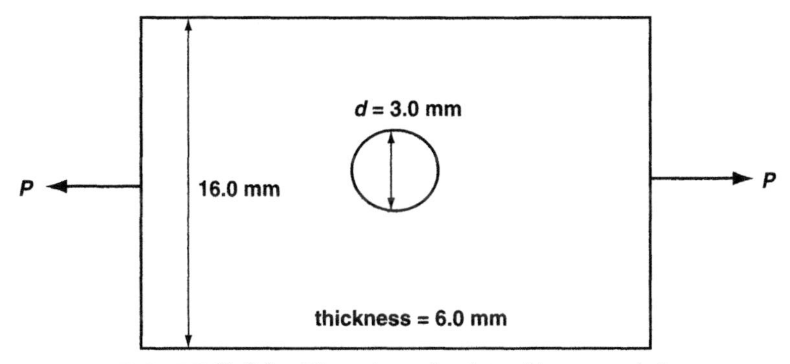

FIGURE 5.31 Dimensions of a plate with a center hole.

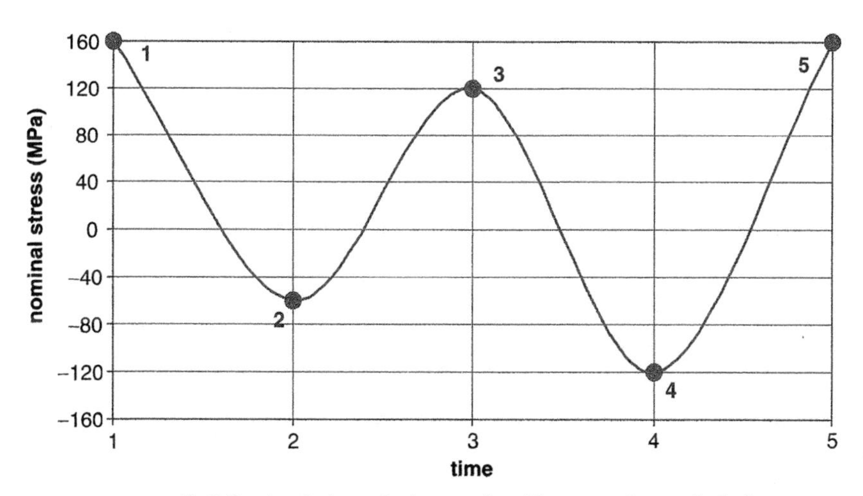

FIGURE 5.32 Applied nominal stress time history to the notched plate.

(1) Determine the fatigue life of the notched plate, using the modified Neuber rule for the notch analysis and the SWT mean stress correction formula.
(2) Repeat Part (1), using the strain energy density method by Molsky and Glinka for the notch analysis.

Solution. (1) To determine the fatigue life of the component by using the modified Neuber rule, the notch sensitivity factor must first be determined. Because $S_u = 320$ MPa < 560 MPa, the empirical Peterson formula for a notch sensitivity factor (q) is no longer valid and has to be estimated from Peterson's empirical notch sensitivity curves for steels. Based on $S_u = 320$ MPa, q is estimated as 0.65.

$$K_f = 1 + (K_t - 1)q = 1 + (2.53 - 1)0.65 = 1.99$$

With all the known material constants and the notch effects to the plate, the local notch stress–strain responses can be simulated by using the modified Neuber rule and the Massing hypothesis.

State of Stress and Strain at Point 1 Based on the Initial Loading Path from Time 0 to 1

First, it is given that the initial nominal stress starts from $S_0 = 0$ MPa to $S_1 = 160$ MPa. Neuber's constant can be calculated by using the nominal stress and the fatigue strength reduction factor:

$$\frac{(K_f S_1)^2}{E} = \frac{(1.99 \times 160)^2}{207000} = 0.48975$$

The true stress, σ_1, can be determined by a nonlinear equation solver in MS Excel or by other means for the following Neuber equation:

$$\frac{(\sigma_1)^2}{207000} + \sigma_1 \left(\frac{\sigma_1}{1240}\right)^{\frac{1}{0.27}} = 0.48975$$

$$\sigma_1 = 208.2\,\text{MPa}$$

Substituting the true stress back into the cyclic stress–strain equation gives

$$\varepsilon_1 = \frac{\sigma_1}{E} + \left(\frac{\sigma_1}{K'}\right)^{\frac{1}{n'}} = \frac{208.2}{207000} + \left(\frac{208.2}{1240}\right)^{\frac{1}{0.27}} = 0.002353 = 2353\,\mu\varepsilon$$

State of Stress and Strain at Point 2 Based on the First Loading Reversal from 1 to 2

For the first reversal, $S_1 = 160$ MPa to $S_2 = -60$ MPa, the nominal stress range is calculated as $\Delta S = 220$ MPa. The Neuber constant can be calculated by using the nominal stress delta and the fatigue strength reduction factor:

$$K_f^2 \Delta S \Delta e = \frac{(K_f \Delta S)^2}{E} = \frac{(1.99 \times 220)^2}{207000} = 0.92594$$

The local stress delta, $\Delta\sigma$, can be determined by a nonlinear equation solver in MS Excel for the following Neuber formula:

$$\frac{(\Delta\sigma)^2}{207000} + 2\Delta\sigma\left(\frac{\Delta\sigma}{2 \times 1240}\right)^{\frac{1}{0.27}} = 0.92594$$

$$\Delta\sigma = 332.7\,\text{MPa}$$

Substituting the true stress delta back into the hysteresis stress–strain equation, the true strain delta, $\Delta\varepsilon$, is determined:

$$\Delta\varepsilon = \frac{\Delta\sigma}{E} + 2\left(\frac{\Delta\sigma}{2K'}\right)^{\frac{1}{n'}} = \frac{332.7}{207000} + 2\left(\frac{332.7}{2 \times 1240}\right)^{\frac{1}{0.27}} = 0.002783 = 2783\,\mu\varepsilon$$

From the true stress and strain deltas, the true stress and strain at Time 2 can be determined by comparing the deltas with the reference true stress and true strain, which means

$$\sigma_2 = \sigma_1 - \Delta\sigma = 208.2 - 332.7 = -124.5\,\text{MPa}$$

$$\varepsilon_2 = \varepsilon_1 - \Delta\varepsilon = 2353 - 2783 = -430\,\mu\varepsilon$$

Following the methodology and calculation for the first loading reversal, the true stress and strain for each remaining reversals can be calculated. Table 5.17 shows the summary of the results for each reversal. Because of the material memory effect, the state of stress and strain at Point 4 follows the path 1–4. Figure 5.33 shows the complete hysteresis loops at the notch root based on the modified Neuber rule.

Next, it is necessary to estimate the fatigue life of the notched plate by using the rainflow cycle counting technique and the SWT method. The damage calculation for each extracted cycle is presented in Table 5.18. Then, the total damage and fatigue life prediction are obtained as follows:

$$\Sigma d_i = 5.28 \times 10^{-6} + 3.30 \times 10^{-5} = 3.83 \times 10^{-5}$$

TABLE 5.17 True Stress and Strain Results for Each Reversal Based on the Modified Neuber Rule

Point No.	Path	S_{Ref} (MPa)	ΔS (MPa)	Neuber Constant	$\Delta\sigma$ (MPa)	$\Delta\varepsilon$ ($\mu\varepsilon$)	σ_{Ref} (MPa)	ε_{Ref} ($\mu\varepsilon$)	σ (MPa)	ε ($\mu\varepsilon$)
1	0–1	0	160	0.48975	208.2	2353	0	0	208.2	2353
2	1–2	160	220	0.92594	332.7	2783	208.2	2353	−124.5	−430
3	2–3	−60	180	0.61984	291.4	2127	−124.5	−430	166.9	1697
4	1–4	160	280	1.49986	385.5	3890	208.2	2353	−177.3	−1537
5	4–5	−120	280	1.49986	385.5	3890	−177.3	−1537	208.2	2353

$$\text{Predicted life} = \frac{1}{\Sigma d_i} = \frac{1}{3.83 \times 10^{-5}} = 26100 \text{ blocks}$$

(2) A very similar process is used to determine the fatigue life, using the strain energy density method. The following steps explain the analysis.

State of Stress and Strain at Point 1 Based on the Initial Loading Path from Time 0 to 1

First it is given that the initial nominal stress starts from $S_0 = 0\,\text{MPa}$ to $S_1 = 160\,\text{MPa}$. Neuber's constant can be calculated by using the nominal stress and the K_t factor:

$$\frac{(K_t S_1)^2}{E} = \frac{(2.53 \times 160)^2}{207000} = 0.79161$$

The true stress σ_1 can be determined by a nonlinear equation solver in MS Excel for the Molsky and Glinka equation:

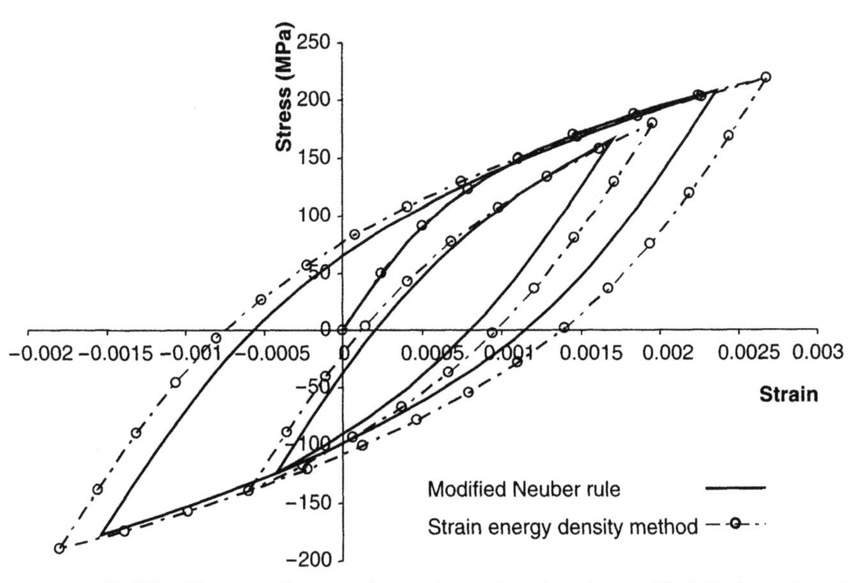

FIGURE 5.33 Hysteresis loops at the notch root based on the modified Neuber rule and the strain energy density method.

TABLE 5.18 SWT Mean Stress Correction Results Based on the Modified Neuber's Rule

n_i (cycles)	From–To	ε_{max} (με)	ε_{min} (με)	σ_{max} (MPa)	σ_{min} (MPa)	ε_a (με)	σ_{max} (MPa)	N_f (cycles)	d_i
1	2 to 3	1697	−430	166.9	−124.5	1064	166.9	189346	5.28E-6
1	1 to 4	2353	−1537	208.2	−177.3	1945	208.2	30273	3.30E-5

$$\frac{(\sigma_1)^2}{207000}+\frac{2\sigma_1}{1+0.27}\left(\frac{\sigma_1}{1240}\right)^{\frac{1}{0.27}}=0.79161$$

$$\sigma_1 = 218.9\,\text{MPa}$$

Substituting the true stress back into the cyclic stress–strain equation gives

$$\varepsilon_1=\frac{\sigma_1}{E}+\left(\frac{\sigma_1}{K'}\right)^{\frac{1}{n'}}=\frac{218.9}{207000}+\left(\frac{218.9}{1240}\right)^{\frac{1}{0.27}}=0.002682=2682\,\mu\varepsilon$$

State of Stress and Strain at Point 2 Based on the First Loading Reversal from 1 to 2

For the first reversal, $S_1 = 160\,\text{MPa}$ to$S_2 = -60\,\text{MPa}$, the nominal stress range is calculated as $\Delta S = 220\,\text{MPa}$.

The Neuber constant can be calculated by using the nominal stress delta and the K_t factor:

$$\frac{(K_t\Delta S)^2}{E}=\frac{(2.53\times 220)^2}{207000}=1.49664$$

The local stress delta, $\Delta\sigma$, can be determined by a nonlinear equation solver in MS Excel for the Molsky and Glinka equation:

$$\frac{(\Delta\sigma)^2}{207000}+\frac{4\Delta\sigma}{1+0.27}\left(\frac{\Delta\sigma}{2\times 1240}\right)^{\frac{1}{0.27}}=1.49664$$

$$\Delta\sigma = 358.6\,\text{MPa}$$

Substituting the true stress delta back into the hysteresis stress–strain equation, the true strain delta, $\Delta\varepsilon$, is determined:

$$\Delta\varepsilon=\frac{\Delta\sigma}{E}+2\left(\frac{\Delta\sigma}{2K'}\right)^{\frac{1}{n'}}=\frac{358.6}{207000}+2\left(\frac{358.6}{2\times 1240}\right)^{\frac{1}{0.27}}=0.003283=3283\,\mu\varepsilon$$

From the true stress and strain deltas, the true stress and strain at Time 2 can be determined by comparing the deltas with the reference true stress and true strain, which means

$$\sigma_2=\sigma_1-\Delta\sigma=218.9-358.6=-139.7\,\text{MPa}$$

$$\varepsilon_2=\varepsilon_1-\Delta\varepsilon=2682-3283=-601\,\mu\varepsilon$$

Following the methodology and calculation for the first loading reversal, the true stress and strain for each remaining reversals can be calculated. Table 5.19 shows the summary of the results for each reversal. Because of the material memory effect, the state of stress and strain at Point 4 follows the path 1–4. Figure 5.33 shows the complete hysteresis loops at the notch root based on the strain energy density method.

TABLE 5.19 True Stress and Strain Results for Each Reversal Based on the Strain Energy Density Method

Point No.	Path	S_{Ref} (MPa)	ΔS (MPa)	Neuber Constant	$\Delta\sigma$ (MPa)	$\Delta\varepsilon$ ($\mu\varepsilon$)	σ_{Ref} (MPa)	ε_{Ref} ($\mu\varepsilon$)	σ (MPa)	ε ($\mu\varepsilon$)
1	0–1	0	160	0.79161	218.9	2682	0	0	218.9	2682
2	1–2	160	220	1.49664	358.6	3283	218.9	2682	−139.7	−601
3	2–3	−60	180	1.00188	319.5	2555	−139.7	−601	179.8	1954
4	1–4	160	280	2.42430	408.6	4488	218.9	2682	−189.7	−1806
5	4–5	−120	280	2.42430	408.6	4488	−189.7	−1806	218.9	2682

TABLE 5.20 SWT Mean Stress Correction Results Based on the Strain Energy Density Method

n_i (cycles)	From–To	ε_{max} ($\mu\varepsilon$)	ε_{min} ($\mu\varepsilon$)	σ_{max} (MPa)	σ_{min} (MPa)	ε_a ($\mu\varepsilon$)	σ_{max} (MPa)	N_f (cycles)	d_i
1	2 to 3	1954	−601	179.8	−139.7	1278	179.8	104084	9.61E-6
1	1 to 4	2682	−1806	218.9	−189.7	2244	218.9	20314	4.92E-5

Next, it is necessary to estimate the fatigue life of the notched plate by using the rainflow cycle counting technique and the SWT method. The damage calculation for each extracted cycle is presented in Table 5.20. The total damage accumulation and life prediction are given below.

$$\Sigma d_i = 9.61 \times 10^{-6} + 4.92 \times 10^{-5} = 5.88 \times 10^{-5}$$

$$\text{Predicted life} = \frac{1}{\Sigma d_i} = \frac{1}{5.88 \times 10^{-5}} = 17000 \text{ blocks}$$

REFERENCES

ASTM, Standard practice for statistical analysis of linear or linearized stress-life (S-N) and strain-life (e-N) fatigue data, ASTM Standard E739, ASTM Designation E 739–91 (reapproved 1998), 1998b, pp. 1–7.

ASTM, Standard practice for strain-controlled fatigue testing, ASTM Standard E 606, ASTM Designation: E 606–92 (reapproved 1998), 1998a, pp. 1–7.

ASTM, Standard practice for verification of specimen alignment under tensile loading, ASTM Standard E1012, ASTM Designation E 1012–99, 1999, pp. 1–8.American Society for Testing and Materials (ASTM), Standard test methods for tension testing of metallic materials [metric], ASTM Standard E 8M, ASTM Designation E 8M-01, 2001, pp. 1–7.

Bannantine, J. A., Comer, J. J., and Handrock, J. L., *Fundamentals of Metal Fatigue Analysis*, Prentice Hall, New York, 1990.

Barkey, M. E., Socie, D. F., and Hsia, K. J. A yield surface approach to the estimation of notch strains for proportional and nonproportional cyclic loading. *Journal of Engineering Materials and Technology*, Vol. 116, 1994, pp. 173–180.

Basquin, O. H., The exponential law of endurance tests, *American Society for Testing and Materials Proceedings*, Vol. 10, 1910, pp. 625–630.

Baumel, A., Jr. and Seeger, T., *Material Data for Cyclic Loading,* Supplemental Volume, Erscheint Anfang Bei, Elsevier, Amsterdam, 1990.

Baumel, A., Jr. and Seeger, T., Thick surface layer model: life calculations for specimens with residual stress distribution and different material zones. In *Proceedings of International Conference on Residual Stress* (ICRS2), London, 1989, pp. 809–914.

Boardman, B. E. Crack initiation fatigue: data, analysis, trends and estimations, SAE 820682, 1982.

Bridgman, P. W. Stress distribution at the neck of tension specimen *Transactions of the American Society for Metals*, Vol. 32, 1944, pp. 553–572.

Coffin, L. F., Jr., A study of the effect of cyclic thermal stresses on a ductile metal, *Transactions of ASME*, Vol. 76, 1954, pp. 931–950.

Dowling, N. E., *Mechanical Behavior of Materials: Engineering Methods for Deformation, Fracture, and Fatigue*, 2nd ed., Prentice Hall, New York, 1998.

Fash, J. and Socie, D. F., Fatigue behavior and mean effects in grey cast iron, *International Journal of Fatigue*, Vol. 4, No. 3, 1982, pp. 137–142.

Forsetti, P. and Blasarin, A., Fatigue behavior of microalloyed steels for hot forged mechanical components, *International Journal of Fatigue*, Vol. 10, No. 3, 1988, pp. 153–161.

Gu, R. and Lee, Y., A new method for estimating nonproportional notch-root stresses and strains, *Journal of Engineering Materials and Technology*, Vol. 119, 1997, pp. 40–45.

Hoffmann, M. and Seeger, T., A generalized method doe estimating multiaxial elastic-plastic notch stresses and strains. Part 1: Theory, *Journal of Engineering Materials and Technology*, Vol. 107, 1985, pp. 250–254.

Koettgen, V. B., Barkey, M. E., and Socie, D. F., Pseudo stress and pseudo strain based approaches to multiaxial notch analysis, *Fatigue and Fracture of Engineering Materials and Structures*, Vol. 18, No. 9, 1995, pp. 981–1006.

Koh, S. K. and Stephens, R. I., Mean stress effects on low cycle fatigue for a high strength steel, *Fatigue and Fracture Engineering Materials and Structures*, Vol. 14, No. 4, 1991, pp. 413–428.

Langlais, T. E. and Vogel, J. H., Overcoming limitations of the conventional strain-life fatigue damage model, *Journal of Engineering Materials and Technology*, Vol. 117, 1995, pp. 103–108.

Landgraf, R. W., *High Fatigue Resistance in Metals and Alloys*, ASTM STP467, American Society for Testing and Materials, 1970, pp. 3–36.

Lawrence, F. V., Burk, J. V., and Yung, J. Y., Influence of residual stress on the predicted fatigue life of weldments. In *Residual Stress in Fatigue*, ASTM STP 776, ASTM, West Conshohocken, PA, pp. 33–43, 1982.

Lee, Y., Chiang, Y., and Wong, H., A constitutive model for estimating multiaxial notch strains, *Journal of Engineering Materials and Technology*, Vol. 117, 1995, pp. 33–40.

Manson, S. S. Behavior of materials under conditions of thermal stress. In *Heat Transfer Symposium,* University of Michigan Engineering Research Institute, MI, 1953, pp. 9–75.

Masing, G. Eigerspannungen and Verfestigung bein Messing, Proceeding of the 2nd International Congress of Applied Mechanics, Zurich, 1976, pp. 332–335.

Moftakhar, A., Buczynski, A., and Glinka, G., Elastic-plastic stress–strain calculation in notched bodies subjected to nonproportional loading, *International Journal of Fracture*, Vol. 70, 1995, pp. 357–373.

Molsky, K. and Glinka, G., A method of elastic-plastic stress and strain calculation at a notch root, *Materials Science and Engineering*, Vol. 50, 1981, pp. 93–100.

Morrow, J. D., Cyclic plastic strain energy and fatigue of metals. In *Internal Friction, Damping, and Cyclic Plasticity*, ASTM, West Conshohocken, PA, 1965, pp. 45–86.

Morrow, J. D. *Fatigue Design Handbook*, Section 3.2, *SAE Advances in Engineering*, Vol. 4, Society for Automotive Engineers, Warrendale, PA, 1968, pp. 21–29.

Muralidharan, U. and Manson, S. S. A modified universal slopes equation for estimating of fatigue characteristics of metals, *Journal of Engineering Materials and Technology*, Vol. 110, 1988, pp. 55–58.

Neuber, H. Theory of stress concentration for shear-strained prismatical bodies with arbitrary nonlinear stress–strain law, *Journal of Applied Mechanics* (*Transactions of ASSM, Section E*), Vol. 28, 1961, pp. 544–550.

Park, J. -H. and Song, J. -H., New estimation method of fatigue properties of aluminum alloys, *Journal of Engineering Materials and Technology*, Vol. 125, 2003, pp. 208–214.

Ramberg, W. and Osgood, W. R., Description of stress–strain curves by three parameters, NACA Technical Note No. 902, 1943.

Reemsnyder, H. S., Evaluating the effect of residual stresses on notched fatigue resistance. In *Materials, Experimentation and Design in Fatigue,* Society of Environmental Engineers, Buntingford, U.K.,1981.

Roessle, M. L., and Fatemi, A., Strain-controlled fatigue properties of steels and some simple approximations, *International Journal of Fatigue*, Vol. 22, 2000, pp. 495–511.

Seeger T. and Heuler P., Generalized application of Neuber's rule, *Journal of Testing and Evaluation*, Vol. 8, No. 4, 1980, pp. 199–204.

Shen, C. L. and Wirsching, P. H., Design curve to characterize fatigue strength, *Journal of Engineering Materials and Technology*, Vol. 118, 1996, pp. 535–541.

Sines, G. and Waisman, J. L., *Metal Fatigue*, McGraw-Hill, New York, 1959.

Smith, K. N., Watson, P., and Topper, T. H., A stress–strain function for the fatigue of metals, *Journal of Materials,* Vol. 5, No. 4, 1970, pp. 767–778.

Topper, T. H., Wetzel, R. M., and Morrow, J., Neuber's rule applied to fatigue of notched specimens, *Journal of Materials*, Vol. 4, No. 1, 1969, pp. 200–209.

Yang, L., Mechanical Behavior and cumulative fatigue damage of vanadium-based micro-alloyed forging steel, Ph.D. Thesis, University of Toledo, Toledo, OH, 1994.

Yan, X., Cordes, T. S., Vogel, J. H., and Dindinger, P. M., A property fitting approach for improved estimates of small cycle fatigue damage, SAE Paper No. 920665, 1992.

Walcher, J., Gary, D., and Manson, S. S. Aspects of cumulative fatigue damage analysis of cold end rotating structures, AIAA Paper 79–1190, 1979.

Wehner T., and Fateni, A. Effect of mean stress on fatigue behavior of a hardened carbon steel, *International Journal of Fatigue*, Vol. 13, No. 3, 1991, pp. 241–248.

Wetzel, R. M., Smooth specimen simulation of the fatigue behavior of notches, *Journal of Materials*, Vol. 3, No. 3, 1968, pp. 646–657.

Williams, C. R., Lee, Y., and Rilly, J. T., A practical method for statistical analysis of strain-life fatigue data, *International Journal of Fatigue*, Vol. 25, No. 5, 2003, pp. 427–436.

6

FRACTURE MECHANICS AND FATIGUE CRACK PROPAGATION

JWO PAN AND SHIH-HUANG LIN
UNIVERSITY OF MICHIGAN

6.1 INTRODUCTION

Use of crack propagation laws based on stress intensity factor ranges is the most successful engineering application of fracture mechanics. This chapter gives a review of the basic concepts of fracture mechanics. In contrast to the traditional stress–life and strain–life approaches to fatigue, cracks are assumed to exist in materials and structures within the context of fracture mechanics. Fracture parameters such as K and J can be used to characterize the stresses and strains near the crack tips. A fundamental understanding of fracture mechanics and the limit of using the fracture parameters is needed for appropriate applications of fracture mechanics to model fatigue crack propagation.

In this chapter, the concept of stress concentration of linear elasticity is first introduced. Griffith's fracture theory and energy release rate are discussed. Different fracture modes and asymptotic crack-tip fields are then presented. Linear elastic stress intensity factors are introduced. Stress intensity factor solutions for simple cracked geometries and loading conditions are given. Also, stress intensity factor solutions for circular and elliptical cracks are given. Plastic zones and requirements of linear elastic fracture mechanics are discussed. Finally, fatigue crack propagation laws based on linear elastic fracture mechanics are presented.

6.2 STRESS CONCENTRATION BASED ON LINEAR ELASTICITY

6.2.1 ELLIPTICAL HOLE

Consider an elliptical hole in a two-dimensional infinite solid under remote uniaxial tension. The major and minor axes for the elliptical hole are denoted as c and b, as shown in Figure 6.1. The elliptical hole surface can be described by the equation

$$\frac{x^2}{c^2} + \frac{y^2}{b^2} = 1 \tag{6.2.1}$$

The radius of curvature at $x = \pm c$ and $y = 0$ is denoted as ρ, as shown in the figure. The radius of curvature ρ can be related to c and b as

$$\rho = \frac{b^2}{c} \tag{6.2.2}$$

The two-dimensional infinite solid is subjected to a uniformly normal stress σ_∞ in the y direction at $y = \pm\infty$. The linear elasticity solution gives the stress solution at $x = \pm c$ and $y = 0$ as

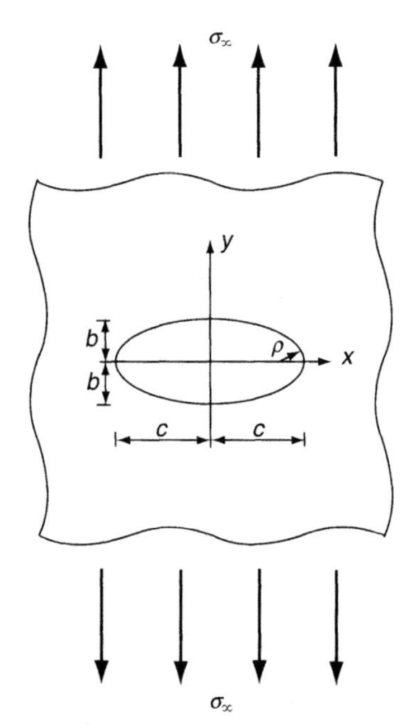

FIGURE 6.1 An elliptical hole in a two-dimensional infinite solid under a remote tension σ_∞.

$$\sigma_{yy}(x = \pm c, y = 0) = \sigma_{\infty}\left(1 + \frac{2c}{b}\right) \quad (6.2.3)$$

Based on Equation 6.2.2, Equation 6.2.3 can be rewritten as

$$\sigma_{yy}(x = \pm c, y = 0) = \sigma_{\infty}\left(1 + 2\sqrt{\frac{c}{\rho}}\right) \quad (6.2.4)$$

We can define the stress concentration factor K_t for the elliptical hole under the remote uniaxial tension as

$$K_t = \frac{\sigma_{yy}(x = \pm c, y = 0)}{\sigma_{\infty}} = 1 + 2\sqrt{\frac{c}{\rho}} \quad (6.2.5)$$

6.2.2 CIRCULAR HOLE

For the special case of a circular hole with a radius of a as shown in Figure 6.2(a), the equations for the elliptical hole will be valid with $\rho = b = c = a$. A polar coordinate system for the circular hole problem is shown in Figure 6.2(a). The solutions for the normalized stresses $\tilde{\sigma}_{rr}$, $\tilde{\sigma}_{\theta\theta}$, and $\tilde{\sigma}_{r\theta}$ for this case are

$$\tilde{\sigma}_{rr} = \frac{\sigma_{rr}}{\sigma_{\infty}} = \frac{1}{2}\left(1 - \frac{a^2}{r^2}\right) - \frac{\cos 2\theta}{2}\left(\frac{3a^4}{r^4} - \frac{4a^2}{r^2} + 1\right) \quad (6.2.6)$$

$$\tilde{\sigma}_{r\theta} = \frac{\sigma_{r\theta}}{\sigma_{\infty}} = \frac{-\sin 2\theta}{2}\left(\frac{3a^4}{r^4} - \frac{2a^2}{r^2} - 1\right) \quad (6.2.7)$$

$$\tilde{\sigma}_{\theta\theta} = \frac{\sigma_{\theta\theta}}{\sigma_{\infty}} = \frac{1}{2}\left(1 + \frac{a^2}{r^2}\right) + \frac{\cos 2\theta}{2}\left(\frac{3a^4}{r^4} + 1\right) \quad (6.2.8)$$

Figure 6.2(b) shows the normalized hoop stress $\tilde{\sigma}_{\theta\theta}$ distribution as a function of the normalized radius r/a along the x axis ($\theta = 0$). As shown in the figure, the normalized hoop stress quickly reduces from 3 to approach to 1 as r/a increases. The hoop stress along the circumference of the circular hole surface is a function of the angular location θ. Figure 6.2(c) shows the normalized hoop stress $\tilde{\sigma}_{\theta\theta}$ as a function of θ along the circumferential surface of the hole ($r = a$). As shown in the figure, the normalized stress $\tilde{\sigma}_{\theta\theta}$ reduces from 3 to -1 and increases to 3 as θ increases from 0° to 90° and then to 180°. The hoop stresses at the locations of $\theta =$ 0°, 90°, 180°, and 270° are schematically shown in Figure 6.2(a). The stress concentration factor K_t for the circular hole in a two-dimensional infinite solid under remote uniaixal tension equals 3. Now consider the case of the circular hole under equal biaxial tension as shown in Figure 6.3. By using the superposition of the elasticity, the stress concentration factor K_t can be obtained as 2. From the example of the circular hole as discussed previously, the stress concentration factor depends not only on geometry but also on loading.

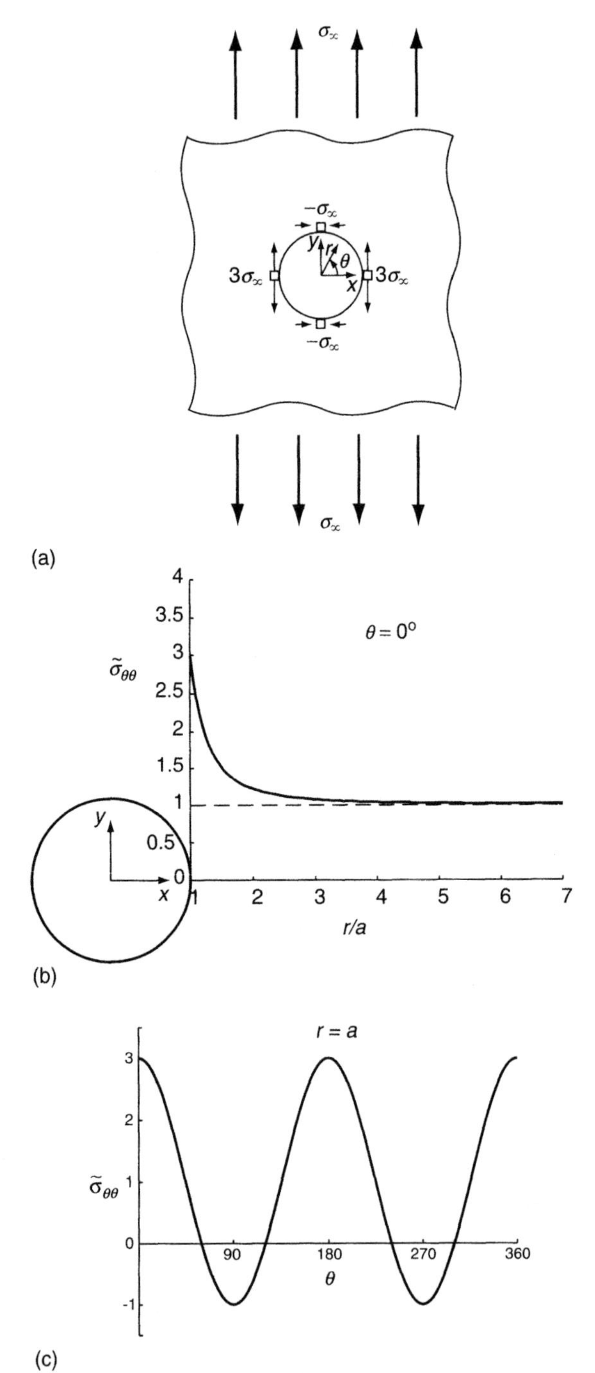

FIGURE 6.2 A circular hole in a two-dimensional infinite solid under a remote tension σ_∞. (a) Cartesian and polar coordinate systems (b) the normalized hoop stress $\tilde{\sigma}_{\theta\theta}$ distribution as a function of the normalized radius r/a along the x axis ($\theta = 0°$), (c) the normalized stress $\tilde{\sigma}_{\theta\theta}$ as a function of θ along the circumferential surface of the hole ($r = a$).

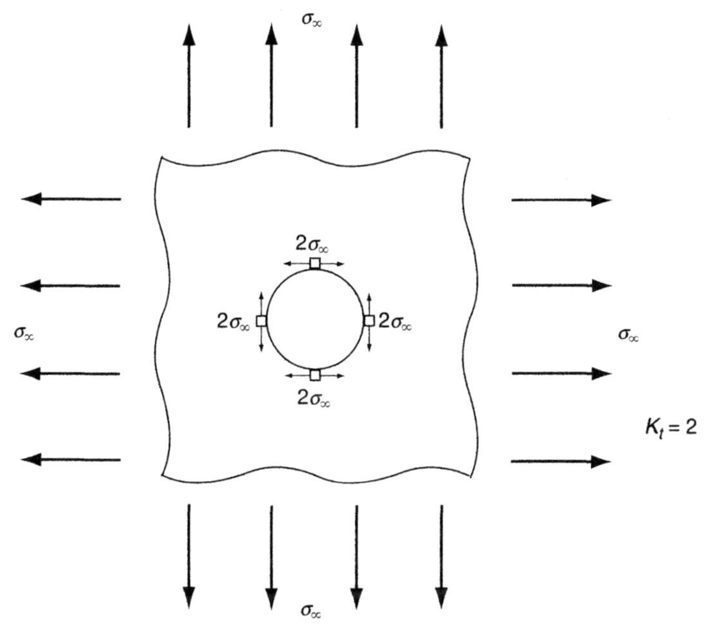

FIGURE 6.3 A circular hole in a two-dimensional infinite solid under an equal biaxial remote tension σ_∞.

6.2.3 LIMIT CASE FOR ELLIPTICAL HOLE

As the elliptical hole becomes increasingly flatter or b/c approaches 0, the geometry of the hole approaches that of a crack. In this case, $b \ll c$ and $\rho \rightarrow 0$. σ_{yy} at $x = \pm c$ and $y = 0$ based on the elastic solution for an elliptical hole becomes unbounded.

6.3 GRIFFITH'S FRACTURE THEORY FOR BRITTLE MATERIALS

As shown in Figure 6.4, we consider a large two-dimensional plate of thickness t under a remote uniform tensile stress σ. A center through-thickness crack with a crack length of $2c$ is shown. This cracked plate is considered under plane-stress loading conditions. According to Griffith (1921), fracture occurs when the decrease of the potential energy due to an increment of crack area equals the surface energy of newly created crack surfaces. For the large plate, the potential energy Φ can be written in terms of σ as

$$\Phi = -\frac{\sigma^2}{2E}V - \pi\frac{\sigma^2 c^2}{E}t \tag{6.3.1}$$

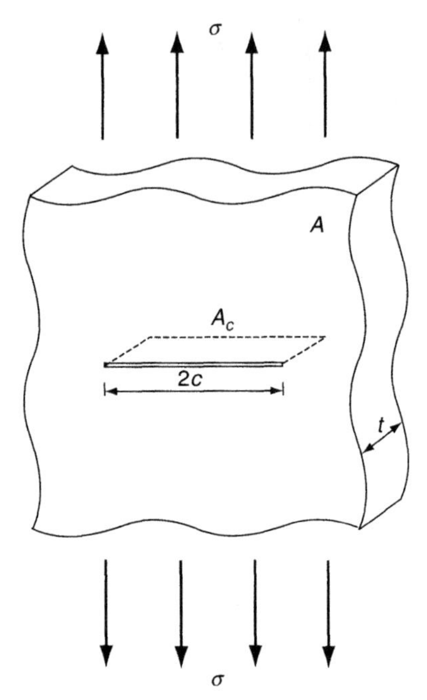

FIGURE 6.4 A large two-dimensional plate of thickness t under a remote uniform tensile stress σ.

where V represents the total volume of the plate.

The energy release rate per unit crack area can be expressed as

$$-\frac{\partial \Phi}{\partial A_c} = -\frac{1}{2t}\frac{\mathrm{d}\Phi}{dc} = \pi\frac{\sigma^2 c}{E} \tag{6.3.2}$$

where A_c represents the crack area. In this case, $A_c = 2ct$. Now, the stress at fracture can be denoted as σ_f. At fracture the energy release per unit crack area growth should equal the surface energy γ of the newly created crack surface. Therefore,

$$-\frac{\partial \Phi}{\partial A_c} = \pi\frac{\sigma_f^2 c}{E} = 2\gamma \tag{6.3.3}$$

In this case, the fracture stress σ_f can be obtained as

$$\sigma_f = \sqrt{\frac{2\gamma E}{\pi c}} \tag{6.3.4}$$

6.4 FRACTURE PARAMETERS BASED ON LINEAR ELASTIC FRACTURE MECHANICS (LEFM)

6.4.1 THREE PRIMARY LOADING MODES

Consider a segment of crack front as shown in Figures 6.5–6.7. Because of different loading conditions, the crack front can be subjected to three primary loading modes and their combinations. A Cartesian coordinate system is assigned such that the crack front is in the z direction. We consider idealized planar crack problems, in which the stresses and strains near the crack tip can be expressed in terms of the in-plane coordinates x and y only. As shown in Figure 6.5, the crack is subject to Mode I, the opening or tensile mode, where the in-plane stresses and strains are symmetric with respect to the x axis. As shown in Figure 6.6, the crack is subject to Mode II, the sliding or in-plane

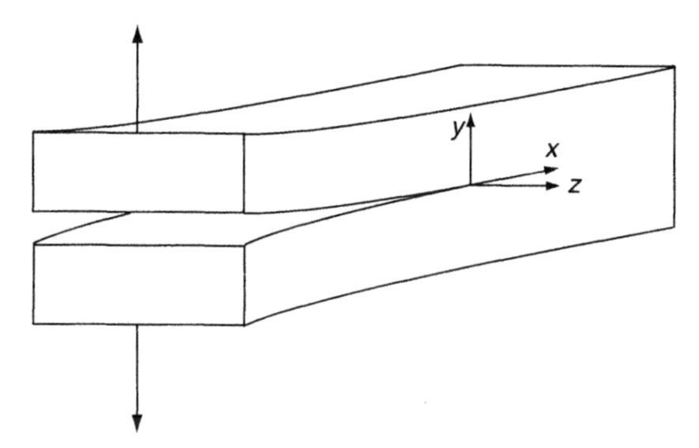

FIGURE 6.5 Mode I or opening mode.

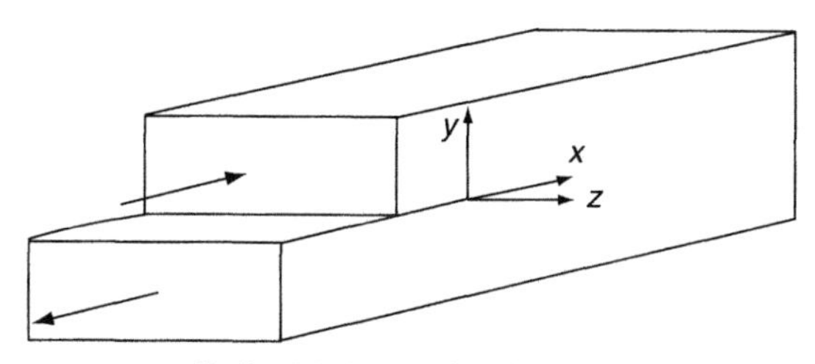

FIGURE 6.6 Mode II or in-plane shearing mode.

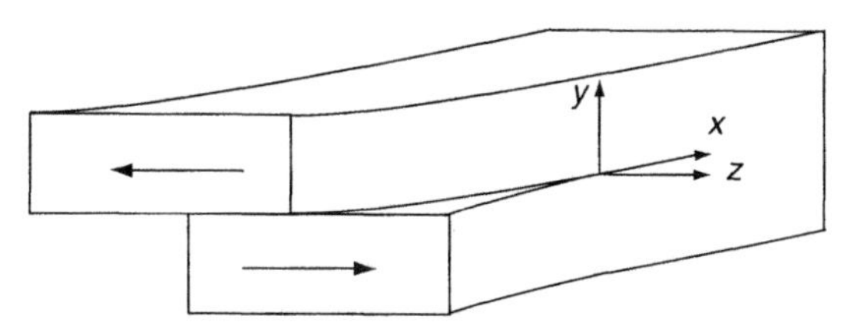

FIGURE 6.7 Mode III or out-of-plane shearing mode.

shearing mode, where the stresses and strains are anti-symmetrical with respect to the x axis. As shown in Figure 6.7, the crack is subject to Mode III, the tearing or anti-plane shearing mode, where the out-of-plane stresses and strains are anti-symmetrical with respect to the x axis.

6.4.2 ASYMPTOTIC CRACK-TIP FIELD FOR LINEAR ELASTIC MATERIALS

Here, the asymptotic crack-tip stresses and strains for different modes of loading are explored for linear elastic isotropic materials. As shown in Figure 6.8, we concentrate on a very small area near a crack tip, where the stresses and strains should be naturally expressed in terms of the polar coordinates r and θ. The three stresses σ_{rr}, $\sigma_{\theta\theta}$, and $\sigma_{r\theta}$ are shown for a material element. The two in-plane equilibrium equations for the plane problem are

$$\frac{\partial \sigma_{rr}}{\partial r} + \frac{1}{r}\frac{\partial \sigma_{r\theta}}{\partial \theta} + \frac{\sigma_{rr} - \sigma_{\theta\theta}}{r} = 0 \tag{6.4.1}$$

$$\frac{\partial \sigma_{r\theta}}{\partial r} + \frac{1}{r}\frac{\partial \sigma_{\theta\theta}}{\partial \theta} + \frac{2\sigma_{rr}}{r} = 0 \tag{6.4.2}$$

As schematically shown in Figure 6.8, the traction on the crack surfaces should vanish. This leads to the traction boundary conditions on the two crack faces as

$$\sigma_{\theta\theta} = \sigma_{r\theta} = 0 \text{ at } \theta = \pm\pi \tag{6.4.3}$$

Combining the compatibility equation, the two in-plane equilibrium equations and the elastic Hooke's law under both plane stress and plane strain conditions leads to

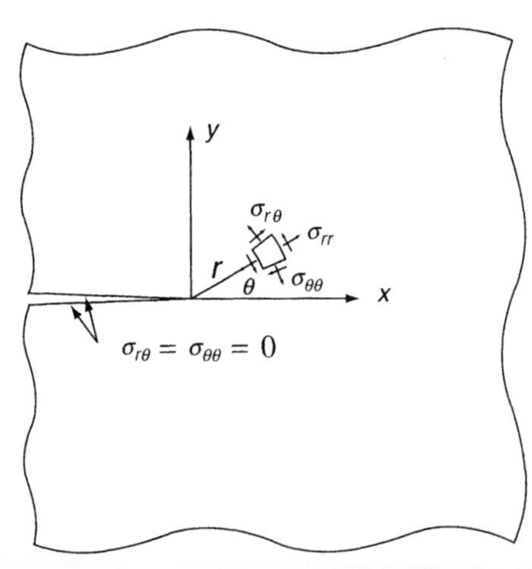

FIGURE 6.8 A crack with the Cartesian and polar coordinate systems centered at the tip.

$$\left(\frac{\partial^2}{\partial r}+\frac{1}{r}\frac{\partial}{\partial r}+\frac{1}{r^2}\frac{\partial^2}{\partial\theta^2}\right)(\sigma_{rr}+\sigma_{\theta\theta})=0 \tag{6.4.4}$$

Solutions are sought for the three governing equations: Equations 6.4.1, 6.4.2, and 6.4.4, with the homogeneous boundary conditions in Equation 6.4.3. The stresses can be assumed to be separable functions of r and θ as

$$\sigma_{rr}=Cr^{\lambda}\tilde{\sigma}_{rr}(\theta) \tag{6.4.5}$$

$$\sigma_{\theta\theta}=Cr^{\lambda}\tilde{\sigma}_{\theta\theta}(\theta) \tag{6.4.6}$$

$$\sigma_{r\theta}=Cr^{\lambda}\tilde{\sigma}_{r\theta}(\theta) \tag{6.4.7}$$

where λ represents an eigenvalue. C is an undetermined constant, which represents the singularity amplitude when λ is negative. $\tilde{\sigma}_{rr}(\theta)$, $\tilde{\sigma}_{r\theta}(\theta)$, and $\tilde{\sigma}_{\theta\theta}(\theta)$ represent the normalized stress functions of θ.

When the boundary-valued problem is solved based on the separable functions in Equations 6.4.5–6.4.7, the allowable singular solution is with $\lambda=-1/2$ by consideration of finite strain energy for a small circular region of size r centered at the crack tip. Under plane strain or plane stress conditions, two primary modes, Modes I and II, are considered. For Mode I, the stresses are symmetrical with respect to the crack line direction ($\theta=0$). Therefore, the stress intensity factor K_{I} for Mode I is defined as the singularity amplitude of the opening stress $\sigma_{\theta\theta}$ directly ahead of the tip ($\theta=0$) as

$$K_{\mathrm{I}}=\lim_{r\to 0}\sqrt{2\pi r}\sigma_{\theta\theta}(r,\,\theta=0) \tag{6.4.8}$$

For Mode II, the stresses are anti-symmetrical with respect to the crack line direction ($\theta=0$). Therefore, the stress intensity factor K_{II} is defined as the singularity amplitude of the shear stress $\sigma_{r\theta}$ directly ahead of the tip ($\theta=0$) as

$$K_{\mathrm{II}}=\lim_{r\to 0}\sqrt{2\pi r}\sigma_{r\theta}(r,\,\theta=0) \tag{6.4.9}$$

Substituting Equations 6.4.5–6.4.7 into the governing equations in Equations 6.4.1, 6.4.2, and 6.4.4 with use of the boundary conditions in Equation 6.4.3 and the definitions in Equation 6.4.8 for K_{I}, the asymptotic stresses under Mode I loading conditions can be obtained as

$$\sigma_{\theta\theta}=\frac{K_{\mathrm{I}}}{\sqrt{2\pi r}}\cos^3\frac{\theta}{2} \tag{6.4.10}$$

$$\sigma_{rr}=\frac{K_{\mathrm{I}}}{\sqrt{2\pi r}}\cos\frac{\theta}{2}\left(1+\sin^2\frac{\theta}{2}\right) \tag{6.4.11}$$

$$\sigma_{r\theta}=\frac{K_{\mathrm{I}}}{\sqrt{2\pi r}}\sin\frac{\theta}{2}\cos^2\frac{\theta}{2} \tag{6.4.12}$$

Similarly, the asymptotic stresses under Mode II loading conditions are

$$\sigma_{\theta\theta} = \frac{K_{\mathrm{II}}}{\sqrt{2\pi r}}\left(-3\sin\frac{\theta}{2}\cos^2\frac{\theta}{2}\right) \tag{6.4.13}$$

$$\sigma_{rr} = \frac{K_{\mathrm{II}}}{\sqrt{2\pi r}}\sin\frac{\theta}{2}\left(1 - 3\sin^2\frac{\theta}{2}\right) \tag{6.4.14}$$

$$\sigma_{r\theta} = \frac{K_{\mathrm{II}}}{\sqrt{2\pi r}}\cos\frac{\theta}{2}\left(1 - 3\sin^2\frac{\theta}{2}\right) \tag{6.4.15}$$

The Mode III stress intensity factor K_{III} can be defined as the singularity amplitude of the shear stress $\sigma_{\theta z}$ directly ahead of the tip ($\theta = 0$) as

$$K_{\mathrm{III}} = \lim_{r \to 0} \sqrt{2\pi r}\sigma_{\theta z}(r, \theta = 0) \tag{6.4.16}$$

Following the similar procedure as summarized above to solve the anti-plane boundary-valued problem with the homogeneous stress boundary conditions, the asymptotic stresses can be obtained as

$$\sigma_{\theta z} = \frac{K_{\mathrm{III}}}{\sqrt{2\pi r}}\cos\frac{\theta}{2} \tag{6.4.18}$$

$$\sigma_{rz} = \frac{K_{\mathrm{III}}}{\sqrt{2\pi r}}\sin\frac{\theta}{2} \tag{6.4.17}$$

For the asymptotic analyses, because the boundary conditions are homogenous ($\sigma_{\theta\theta} = \sigma_{r\theta} = \sigma_{\theta z} = 0$ at $\theta = \pm\pi$), only the angular dependence of the stresses can be determined. The singularity amplitudes K_{I}, K_{II}, and K_{III} cannot be determined by the asymptotic analyses. K_{I}, K_{II}, and K_{III} depend on the crack geometry and loading conditions. In general, the solutions of K_{I}, K_{II}, and K_{III} for relatively simple geometric configurations and loading conditions can be found in stress intensity factor handbooks such as in Tada et al. (2000).

6.4.3 STRESS INTENSITY FACTOR SOLUTIONS FOR SIMPLE GEOMETRIES AND LOADING CONDITIONS

For small cracks in infinite or semi-infinite solids, dimensional analyses can guide the selection of the form of the stress intensity factor solutions. For example, K_{I} has a dimension of

$$[K_{\mathrm{I}}] = \frac{[F]}{[L]^2}[L]^{1/2} \tag{6.4.19}$$

where $[K_{\mathrm{I}}]$ represents the dimension of the stress intensity factor K_{I}, $[F]$ the force dimension, and $[L]$ the length dimension. The unit of stress intensity factor K_{I} can be $\mathrm{MPa} \cdot \mathrm{m}^{1/2}$ or $\mathrm{ksi} \cdot \mathrm{in}^{1/2}$. We now consider three geometric and loading conditions as shown in Figures 6.9–6.11.

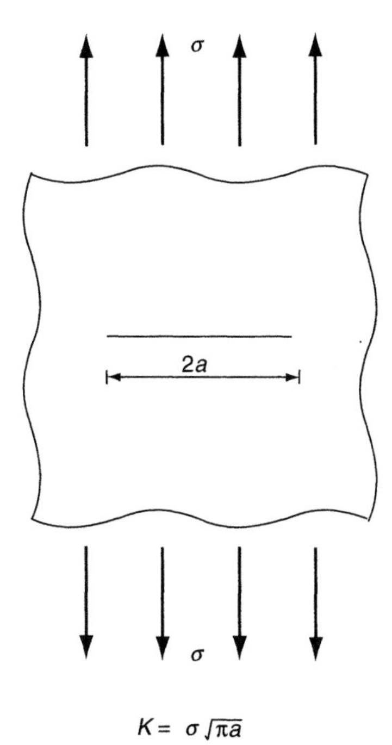

FIGURE 6.9 A center crack in a two-dimensional infinite solid under a remote uniform tensile stress σ.

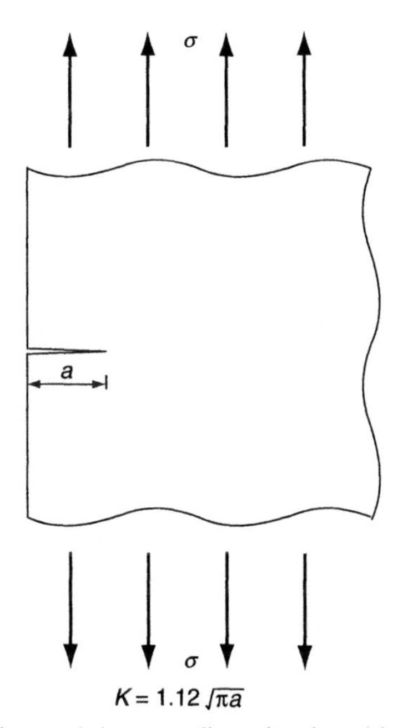

FIGURE 6.10 An edge crack in a two-dimensional semi-infinite solid under a remote uniform tensile stress σ.

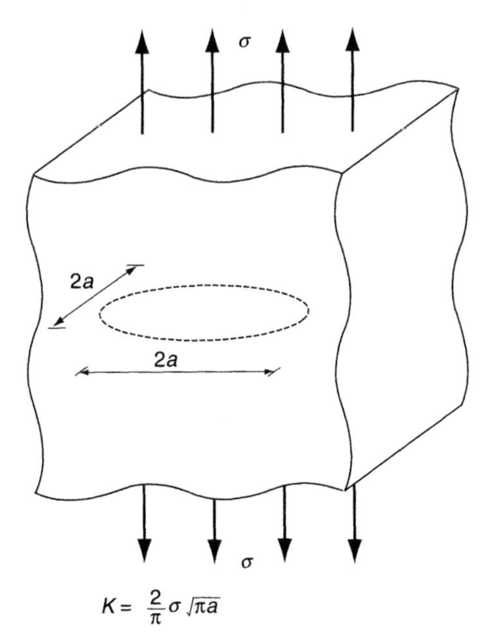

FIGURE 6.11 A penny-shaped crack in a three-dimensional infinite solid under a remote uniform tensile stress σ.

6.4.3.1 Center Crack in a Two-Dimensional Infinite Solid

Figure 6.9 shows a center crack with the crack length $2a$ in a two-dimensional infinite solid under a remote uniform tensile stress σ. For the center crack, the stress intensity factor solution is

$$K_{\mathrm{I}} = \sigma\sqrt{\pi a} \tag{6.4.20}$$

6.4.3.2 Edge Crack in a Two-Dimensional Semi-infinite Solid

Figure 6.10 shows an edge crack with the crack length a in a two-dimensional semi-infinite solid under a remote uniform tensile stress σ. For the edge crack, the stress intensity factor solution is

$$K_{\mathrm{I}} = 1.12\sigma\sqrt{\pi a} \tag{6.4.21}$$

6.4.3.3 Penny-Shaped Crack in a Three-Dimensional Infinite Solid

Figure 6.11 shows a penny-shaped crack with the diameter $2a$ in a three-dimensional infinite solid under a remote uniform tensile stress σ. For the penny-shaped crack, the stress intensity factor solution is

$$K_{\mathrm{I}} = \frac{2}{\pi}\sigma\sqrt{\pi a} \tag{6.4.22}$$

The stress intensity factor solutions for these three crack problems have the same forms except the proportionality constants for different geometries and loading conditions. The stress intensity factor solutions must be proportional to the magnitude of the remote uniform tensile stress σ due to the linearity of linear elasticity. The stress intensity factor solutions must be proportional to the square root of the only significant length dimension, which is the crack length a. For relatively simple geometries and loading conditions, finite element or other numerical methods can be used to find the relevant proportionality constants in the stress intensity factor solutions determined by dimensional analyses.

In general, for a center-cracked panel or an edge-cracked panel of finite size, the stress intensity factor can be written in the following form:

$$K_{\mathrm{I}} = F(\alpha) S_g \sqrt{\pi a} \tag{6.4.23}$$

where a represents the crack length; S_g the gross nominal stress, either the nominal stress or the maximum nominal bending stress; and $F(\alpha)$ a geometric function of the normalized crack length α. Here, $\alpha = a/b$, where b represents the width of the panel.

6.4.3.4 Center-Cracked Plate Under Tension

Figure 6.12(a) shows a center-cracked panel subject to a tensile load P. For the center-cracked panel in tension, S_g and $F(\alpha)$ are

$$S_g = \frac{P}{2bt} \tag{6.4.24}$$

$$F(\alpha) = \frac{1 - 0.5\alpha + 0.326\alpha^2}{\sqrt{1-\alpha}} \quad \text{for } h/b \geq 1.5 \tag{6.4.25}$$

where α represents the normalized crack length, h the height of the panel, and b the width of the panel.

6.4.3.5 Edge-Cracked Plate Under Tension

Figure 6.12(b) shows an edge-cracked panel subject to a tensile load P. For the edge-cracked panel in tension, S_g and $F(\alpha)$ are

$$S_g = \frac{P}{bt} \tag{6.4.26}$$

$$F(\alpha) = 0.265(1-\alpha)^4 + \frac{0.857 + 0.265\alpha}{(1-\alpha)^{3/2}} \quad \text{for } h/b \geq 1.0 \tag{6.4.27}$$

where α represents the normalized crack length, h the height of the panel, and b the width of the panel.

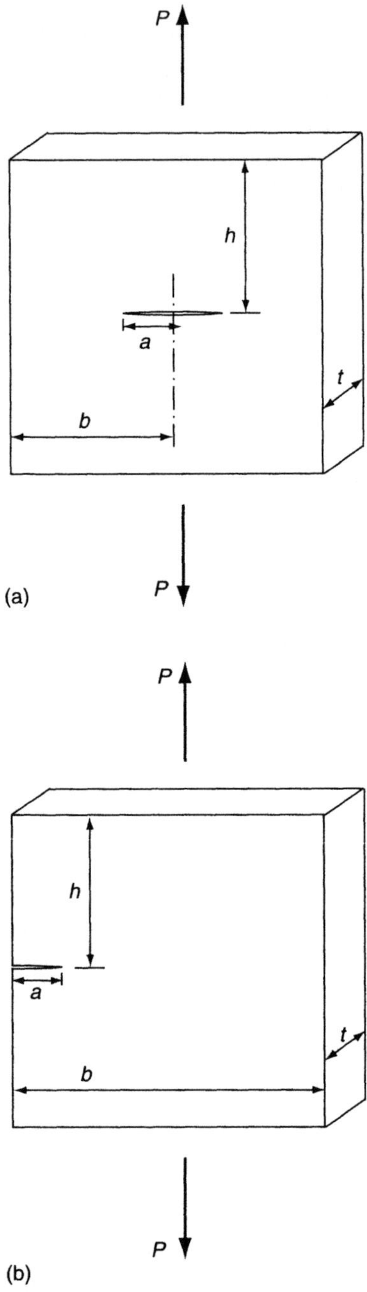

FIGURE 6.12 (a) A center cracked panel subject to a remote tensile load P. (b) An edge cracked panel subject to a tensile load P.

Continued

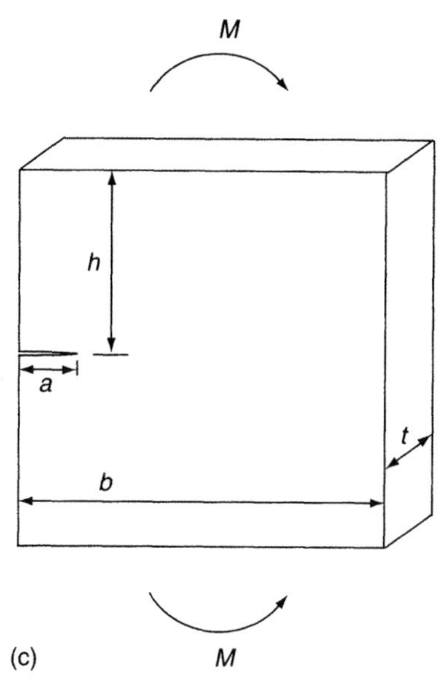

FIGURE 6.12 *Cont'd* (c) An edge cracked panel subject to a bending moment M

6.4.3.6 Edge-Cracked Plate Under Bending

Figure 6.12(c) shows an edge-cracked panel subject to a bending moment M. For the edge-cracked panel in bending, S_g and $F(\alpha)$ are

$$S_g = \frac{6M}{b^2 t} \tag{6.4.28}$$

$$F(\alpha) = \sqrt{\frac{2}{\pi\alpha}\tan\frac{\pi\alpha}{2}}\left[\frac{0.923 + 0.199\left(1 - \sin\frac{\pi\alpha}{2}\right)^4}{\cos\frac{\pi\alpha}{2}}\right] \quad \text{for large } h/b\text{'s} \tag{6.4.29}$$

where α represents the normalized crack length, h the height of the panel, and b the width of the panel.

Figure 6.13 show the geometric function $F(\alpha)$ for the three cases. As shown in the figure, as α becomes smaller, $F(\alpha)$ approaches 1 for the center-cracked panel case and 1.12 for the edge-cracked panel cases in order to be consistent with the solutions for a center crack in an infinite solid and an edge crack in a semi-infinite solid under remote tensile stress. As shown in the figure, when α increases, $F(\alpha)$ increases.

6.4.3.7 Cracks Emanating from a Hole Under Remote Tension

As shown in Figure 6.14, small cracks with the crack length l are emanating from a hole with a radius of c under a remote tensile stress S. When $l \ll c$, the stress intensity factor K_A can be expressed as

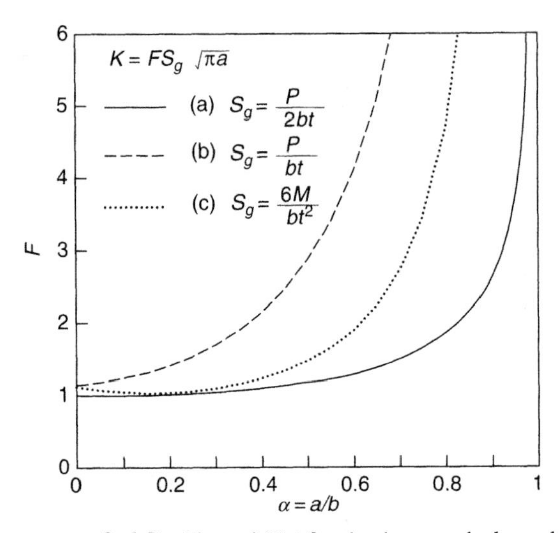

FIGURE 6.13 Plots of $F(\alpha)$ for the three cracked panels.

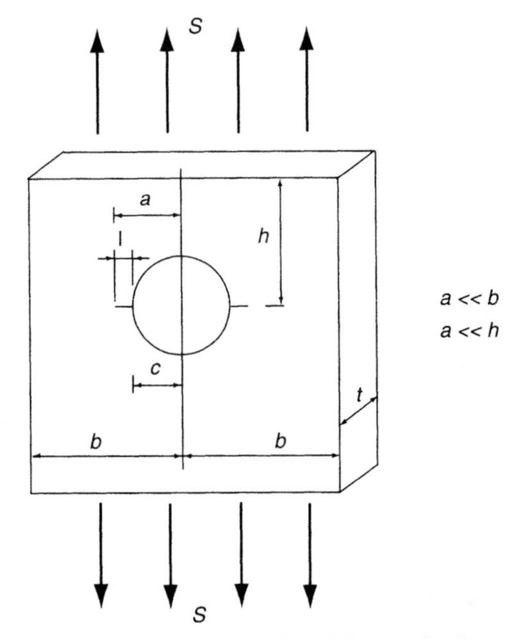

FIGURE 6.14 Small cracks with the crack length l emanating from a hole with a radius of c under a remote tensile stress S.

$$K_A = 1.12K_tS\sqrt{\pi l} \tag{6.4.30}$$

where the stress concentration factor K_t equals 3 in this case. When l is in the same order of c, the stress intensity factor K_B can be expressed as

$$K_B = FS\sqrt{\pi a} \tag{6.4.31}$$

where $a = c + l$. In general, the stress concentration factor K can be expressed as (Tada et al., 2000)

$$K = F(\mathrm{d})S\sqrt{\pi l} \tag{6.4.32}$$

Here, the geometric function $F(d)$ is in terms of the normalized crack length d as

$$F(d) = 0.5(3 - d)(1 + 1.243(1 - d)^3) \tag{6.4.33}$$

where

$$d = \frac{l}{a} = \frac{l}{c + l} \tag{6.4.34}$$

6.4.4 NON-SINGULAR ASYMPTOTIC CRACK-TIP FIELD FOR LINEAR ELASTIC MATERIALS

As discussed earlier, $\lambda = -1/2$ is adopted for the most singular asymptotic crack-tip stress and strain solutions. The next available solution is based on $\lambda = 0$ to satisfy the governing equations and boundary conditions for the asymptotic crack-tip fields. The non-singular stress satisfied the boundary conditions for the in-plane modes is the normal stress σ_{xx} or σ_{11} in the crack line (x or x_1) direction. The non-singular stress that satisfied the boundary conditions for the anti-plane mode is the shear stress σ_{zx} or σ_{31}. Under generalized plane strain conditions, we can write the asymptotic crack-tip fields with the non-singular stresses T for the in-plane modes and S for the anti-plane mode (Rice, 1974) as

$$\begin{aligned}\sigma_{ij} = & \frac{K_{\mathrm{I}}}{\sqrt{2\pi r}}\tilde{\sigma}^{\mathrm{I}}_{ij}(\theta) + \frac{K_{\mathrm{II}}}{\sqrt{2\pi r}}\tilde{\sigma}^{\mathrm{II}}_{ij}(\theta) + \frac{K_{\mathrm{III}}}{\sqrt{2\pi r}}\tilde{\sigma}_{ij}{}^{\mathrm{III}}(\theta) \\ & + \delta_{i1}\delta_{j1}T + \delta_{i3}\delta_{j3}\nu T + \delta_{i1}\delta_{j3}S + \cdots\end{aligned} \tag{6.4.35}$$

where $\tilde{\sigma}^{\mathrm{I}}_{ij}(\theta)$, $\tilde{\sigma}^{\mathrm{II}}_{ij}(\theta)$, and $\tilde{\sigma}^{\mathrm{III}}_{ij}(\theta)$ are the normalized angular functions for Mode I, II, and III asymptotic stress solutions, respectively, as specified earlier. The range of the subscripts i and j is from 1 to 3. Here, δ_{ij} represents the components of the Kronecker delta function. For example, δ_{i1} equals 1 when i equals 1 and δ_{i1} equals 0 when i does not equal 1. The T and S stresses, just as the stress intensity factors K_{I}, K_{II}, and K_{III} are functions of the geometries and loading conditions [e.g., see Sham (1991)].

Figure 6.15 shows a center crack in a two-dimensional infinite solid subject to remote biaxial stressing conditions. Here, σ represents the normal stress perpendicular to the crack and $R\sigma$ represents the normal stress in the crack line direction. The stress intensity factor solution for this case is

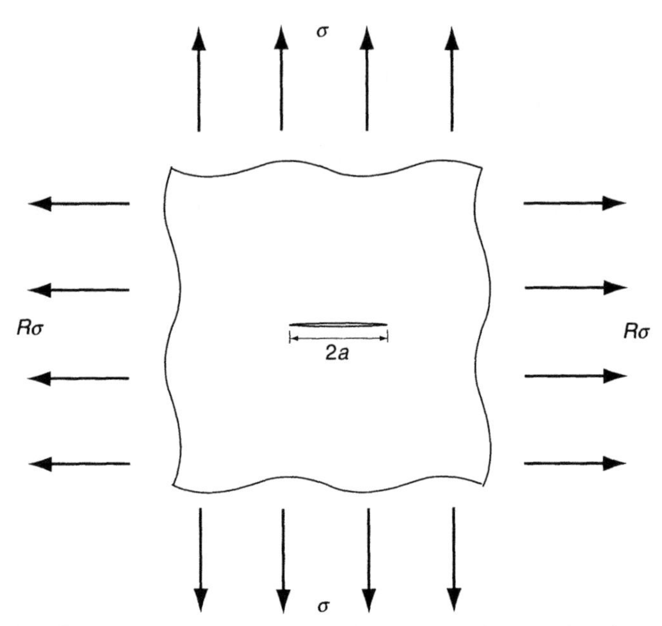

FIGURE 6.15 A center crack in a two-dimensional infinite solid subject to remote biaxial stressing conditions.

$$K_{\mathrm{I}} = \sigma\sqrt{\pi a} \tag{6.4.36}$$

The T stress solution can be expressed as

$$\frac{T}{K_{\mathrm{I}}} = \frac{R-1}{\sqrt{\pi a}} \tag{6.4.37}$$

Substituting the stress intensity solution into the T stress solution gives a simple solution for the T stress as

$$T = \sigma(R-1) \tag{6.4.38}$$

Therefore, under uniaxial tension, $R = 0$ and $T = -\sigma$. Under equal biaxial tension, $R = 1$ and $T = 0$. When $R > 1$, $T > 0$ and when $R < 1$, $T < 0$. When $T < 0$, there should be stable Mode I crack growth. When $T > 0$, there should be unstable Mode I crack growth (Cotterell and Rice, 1980). The effects of the T stress on the stability of Mode I crack growth is schematically shown in Figure 6.16.

6.4.5 CIRCULAR AND ELLIPTICAL CRACKS

6.4.5.1 Circular Cracks

In practical situations, circular and semi-circular cracks are usually found in engineering structures. The stress intensity factor solution for a

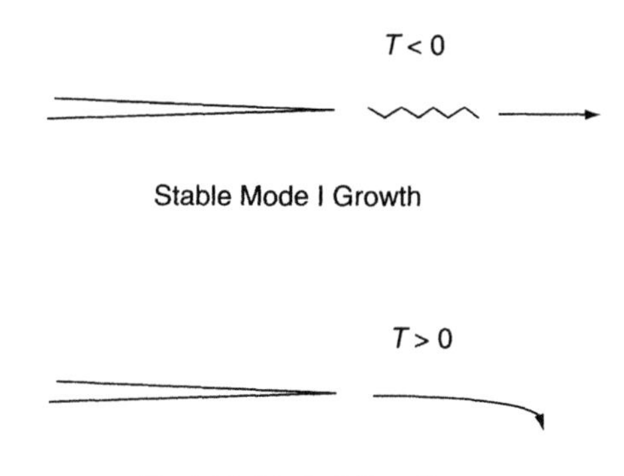

FIGURE 6.16 A schematic plot of the effects of the T stress on the stability of mode I crack growth.

penny-shaped crack in a three-dimensional infinite solid under a remote tensile stress S is available as discussed earlier:

$$K_{\mathrm{I}} = \frac{2}{\pi} S\sqrt{\pi a} \tag{6.4.39}$$

where a represents the crack length. When the penny-shaped crack is in a finite body, the size of the body can affect the stress intensity factor solution. When the crack length a is small compared with the width or the thickness of the cracked structures, Equation 6.4.39 is a good approximation. Now, we will concentrate on surface cracks. For a semi-circular surface crack in a rectangular bar under a tensile force P and a bending moment M as shown in Figure 6.17(a), the stress intensity factor solution varies along the crack front (Newman and Raju, 1986). The maximum values occur at the crack front on the free surface. The gross nominal stresses for the bar due to tension and bending, S_t and S_b, can be expressed as

$$S_t = \frac{P}{2bt} \tag{6.4.40}$$

$$S_b = \frac{3M}{bt^2} \tag{6.4.41}$$

where b represents the width and t the thickness.

For small surface cracks, the maximum values of the stress intensity factor solutions due to tension and bending can be expressed, respectively, as

$$K_{\mathrm{I}} = 0.728 S_t\sqrt{\pi a} \tag{6.4.42}$$

$$K_{\mathrm{I}} = 0.728 S_b\sqrt{\pi a} \tag{6.4.43}$$

The stress intensity factor solution decreases by about 10% at the crack deepest front.

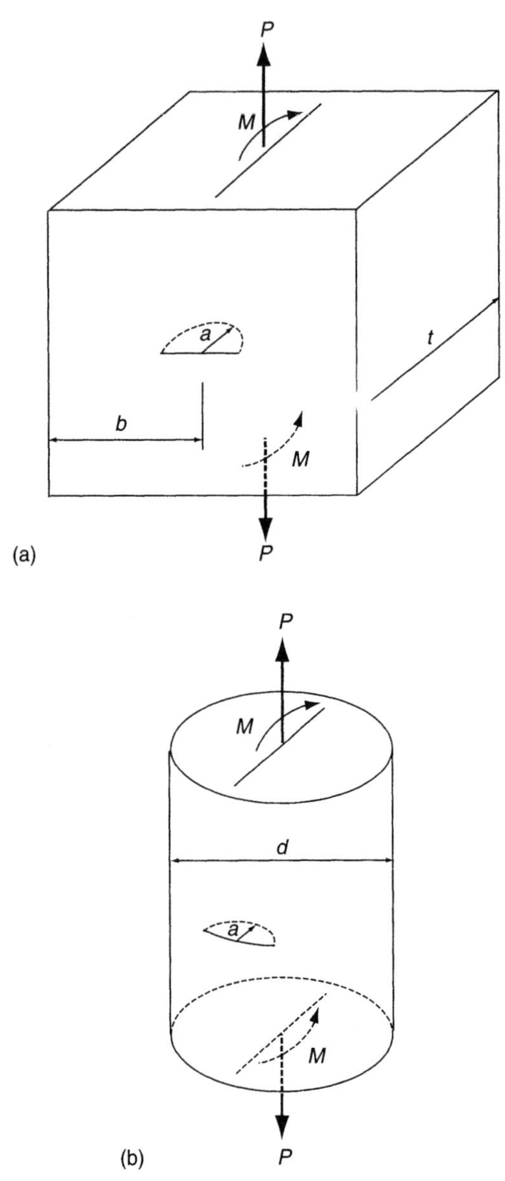

FIGURE 6.17 (a) A semi-circular surface crack (b) in a rectangular bar under tension and bending. (b) A semi-circular surface crack in a circular bar under tension and bending.

For a semi-circular surface crack in a circular bar under a tensile force P and a bending moment M as shown in Figure 6.17(b), the gross nominal stresses for the bar due to tension and bending, S_t and S_b, can be expressed, respectively, as (Newman and Raju, 1986)

$$S_t = \frac{4P}{\pi d^2} \tag{6.4.44}$$

$$S_b = \frac{32M}{\pi d^3} \tag{6.4.45}$$

where d represents the diameter.

For small surface cracks, the maximum values of the stress intensity factor solutions due to tension and bending can be expressed, respectively, as

$$K_\mathrm{I} = 0.728 S_t \sqrt{\pi a} \tag{6.4.46}$$

$$K_\mathrm{I} = 0.728 S_b \sqrt{\pi a} \tag{6.4.47}$$

6.4.5.2 Elliptical Cracks

The stress intensity factor solution for an elliptical crack in an infinite solid under a remote tensile stress S is also available. Figure 6.18 shows an elliptical crack in an infinite solid under a remote tensile stress S. The major axis for the elliptical crack front is c and the minor axis is a. Figure 6.19 shows a top view of the crack and the coordinate system. The stress intensity factor varies along the crack front. The maximum value of K_I occurs at Point D and the minimum value of K_I occurs at Point E. The stress intensity factor solution K_I is expressed in terms of the orientation angle ϕ as shown in Figure 6.19.

The solution K_I is expressed as a function of the orientation angle ϕ as

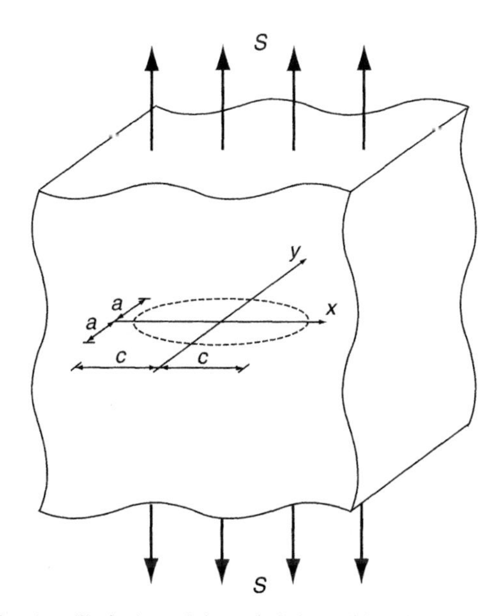

FIGURE 6.18 An elliptical crack in an infinite solid under a remote tensile stress S.

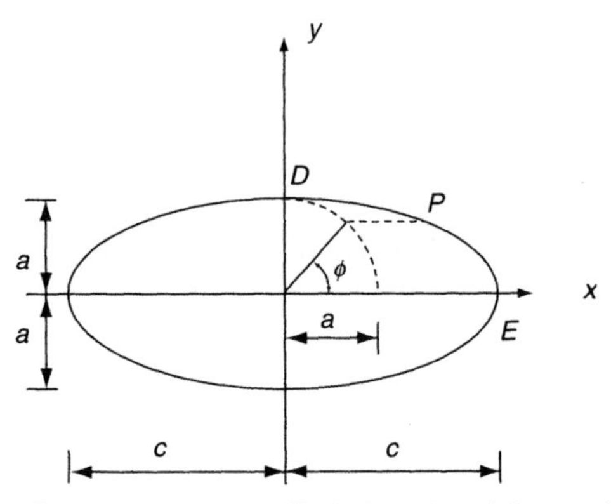

FIGURE 6.19 Top view of the elliptical crack and the coordinate system.

$$K = S\sqrt{\frac{\pi a}{Q}} f_\phi \tag{6.4.48}$$

where

$$f_\phi = \left[\left(\frac{a}{c}\right)^2 \cos^2\phi + \sin^2\phi\right]^{\frac{1}{4}} \text{ for } \frac{a}{c} \leq 1 \tag{6.4.49}$$

and

$$\sqrt{Q} = E(k) = \int_0^{\pi/2} \sqrt{1 - k^2 \sin^2\beta}\, d\beta \tag{6.4.50}$$

where Q represents the flaw shape factor and

$$k^2 = 1 - \left(\frac{a}{c}\right)^2 \tag{6.4.51}$$

Here, $E(k)$ is the elliptical integral of the second kind. $E(k)$ can be approximated as

$$\sqrt{Q} = E(k) \approx \left(1 + 1.464\left(\frac{a}{c}\right)^{1.65}\right)^{1/2} \tag{6.4.52}$$

The maximum value of K_I occurs at Point D where $\phi = 90°$ and $f_\phi = 1$. The minimum value of K_I occurs at Point E where $\phi = 0°$ and $f_\phi = \sqrt{a/c}$. Therefore, the maximum value of K_I at Point D is

$$K_D = S\frac{\sqrt{\pi a}}{E(k)} \tag{6.4.53}$$

The minimum value of K_I at Point E is

$$K_E = S\frac{\sqrt{\pi a}}{E(k)}\sqrt{\frac{a}{c}} \tag{6.4.54}$$

For semi-elliptical cracks, the maximum value of K_I still occurs at Point D when the ratio a/c is small. When the ratio a/c is close to unity, the maximum value of K_I occurs at Point E for the semi-circular surface cracks. For semi-elliptical surface cracks, the stress intensity factor solutions should be based on the embedded elliptical crack solutions multiplied by the free surface factor of 1.12.

6.5 PLASTIC ZONE SIZE AND REQUIREMENT OF LINEAR ELASTIC FRACTURE MECHANICS (LEFM)

6.5.1 ESTIMATION OF PLASTIC ZONE SIZE AHEAD OF A MODE I CRACK

We consider a crack with the Cartesian and the polar coordinate systems centered at the tip as shown in Figure 6.20. Here, z represents the out-of-plane coordinate. For in-plane modes (Modes I and II), the out-of-plane shear stresses are

$$\sigma_{zr} = \sigma_{z\theta} = 0 \tag{6.5.1}$$

6.5.1.1 Plane Stress Conditions

Under plane stress conditions, the out-of-plane normal stress is

$$\sigma_{zz} = 0 \tag{6.5.2}$$

We can estimate the plastic zone size at $\theta = 0$ directly ahead of a crack tip. Based on the asymptotic stress solutions in Equations 6.4.10–6.4.12, the stresses ahead of the crack tip at $\theta = 0$ are

$$\sigma_{rr} = \sigma_{\theta\theta} = \frac{K_I}{\sqrt{2\pi r}} \tag{6.5.3}$$

$$\sigma_{r\theta} = 0 \tag{6.5.4}$$

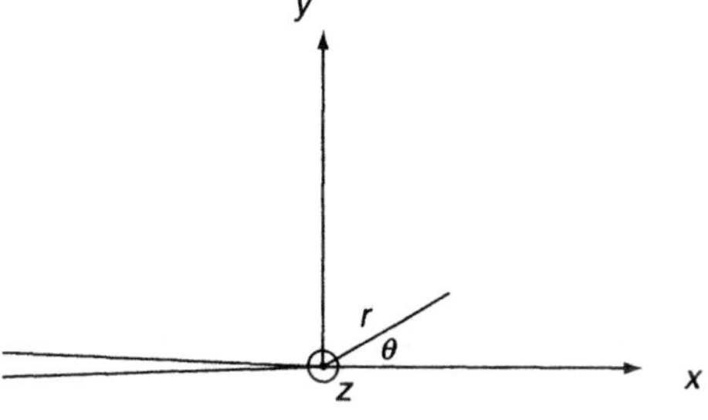

FIGURE 6.20 A crack with the Cartesian and the polar coordinate systems centered at the tip.

Figure 6.21 shows schematically the distribution of $\sigma_{\theta\theta}$ as a function of the radial distance r or x ahead of the crack tip. As shown in the figure, $\sigma_{\theta\theta}$ increases as x decreases. As the stress increases, plastic flow generally occurs for metallic materials. The Mises yield criterion can be expressed in terms of the three principal stresses σ_{rr}, $\sigma_{\theta\theta}$, and σ_{zz} as

$$\sigma_0^2 = \frac{1}{2}\left[(\sigma_{rr} - \sigma_{\theta\theta})^2 + (\sigma_{rr} - \sigma_{zz})^2 + (\sigma_{\theta\theta} - \sigma_{zz})^2\right] \tag{6.5.5}$$

where σ_0 is the tensile yield stress.

Under the plane stress conditions with Equation 6.5.2, the yield criterion for the material element at $\theta = 0$ becomes

$$\sigma_0^2 = \sigma_{\theta\theta}^2 \tag{6.5.6}$$

For Mode I cracks, this leads to

$$\sigma_{\theta\theta} = \sigma_0 \tag{6.5.7}$$

As shown in Figure 6.21, when the yield criterion is satisfied, $\sigma_0 = \sigma_{\theta\theta}$ at $r = r_{0\sigma}$. Therefore, from Equation 6.5.3,

$$\sigma_0 = \sigma_{\theta\theta} = \frac{K_\mathrm{I}}{\sqrt{2\pi r_{0\sigma}}} \tag{6.5.8}$$

Rearranging Equation 6.5.8 gives the plastic zone size ahead of the crack tip as

$$r_{0\sigma} = \frac{1}{2\pi}\left(\frac{K_\mathrm{I}}{\sigma_0}\right)^2 \tag{6.5.9}$$

Now we consider the elastic perfectly plastic material behavior. The tensile stress–strain curve for an elastic perfectly plastic material is shown in Figure 6.22. For the elastic perfectly plastic material, the plastic zone should be larger than the size of $r_{0\sigma}$ estimated from the linear elastic asymptotic stress solution. An estimated plastic zone size for the elastic perfectly plastic material as schematically shown in Figure 6.21 can be expressed as

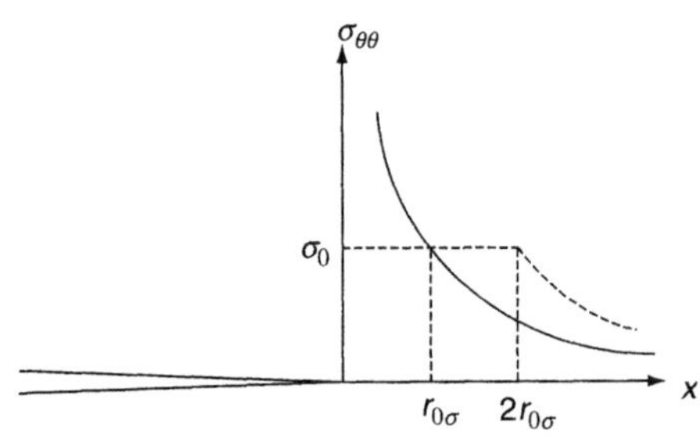

FIGURE 6.21 The distribution of $\sigma_{\theta\theta}$ as a function of the radial distance r or x ahead of the crack tip.

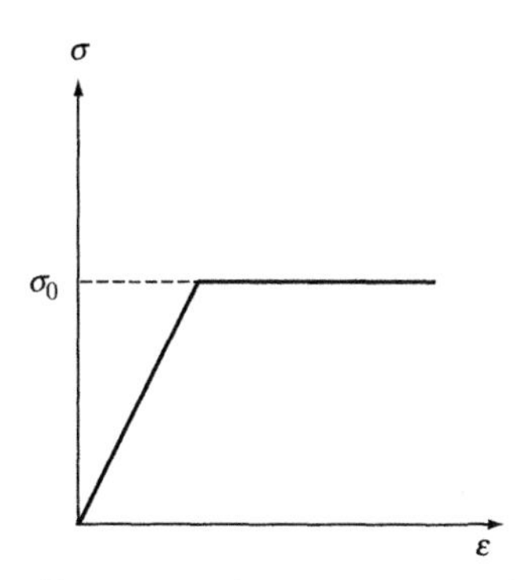

FIGURE 6.22 The tensile stress–strain curve of an elastic perfectly plastic material.

$$2r_{0\sigma} = \frac{1}{\pi}\left(\frac{K_{\mathrm{I}}}{\sigma_0}\right)^2 \tag{6.5.10}$$

6.5.1.2 Plane Strain Conditions

Under plane strain conditions, $\varepsilon_{zz} = 0$. This leads to

$$\sigma_{zz} = \nu(\sigma_{xx} + \sigma_{yy}) = \nu(\sigma_{rr} + \sigma_{\theta\theta}) \tag{6.5.11}$$

Directly ahead of the crack tip at $\theta = 0$, $\sigma_{r\theta} = 0$ and

$$\sigma_{rr} = \sigma_{\theta\theta} = \frac{K_{\mathrm{I}}}{\sqrt{2\pi r}} \tag{6.5.12}$$

Based on Equations 6.5.11 and 6.5.12, σ_{zz} is

$$\sigma_{zz} = 2\nu\sigma_{\theta\theta} \tag{6.5.13}$$

The yield criterion for the material element at $\theta = 0$ now becomes

$$\sigma_0^2 = [(1 - 2\nu)\sigma_{\theta\theta}]^2 \tag{6.5.14}$$

For Mode I cracks, when the yield criterion is satisfied,

$$\sigma_0 = (1 - 2\nu)\sigma_{\theta\theta} \tag{6.5.15}$$

or

$$\sigma_{\theta\theta} = \frac{1}{1 - 2\nu}\sigma_0 \tag{6.5.16}$$

For example, when $\nu = 0.3$, $\sigma_{\theta\theta}$ has to be equal to 2.5 σ_0 to satisfy the yield criterion under plane strain conditions. Therefore, the plastic zone size ahead of the crack tip should be smaller under plane strain conditions when compared to that under plane stress conditions. From Equation 6.5.12,

$$\frac{\sigma_0}{1-2\nu} = \sigma_{\theta\theta} = \frac{K_{\mathrm{I}}}{\sqrt{2\pi r_{0\varepsilon}}} \tag{6.5.17}$$

at $r = r_{0\varepsilon}$. This leads to

$$r_{0\varepsilon} = \frac{(1-2\nu)^2}{2\pi}\left(\frac{K_{\mathrm{I}}}{\sigma_0}\right)^2 \tag{6.5.18}$$

Similar to the plane stress case, the plastic zone size estimated for elastic plastic materials is

$$2r_{0\varepsilon} = \frac{(1-2\nu)^2}{\pi}\left(\frac{K_{\mathrm{I}}}{\sigma_0}\right)^2 \tag{6.5.19}$$

For example, when $\nu = 0.3$, $(1-2\nu)^2 = 0.16$. Therefore, the plastic zone ahead of the crack tip under plane strain conditions is much smaller than that under plane stress conditions. Estimation of the plastic zone from Irwin can be written as

$$r_{0\varepsilon} = \frac{1}{6\pi}\left(\frac{K_{\mathrm{I}}}{\sigma_0}\right)^2 \tag{6.5.20}$$

$$2r_{0\varepsilon} = \frac{1}{3\pi}\left(\frac{K_{\mathrm{I}}}{\sigma_0}\right)^2 \tag{6.5.21}$$

6.5.2 ESTIMATION OF PLASTIC ZONE SIZE AND SHAPE UNDER MODE I LOADING CONDITIONS

Plastic zone size and shape for Mises materials can be estimated based on the asymptotic linear elastic crack-tip stress fields. The Mises yield criterion in the cylindrical coordinate system for the in-plane modes can be expressed as

$$\sigma_0^2 = \frac{1}{2}\left[(\sigma_{rr}-\sigma_{\theta\theta})^2 + (\sigma_{\theta\theta}-\sigma_{zz})^2 + (\sigma_{zz}-\sigma_{rr})^2\right] + 3\sigma_{r\theta}^2 \tag{6.5.22}$$

The asymptotic stress fields in Equations 6.4.10–6.4.12 can be substituted into the Mises yield criterion in Equation 6.5.22.

Under plane stress conditions, the plastic zone size $r_p(\theta)$ as a function of θ can be obtained as

$$r_p(\theta) = \frac{1}{2\pi}\left(\frac{K_{\mathrm{I}}}{\sigma_0}\right)^2\left[\frac{1}{2}(1+\cos\theta) + \frac{3}{4}\sin^2\theta\right] \tag{6.5.23}$$

Under plane strain conditions, the plastic zone size $r_p(\theta)$ as a function of θ can be obtained as

$$r_p(\theta) = \frac{1}{2\pi}\left(\frac{K_{\mathrm{I}}}{\sigma_0}\right)^2\left[\frac{1}{2}(1-2\nu)^2(1+\cos\theta) + \frac{3}{4}\sin^2\theta\right] \tag{6.5.24}$$

The size and shape of the plastic zones based on Equations 6.5.23 and 6.5.24 in the normalized coordinate system under plane stress and plane strain conditions for $\nu = 0.3$ are shown in Figure 6.23. As shown in the figure, the size of the plastic zone under plane strain conditions is much smaller than that under plane stress conditions.

The non-singular stress T can also affect the plastic zone size and shape. However, the size and shape based on the linear elastic asymptotic crack-tip fields do not in general agree well with the computational results for elastic plastic materials [e.g., see Ben-Aoun and Pan (1996)]. Figures 6.24 and 6.25 show the general trends of plastic zone size and shape for different T stresses for elastic perfectly plastic materials under small-scale yielding plane strain and plane stress conditions.

6.5.3 REQUIREMENT OF LINEAR ELASTIC FRACTURE MECHANICS

The most important requirement of linear elastic fracture mechanics (LEFM) is that the size of the plastic zone near the crack tip should be much smaller than any relevant length dimensions for the crack problem of interest. In this way, the stress intensity factor K controls the plastic deformation near the crack tip, and, in turn, the fracture process near the crack tip. Figure 6.26 shows a schematic representation of the K field and the plastic zone near the tip of a crack with the length a in a specimen. The requirement of LEFM is

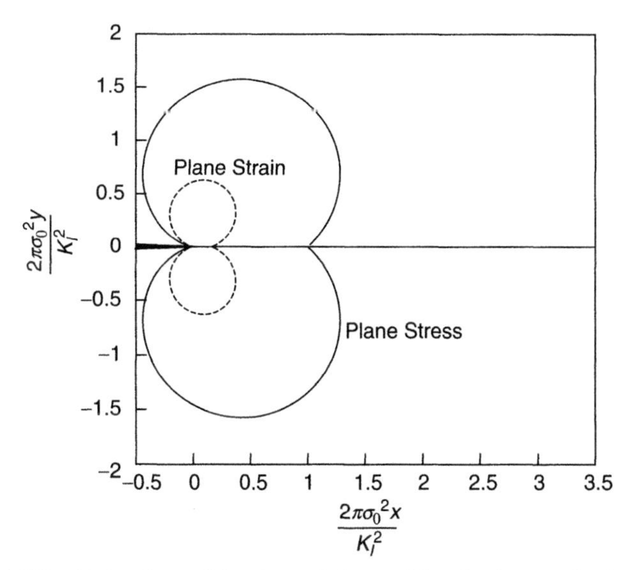

FIGURE 6.23 Estimations of the size and shape of the plastic zones in the normalized coordinate system under plane stress and plane strain conditions for $\nu = 0.3$.

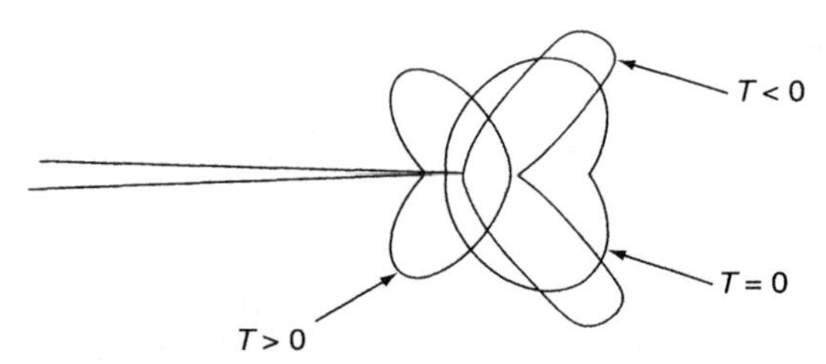

FIGURE 6.24 The general trends of plastic zone size and shape for different T stresses under small-scale yielding plane strain conditions.

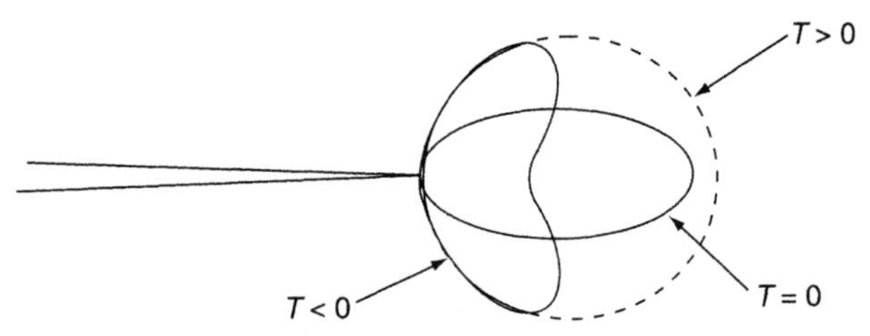

FIGURE 6.25 The general trends of plastic zone size and shape for different T stresses under small-scale yielding plane stress conditions.

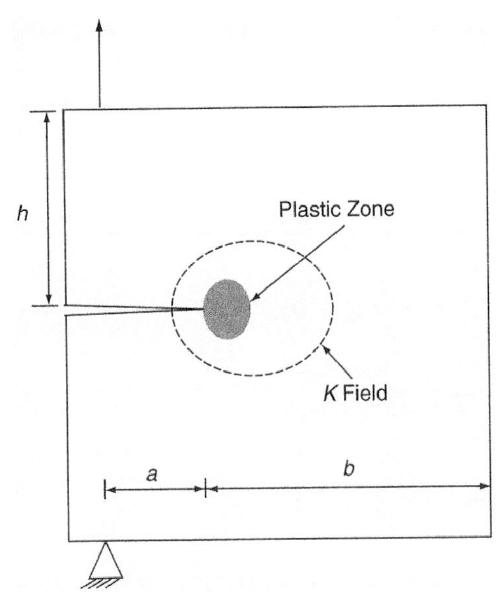

FIGURE 6.26 A schematic representation of the K field and the plastic zone near the tip of a crack in a specimen.

$$a,\ b,\ h \geq \frac{4}{\pi}\left(\frac{K_{\mathrm{I}}}{\sigma_0}\right)^2 \tag{6.5.25}$$

where b and h are the remaining ligament and height of the specimen, respectively.

The definition of plane stress and plane strain conditions depends on the thickness of the specimens or structures and the load amplitude. For example, consider a cracked specimen with the thickness t and the plastic zone $2r_0$ on the lateral surface as shown in Figure 6.27. A Cartesian coordinate system is shown in the figure. The crack front is in the z direction. The plastic zone size is larger on the specimen lateral surface, because the surface material element has no constraint and is under plane stress conditions. However, for the crack front in the middle of the specimen, the plastic zone becomes smaller because the material element has the constraint in the z direction. This concept is illustrated in Figure 6.28. Figure 6.28 shows the side views of the specimen with the plastic zones shown as the shaded regions. When the load P is small, most of the plastic zone along the crack front is small compared to the thickness of the specimen. The specimen is subjected to mostly plane strain conditions. When the load P is large, the plastic zone along the crack front becomes large compared to the thickness of the specimen and the specimen is subjected to plane stress conditions. The requirement of plane strain conditions is

$$t,\ a,\ h,\ b \geq 2.5\left(\frac{K_{\mathrm{I}}}{\sigma_0}\right)^2 \tag{6.5.26}$$

6.5.4 FRACTURE TOUGHNESS TESTING

Figure 6.29 shows a three-point bend specimen and Figure 6.30 a compact tension specimen. These two types of specimens are used for fracture toughness testing according to the ASTM standard E399 (ASTM, 1997). The stress intensity factor K_Q can be related to the load P_Q by

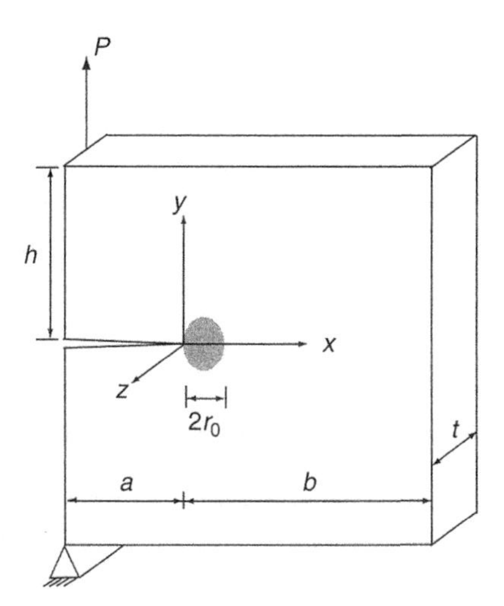

FIGURE 6.27 A cracked specimen with the thickness t and the plastic zone $2r_0$.

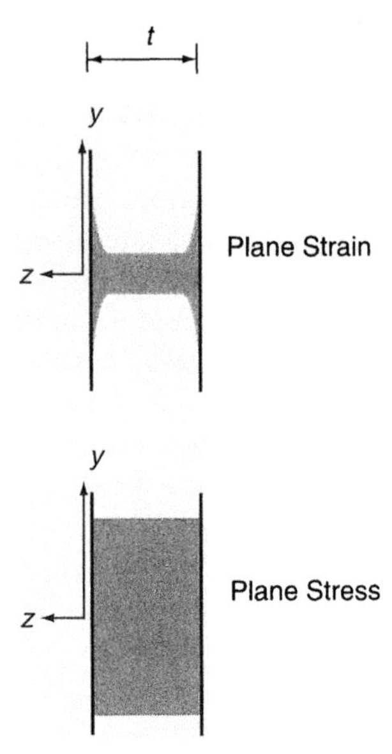

FIGURE 6.28 Side views of the plastic zone size distribution along the crack front. Shaded regions represent the plastic zones.

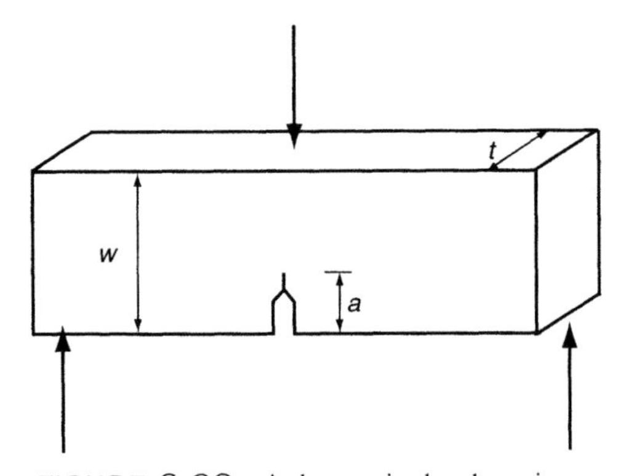

FIGURE 6.29 A three-point bend specimen.

$$K_Q = P_Q \frac{1}{t\sqrt{W}} F\left(\frac{a}{W}\right) \tag{6.5.27}$$

where $F(a/W)$ represents the geometric function for the three-point bend specimen and the compact tension specimen.

There are several ways to determine the load P_Q from different types of the load–displacement curves. Figure 6.31 shows a simple rule for a Type I load–displacement curve. In the figure, P represents the load and v the displacement. The initial slope of the load–displacement curve is denoted as

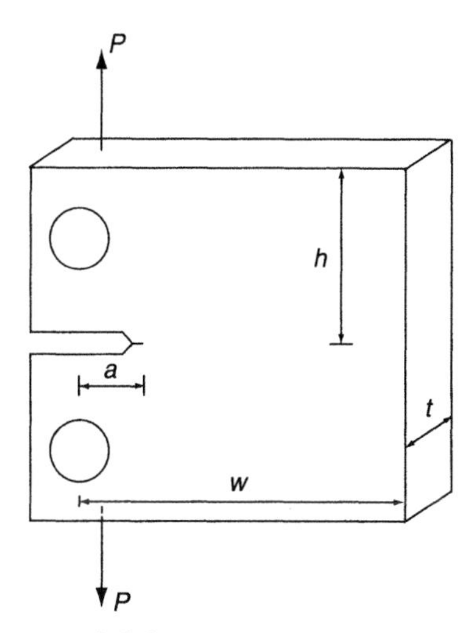

FIGURE 6.30 A compact tension specimen.

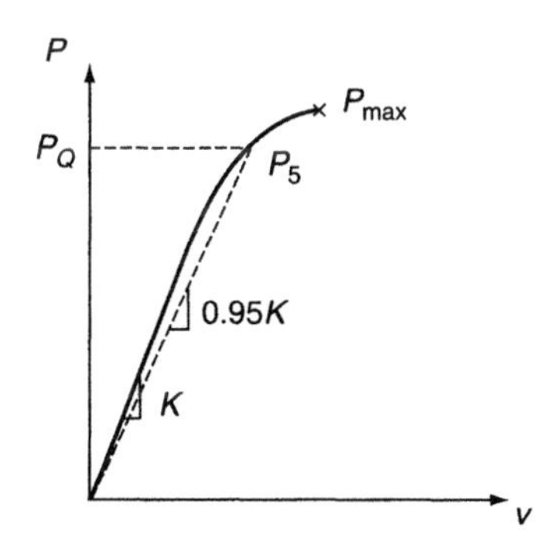

FIGURE 6.31 A simple rule to determine the load P_Q for a typical Type I load-displacement response.

K. The intersection of the load–displacement curve and a line with the slope $0.95\,K$ from the origin defines the load P_5. P_Q is the smaller value of P_5 or the maximum load P_{max}.

Figure 6.32 shows a schematic plot of K_Q as a function of the thickness t. In general, as the thickness t increases, the value of K_Q decreases. The fracture toughness K_{Ic} is the asymptotic value of K_Q. Under the plane strain conditions, the thickness and other dimensions of the specimens must satisfy Equation 6.5.26. The values of K_{Ic} are in the range of 20–200 $\mathrm{MPa}\sqrt{\mathrm{m}}$ for metals, 1–5 $\mathrm{MPa}\sqrt{\mathrm{m}}$ for polymers, and 1–5 $\mathrm{MPa}\sqrt{\mathrm{m}}$ for ceramics. Typical values of the fracture toughness K_{Ic} for several metals are listed in Table 6.1 (Dieter, 1988). In general, as the yield stress increases, the value of K_{Ic} usually decreases.

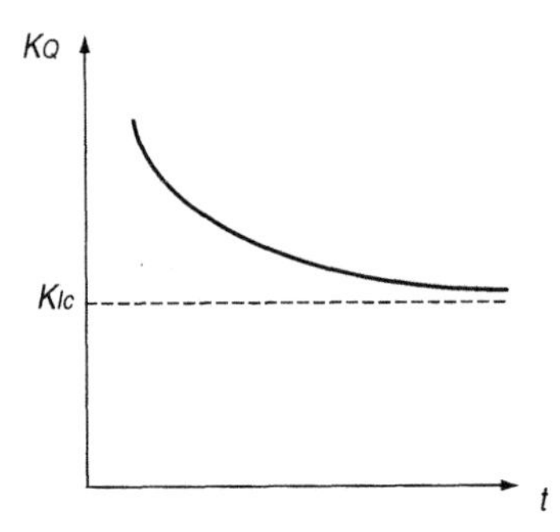

FIGURE 6.32 A schematic plot of K_Q as a function of the thickness t. The asymptotic value represents the fracture toughness K_{Ic}.

TABLE 6.1 Typical Values of K_{Ic}

Material	Yield Strength (MPa)	Fracture Toughness K_{Ic} (MPa $\sqrt{m}$)
Maraging steel	1730	90
4340 steel	1470	46
Ti-6Al-4V	900	57
7075-T6 Al alloy	500	24
2024-T3 Al alloy	385	26

6.5.5 PLASTIC ZONE UNDER MONOTONIC AND CYCLIC LOADING CONDITIONS

As discussed earlier, the plastic zone sizes ahead of the crack tip under Mode I plane stress and plane strain loading conditions can be expressed as

$$2(r_{0\sigma})_{\text{monotonic}} = \frac{1}{\pi}\left(\frac{K_I}{\sigma_0}\right)^2 \tag{6.5.28}$$

$$2(r_{0\varepsilon})_{\text{monotonic}} = \frac{1}{3\pi}\left(\frac{K_I}{\sigma_0}\right)^2 \tag{6.5.29}$$

where σ_0 represents the uniaxial tensile yield stress under monotonic loading conditions.

However, under cyclic loading conditions, if we adopt the elastic perfectly plastic material behavior with an anisotropic kinematic hardening law, the effective Mises yield stress for the fully reversed plastic reloading is $2\sigma_0$. Therefore, the estimations of the plastic zone sizes under cyclic loading conditions are one quarter of those under monotonic loading conditions.

$$2(r_{0\sigma})_{\text{cyclic}} = \frac{1}{\pi}\left(\frac{K_I}{2\sigma_0}\right)^2 = \frac{1}{4}2(r_{0\sigma})_{\text{monotonic}} \tag{6.5.30}$$

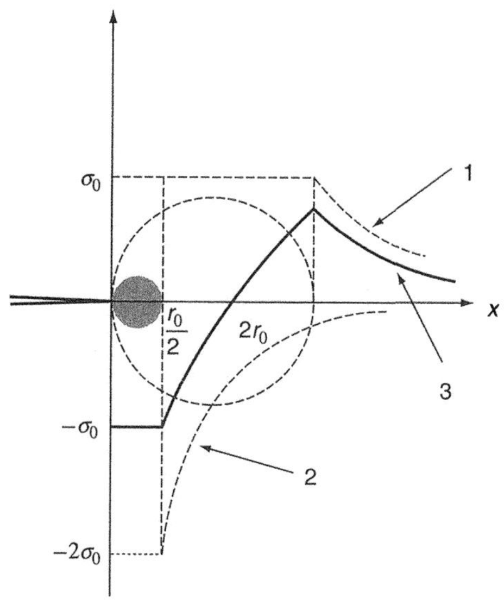

FIGURE 6.33 Estimation of the plastic zone size under cyclic loading conditions based on the concept of elastic unloading and plastic reloading near the crack tip.

$$2(r_{0\varepsilon})_{\text{cyclic}} = \frac{1}{3\pi}\left(\frac{K_{\text{I}}}{2\sigma_0}\right)^2 = \frac{1}{4}2(r_{0\varepsilon})_{\text{monotonic}} \qquad (6.5.31)$$

Figure 6.33 illustrates the concept of elastic unloading and plastic reloading near the crack tip. Curve 1 represents the distribution of σ_{yy} due to the initial tensile loading. The plastic zone size is denoted as $2r_0$. Once the load is released, the unloading process introduces a compressive crack-tip stress field. This will cause the plastic reloading close to the crack tip and elastic unloading outside the plastic reloading zone. The distribution of σ_{yy} due to the unloading can be represented by Curve 2. The combination of Curves 1 and 2 gives Curve 3, which represents the resulting distribution of σ_{yy} ahead of the crack tip.

6.5.6 ENERGY RELEASE RATE UNDER DISPLACEMENT-CONTROLLED CONDITIONS

Consider a nonlinear elastic cracked solid as shown of thickness t in Figure 6.34 under displacement-controlled conditions. Figure 6.35 shows the load–displacement curve of a cracked solid. In the figure, P represents the load and v represents the displacement. The area under the loading–displacement curve represents the strain energy U stored in the cracked solid. The potential energy Φ of the cracked solid is

$$\Phi = U \qquad (6.5.32)$$

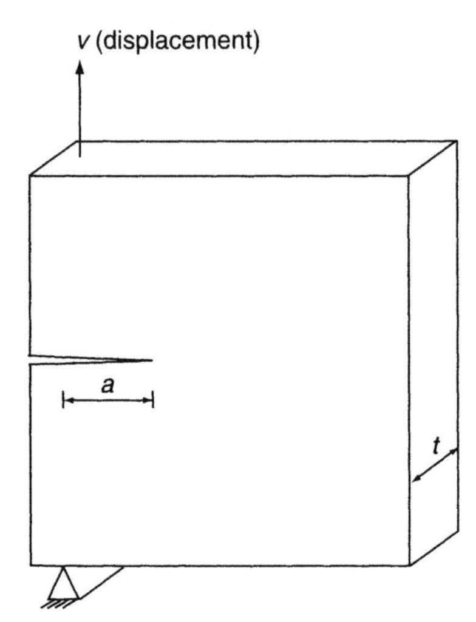

FIGURE 6.34 A nonlinear elastic cracked solid under displacement-controlled conditions.

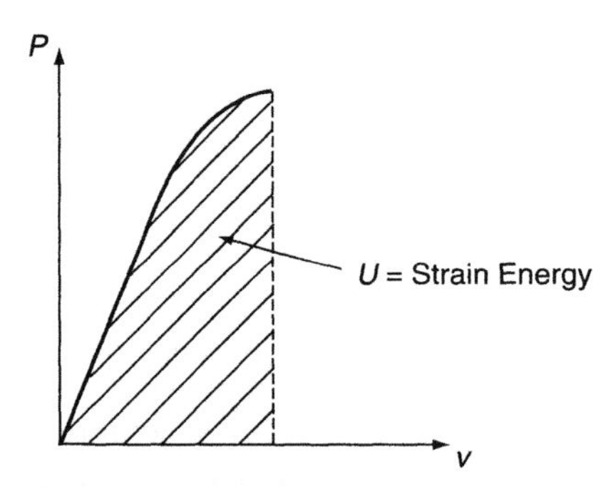

FIGURE 6.35 The load–displacement curve of a nonlinear elastic cracked solid. The area under the loading–displacement curve represents the strain energy U.

The energy release rate $\mathcal{G}$ is defined as

$$\mathcal{G} = -\frac{1}{t}\frac{\partial \Phi}{\partial a} = -\frac{1}{t}\frac{\partial U}{\partial a}\bigg|_{v} \tag{6.5.33}$$

Estimation of the energy release rate $\mathcal{G}$ based on the load–displacement curve is illustrated in Figure 6.36. For a small amount of crack growth Δa, the change of strain energy, ΔU, can be used to estimate the energy release rate $\mathcal{G}$ as

$$\mathcal{G} \approx -\frac{1}{t}\frac{\Delta U}{\Delta a} \tag{6.5.34}$$

For linear elastic materials, the load–displacement curve is linear, as shown in Figure 6.37. The slope of the curve represents the stiffness of the

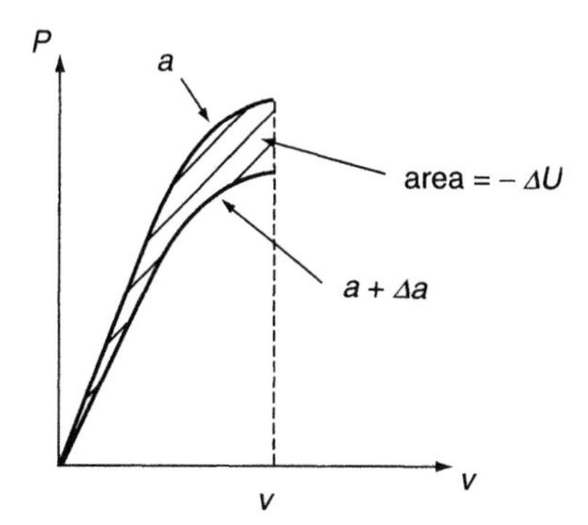

FIGURE 6.36 Estimation of the energy release rate $\mathcal{G}$ based on the change of the strain energy, ΔU, for a small amount of crack growth Δa.

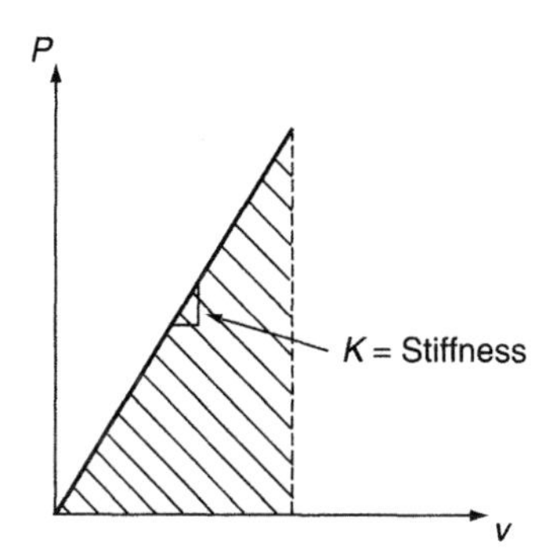

FIGURE 6.37 The linear load–displacement curve of a linear elastic-cracked solid.

cracked solid as shown. The stiffness of the cracked solid, $K(a)$, depends on the crack length a. Under displacement-controlled conditions, the potential energy Φ of the cracked solid can be expressed as

$$\Phi = U = \frac{1}{2}Pv = \frac{1}{2}K(a)v^2 \tag{6.5.35}$$

The energy release rate $\mathcal{G}$ is defined as

$$\mathcal{G} = -\frac{1}{t}\frac{\partial U}{\partial a}\bigg|_v = -\frac{1}{t}K'(a)v^2 \tag{6.5.36}$$

Estimation of $\mathcal{G}$ based on the change of the strain energy, ΔU, is illustrated in Figure 6.38.

$\mathcal{G}$ can be related to stress intensity factors for linear elastic materials. For a crack front under combined Modes I, II, and III loading conditions, the local energy release rate $\mathcal{G}$ (Irwin, 1957) is

$$\mathcal{G} = \frac{1-v^2}{E}(K_{\mathrm{I}}^2 + K_{\mathrm{II}}^2) + \frac{1+v}{E}K_{\mathrm{III}}^2 \tag{6.5.37}$$

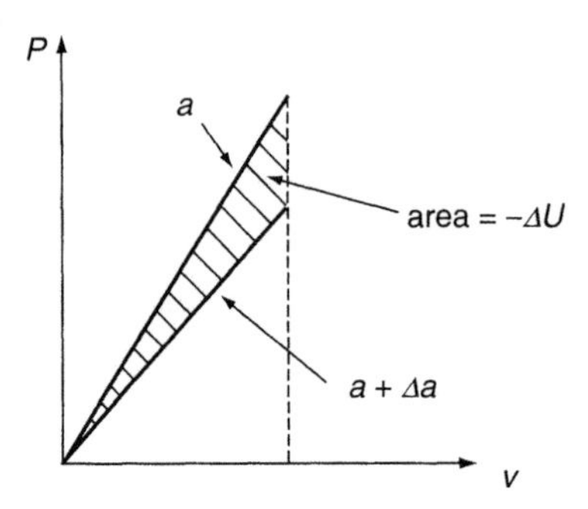

FIGURE 6.38 Estimation of the energy release rate $\mathcal{G}$ based on the change of the strain energy, ΔU, for a linear elastic-cracked solid.

Under plane stress conditions,

$$\mathcal{G} = \frac{1}{E}(K_{\mathrm{I}}^2 + K_{\mathrm{II}}^2) \tag{6.5.38}$$

For nonlinear elastic power-law hardening material, $\mathcal{G}$ can be related to the path-independent J integral (Rice, 1968), which represents the singularity amplitude of the HRR crack-tip singular stress and strain fields (Hutchinson, 1968; Rice and Rosengren, 1968)

6.6 FATIGUE CRACK PROPAGATION BASED ON LINEAR ELASTIC FRACTURE MECHANICS

6.6.1 EXTENDED SERVICE LIFE

Different inspection methods that use visual, X-ray, ultrasonic waves, and electric potential drop methods can be employed to detect cracks. Once cracks are detected, fracture mechanics methods can be used to determine the extended life of the structures or components.

Figure 6.39 shows the concept by plotting the crack length a as a function of the number of cycles, N. In the figure, a_i represents the initial crack length within a structure or component, a_d a detectable crack length, and a_c the critical crack length when the stress intensity factor of the crack reaches the critical value or the structure or component reaches fully plastic conditions. The number of cycles when a crack is detected is denoted as N_d. The number of cycles when the crack length reaches the critical length a_c is denoted as N_c. The extended life N_{cd} equals $N_c - N_d$. The damage-tolerant approach used in the aerospace industry is based on the worst-case analysis with the assumption that the largest cracks missed by each inspection exist in the structures or components.

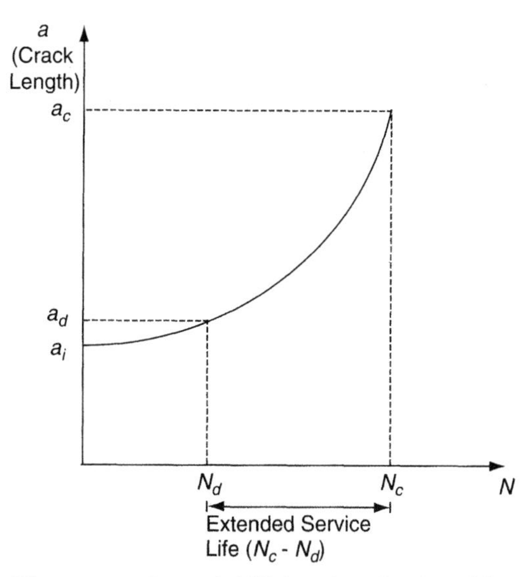

FIGURE 6.39 The concept of extended life based on the plot of the crack length a as a function of the number of cycles, N.

6.6.2 CRACK GROWTH BEHAVIOR

Linear elastic fracture mechanics has been successfully used to model the fatigue crack growth behavior. Consider a cracked structure with a crack length a under cyclic stressing conditions, as shown in Figure 6.40(a). The applied stress S as a function of time t is shown in Figure 6.40(b). The maximum and minimum stress intensity factors K_{max} and K_{min} are linearly related to the maximum and minimum applied stresses S_{max} and S_{min}, respectively, according to the linear elastic fracture mechanics.

$$K_{max} = FS_{max}\sqrt{\pi a} \tag{6.6.1}$$

$$K_{min} = FS_{min}\sqrt{\pi a} \tag{6.6.2}$$

where F represents the function depending on the geometries and loading conditions.

The stress intensity factor range ΔK is defined as

$$\Delta K = K_{max} - K_{min} \tag{6.6.3}$$

The load ratio R in the stress–life approach is defined as

$$R = \frac{S_{min}}{S_{max}} \tag{6.6.4}$$

Due to the linearity, R also represents the stress intensity factor ratio

$$R = \frac{K_{min}}{K_{max}} \tag{6.6.5}$$

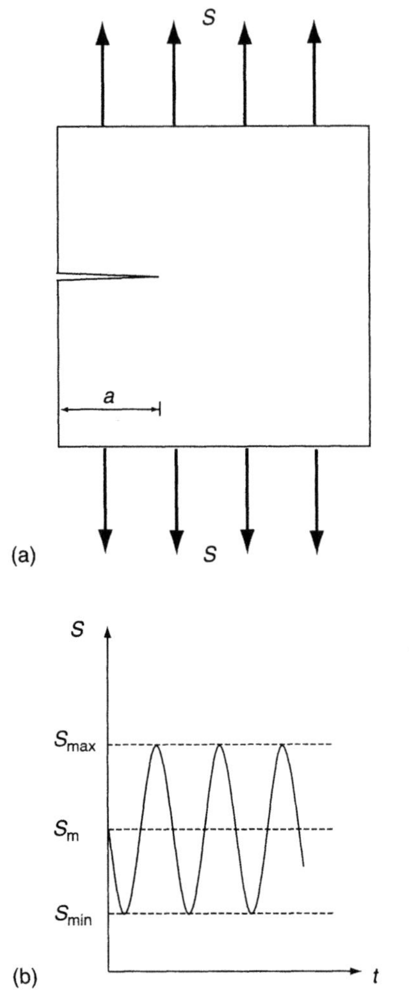

FIGURE 6.40 (a) A cracked structure with a crack length a under cyclic stressing conditions. (b) The applied stress S varies as a function of time t.

Based on experimental results, Figure 6.41 shows schematically the crack growth rate da/dN as a function of the stress intensity factor range ΔK in a log–log scale. Note that 10^{-7} mm represents the spacing between atoms. In Region I, when ΔK decreases, the crack growth rate drops significantly. The asymptote ΔK_{th} is the threshold of the stress intensity factor range below which no fatigue crack growth should occur. The values of ΔK_{th} are 5–16 MPa $\sqrt{\text{m}}$ for steels and 3–6 MPa $\sqrt{\text{m}}$ for aluminum alloys, depending on the value of R. In Region III, when ΔK is large, the crack growth rate accelerates significantly. This can happen when $K_{\max}$ approaches K_c, which represents the critical K for crack initiation for a given thickness.

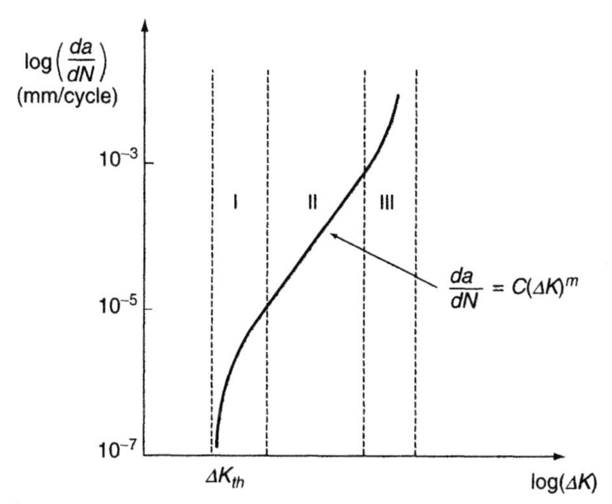

FIGURE 6.41 Crack growth rate *da/dN* schematically plotted as a function of the stress intensity factor range ΔK in a log–log scale.

In Region II, the crack growth rate da/dN can be approximately linear related to the stress intensity factor range ΔK in the log–log scale plot. The Paris law for crack growth (Paris et al., 1961; Paris and Erdogan, 1963) is

$$\frac{\mathrm{d}a}{\mathrm{d}N} = C(\Delta K)^m \tag{6.6.6a}$$

where C and m represent the material constants. To obtain the values of C and m, a least-square method can usually be adopted to fit the fatigue data in the log–log scale plot of da/dN as a function of ΔK as

$$Y = A + BX \tag{6.6.6b}$$

where

$$Y = \log\left(\frac{\mathrm{d}a}{\mathrm{d}N}\right), \; X = \log(\Delta K), \; A = \log C, \text{ and } B = m \tag{6.6.6c}$$

The typical value of m is 3.0 for ferrite-pearlite steel, 2.25 for matensitic steel, and 3.25 for austenitic steel. The load ratio R can affect the fatigue crack growth rate. Typical values of material constants C and m for steels are listed in Table 6.2 (Barsom and Rolfe, 1987). Two approaches with consideration of the load ratio effects are discussed in the following section.

6.6.3 WALKER EQUATION

Load ratio effects can be accounted for by an equivalent zero-to-maximum stress intensity factor range $\Delta\bar{K}$ introduced by Walker (1970):

TABLE 6.2 Typical Values of Material Constants C and m in the Paris Law for $R \approx 0$

Material	C[a]	m
Martensitic steel	1.36×10^{-7}	2.25
Ferritic-pearlitic steel	6.89×10^{-9}	3
Austentic steel	5.61×10^{-9}	3.25

[a] The values of C are obtained with the units of MPa $\sqrt{\text{m}}$ for ΔK and mm/cycle for da/dN.

$$\Delta\bar{K} = K_{\text{max}}(1 - R)^{\gamma} \tag{6.6.7}$$

where γ is a constant in the range of 0.4 to 0.9. The Walker equation can be rewritten as

$$\Delta\bar{K} = \Delta K(1 - R)^{\gamma-1} \tag{6.6.8}$$

or

$$\Delta\bar{K} = \frac{\Delta K}{(1 - R)^{1-\gamma}} \tag{6.6.9}$$

The Paris law can be rewritten as

$$\frac{\text{d}a}{\text{d}N} = C_1(\Delta\bar{K})^{m_1} \tag{6.6.10}$$

where C_1 and m_1 are material constants. For $R = 0$, Equation 6.6.9 becomes

$$\Delta\bar{K} = \Delta K \tag{6.6.11}$$

Therefore,

$$\frac{\text{d}a}{\text{d}N} = C_1(\Delta K)^{m_1} \tag{6.6.12}$$

For $R > 0$, substituting Equation 6.6.9 into 6.6.10 gives

$$\frac{\text{d}a}{\text{d}N} = C_1\left(\frac{\Delta K}{(1 - R)^{1-\gamma}}\right)^{m_1} \tag{6.6.13}$$

Rearranging Equation 6.6.13 gives

$$\frac{\text{d}a}{\text{d}N} = \frac{C_1}{(1 - R)^{m_1(1-\gamma)}}(\Delta K)^{m_1} \tag{6.6.14}$$

Comparing Equation 6.6.13 with 6.6.6(a) gives

$$C = \frac{C_1}{(1 - R)^{m_1(1-\gamma)}} \tag{6.6.15}$$

and

$$m = m_1 \tag{6.6.16}$$

For $R < 0$, $\gamma = 0$. Therefore,

$$\Delta\overline{K} = \frac{\Delta K}{(1-R)^{1-\gamma}} = \frac{\Delta K}{(1-R)} = K_{\max} \tag{6.6.17}$$

Therefore,

$$\frac{\mathrm{d}a}{\mathrm{d}N} = C_1 K_{\max}^{m_1} \tag{6.6.18}$$

If $\gamma = 1$,

$$\Delta\overline{K} = \frac{\Delta K}{(1-R)^{1-\gamma}} = \Delta K \tag{6.6.19}$$

Therefore, there is no effect of R on the crack growth rate.

6.6.4 FORMAN EQUATION

The crack growth depends on K_c, which is the critical stress intensity factor for a material and a given thickness. The crack growth rate is written as (Forman et al., 1967)

$$\frac{\mathrm{d}a}{\mathrm{d}N} = \frac{C_2(\Delta K)^{m_2}}{(1-R)K_c - \Delta K} \tag{6.6.20}$$

or

$$\frac{\mathrm{d}a}{\mathrm{d}N} = \frac{C_2(\Delta K)^{m_2}}{(1-R)(K_c - K_{\max})} \tag{6.6.21}$$

Typical values of material constants for the Walker and Forman equations are listed in Table 6.3 (Dowling, 1999). When the crack-tip plastic zone becomes large compared with the relevant physical length of the cracked solid, linear elastic fracture mechanics should not be used. Use of the crack-tip opening displacement (CTOD) or the J integral to model the fatigue crack growth is possible.

6.6.5 LIFE ESTIMATION

Life estimation under constant load amplitude loading conditions usually needs numerical integration. In general, ΔK increases as a increases with ΔS being kept as constant. The crack growth rate $\mathrm{d}a/\mathrm{d}N$ can be expressed as

$$\frac{\mathrm{d}a}{\mathrm{d}N} = f(\Delta K, R) \tag{6.6.22}$$

TABLE 6.3 Typical Values of Material Constants in the Walker and Forman Equations

	Walker Equation				Forman Equation		
Material	C_1^a	m_1	$\gamma(R \geq 0)$	$\gamma(R<0)$	C_2^a	m_2	K_c^b
4340 Steel	5.11×10^{-10}	3.24	0.42	0	n/a	n/a	n/a
2024-T3 Al alloy	1.42×10^{-8}	3.59	0.68	n/a	2.31×10^{-6}	3.38	110
7075-T6 Al alloy	2.71×10^{-8}	3.70	0.64	0	5.29×10^{-6}	3.21	78.7

[a] The values of C_1 and C_2 are obtained with the units of MPa $\sqrt{\text{m}}$ for ΔK and mm/cycle for da/dN.
[b] The thickness of the material is 2.3 mm.

Integration of Equation 6.6.22 gives

$$\int_{N_i}^{N_f} \mathrm{d}N = \int_{a_i}^{a_f} \frac{\mathrm{d}a}{f(\Delta K, R)} \tag{6.6.23}$$

where a_i represents the initial crack length, a_f the final crack length, N_i the number of cycles at the initial crack length, and N_f the number of cycles at the final crack length. The left hand side of the equation can be integrated as

$$N_{if} = N_f - N_i = \int_{N_i}^{N_f} \mathrm{d}N \tag{6.6.24}$$

where N_{if} represents the number of cycles for crack propagation from the initial crack length a_i to the final crack length a_f.

The Paris law gives a specific form for the crack growth rate $\mathrm{d}a/\mathrm{d}N$ as

$$\frac{\mathrm{d}a}{\mathrm{d}N} = C(\Delta K)^m \tag{6.6.25}$$

Consider the solution for ΔK in the form as

$$\Delta K = F \Delta S \sqrt{\pi a} \tag{6.6.26}$$

where F is a function of geometry and loading.

The crack growth rate $\mathrm{d}a/\mathrm{d}N$ now can be written as

$$\frac{\mathrm{d}a}{\mathrm{d}N} = C(F \Delta S \sqrt{\pi})^m a^{m/2} \tag{6.6.27}$$

Then, the number of cycles N_{if} for crack propagation from the initial crack length a_i to the final crack length a_f can be integrated as

$$N_{if} = \int_{N_i}^{N_f} \mathrm{d}N = \int_{a_i}^{a_f} \left(\frac{\mathrm{d}N}{\mathrm{d}a}\right) \mathrm{d}a = \int_{a_i}^{a_f} \frac{1}{C(F \Delta S \sqrt{\pi})^m} a^{\frac{-m}{2}} \mathrm{d}a \tag{6.6.28}$$

If we assume that F is a constant, Equation 6.6.27 can be integrated explicitly as

$$N_{if} = \frac{\left(1 - \left(\frac{a_i}{a_f}\right)^{\frac{m}{2}-1}\right)}{C(F\Delta S\sqrt{\pi})^m \left(\frac{m}{2} - 1\right)} \frac{1}{a_i^{\frac{m}{2}-1}} \tag{6.6.29}$$

Equation 6.6.29 indicates that N_{if} is very sensitive to the selection of a_i. Because F is a function of geometry, loading, and crack length, a numerical integration is usually needed. The numerical integration is usually carried out by taking F as a constant for a small range of a.

From the fracture mechanics viewpoint, a structure of interest contains initial defects or cracks. Therefore, a dominant crack with an initial crack length can be found to correlate the fatigue life of the structure under cyclic loading conditions. From the strain–life or stress–life viewpoint, the fatigue initiation life for the material near the notches or stress concentrators in the structure can be estimated. Then, the fatigue life due to crack propagation can be estimated by the fracture mechanics methodology based on the Paris law. The total fatigue life of the structure is the combination of the initiation life and propagation life [e.g., see Dowling (1979) and Socie et al. (1984)].

Example 6.6.1. Consider a very large plate with a small center crack. The initial total crack length $2a$ is 20 mm. The plate is subjected to repeated remote stress cycles from $\sigma_{\min} = 0$ MPa to $\sigma_{\max} = 75$ MPa. The initial crack grows as the stress cycles are applied. The critical stress intensity factor K_c for fracture initiation for the plate is 25 MPa $\sqrt{\text{m}}$. The material is assumed to have a very high yield stress such that the plastic zone size is small compared to the crack size under the applied stresses. Assume that the Paris law will be applicable to the plate material to characterize the fatigue crack growth

$$\frac{da}{dN} = C(\Delta K)^m \tag{6.6.25}$$

where $C = 1 \times 10^{-9}$ and $m = 3$. The value of C is obtained with the units of MPa $\sqrt{\text{m}}$ for ΔK and mm/cycle for da/dN.

1. Determine the final half crack length a_f at fracture initiation.
2. Determine the number of cycles before the fracture initiation of the plate.

Solution. For a center crack in a large plate,

$$K_{\text{I}} = \sigma\sqrt{\pi a} \tag{6.4.20}$$

The critical K_c for fracture initiation is 25 MPa $\sqrt{\text{m}}$. Fracture initiation takes place when $K_{\text{I}} = K_c$ at the final half crack length a_f under the maximum applied stress $\sigma_{\max}$. Equation 6.4.20 becomes

$$K_c = \sigma_{\max}\sqrt{\pi a_f} \tag{6.6.30}$$

The final half crack length a_f can be obtained by substituting $K_c = 25$ MPa $\sqrt{m}$ and $\sigma_{max} = 75$ MPa into Equation 6.6.30 as

$$a_f = 35.37\,\text{mm}$$

The initial half crack length a_i is

$$a_i = 10\,\text{mm}$$

The stress range $\Delta\sigma$ is determined as

$$\Delta\sigma = \sigma_{max} - \sigma_{min} = 75\,\text{MPa}$$

For a center crack in a large plate, the geometric parameter F in Equation 6.6.29 is 1. Equation 6.6.29 becomes

$$N_{if} = \frac{\left(1 - \left(\frac{a_i}{a_f}\right)^{\frac{m}{2}-1}\right)}{C(\Delta\sigma\sqrt{\pi})^m\left(\frac{m}{2} - 1\right)} \frac{1}{a_i^{\frac{m}{2}-1}} \tag{6.6.29a}$$

Based on Equation 6.6.29 and the given values of C and m, the number of cycles N_{if} for crack propagation from the initial half crack length a_i to the final half crack length a_f under the stress range $\Delta\sigma$ is obtained as 3.99×10^6.

Example 6.6.2. Consider a spot-weld cup specimen of dual-phase steel as shown in Figure 6.42(a) under cyclic loads of $P_{max} = 200$ N and $P_{min} = 0$. Figure 6.42(a) only shows a half specimen to represent the weld nugget clearly. As shown in the figure, the sheet thickness is represented by t, the nugget diameter by d, and the specimen diameter by D. Here, t is 1 mm, d is 5 mm, and D is 40 mm. Figure 6.42(b) shows a symmetric cross section of the cup specimen containing the weld nugget before testing. The notch near the weld nugget can be considered as an original crack. When the cup specimen is under pure opening loading conditions, the stress intensity factor K_I for the original crack can be obtained as (Pook, 1975; Wang et al., 2004)

$$K_I = \frac{P}{2\pi}\left(\frac{3}{t^3}\right)^{1/2}\left[\frac{2\left(\frac{D}{d}\right)^2 \ln\left(\frac{D}{d}\right)}{\left(\frac{D}{d}\right)^2 - 1} - 1\right] \tag{6.6.31}$$

The stress intensity factor K_{II} is 0 for the specimen under pure opening loading conditions. During the testing, a kinked crack is initiated near the weld nugget and then propagates through the thickness with a kink angle α as shown in Figure 6.42(c).

Figure 6.43 shows a schematic plot of a main crack and a kinked crack with the kink length a and the kink angle α. Here, K_I and K_{II} represent the global stress intensity factors for the main crack, and k_I and k_{II} represent

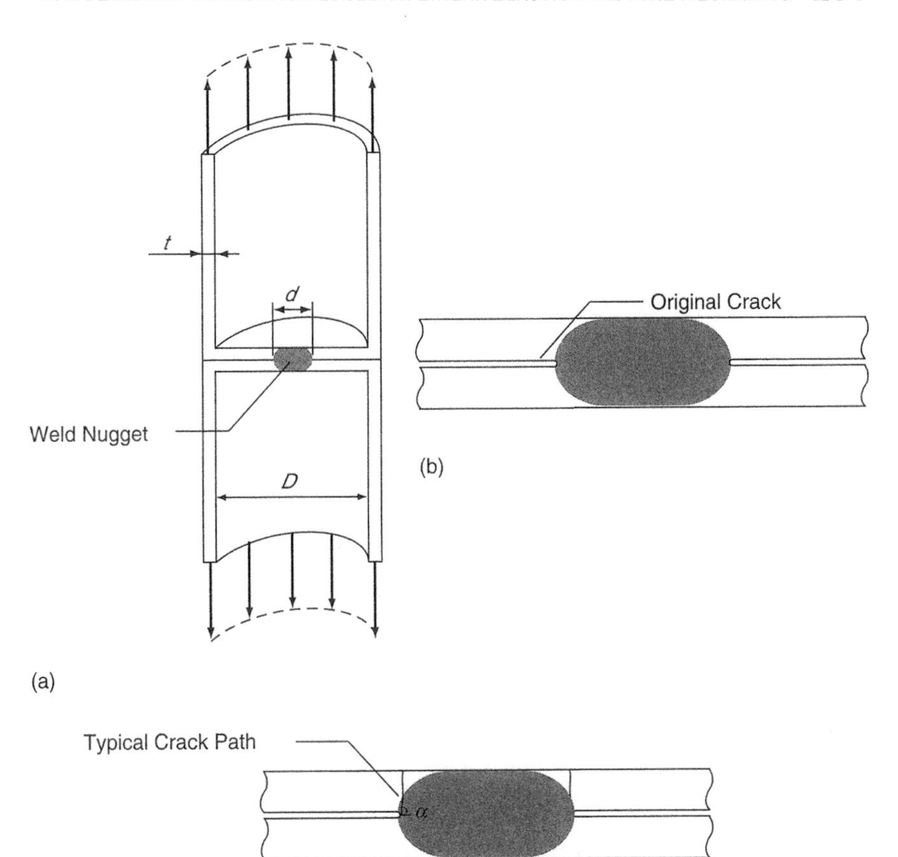

FIGURE 6.42 (a) A schematic plot of a half specimen with the sheet thickness t, nugget diameter d, and the specimen diameter D under the applied force (shown as the bold arrows). (b) A schematic plot of the symmetry cross section near the spot weld with the notch considered as an original crack. (c) The typical fatigue crack propagation path shown with a kink angle α.

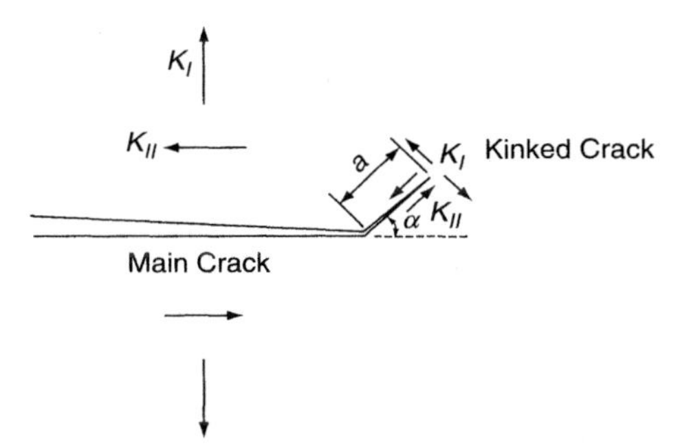

FIGURE 6.43 A schematic plot of a main crack and a kinked crack with the kink length a and the kink angle α.

the local stress intensity factors for the kinked crack. The arrows in the figure represent the positive values of the global and local stress intensity factors K_I, K_{II}, k_I, and k_{II}. In general, the local stress intensity factors of the kinked crack, k_I and k_{II}, are different from the global stress intensity factors of the original crack, K_I and K_{II}. When the kink length a approaches 0, the local stress intensity factors k_I and k_{II} can be obtained from the global stress intensity factors K_I and K_{II} as (Bibly et al., 1978; Cotterell and Rice, 1980)

$$k_I = \frac{1}{4}\left(3\cos\frac{\alpha}{2} + \cos\frac{3\alpha}{2}\right)K_I + \frac{3}{4}\left(\sin\frac{\alpha}{2} + \sin\frac{3\alpha}{2}\right)K_{II} \tag{6.6.32}$$

$$k_{II} = -\frac{1}{4}\left(\sin\frac{\alpha}{2} + \sin\frac{3\alpha}{2}\right)K_I + \frac{1}{4}\left(\cos\frac{\alpha}{2} + 3\cos\frac{3\alpha}{2}\right)K_{II} \tag{6.6.33}$$

For the cup specimen subjected to pure opening loading conditions, the stress intensity factor K_{II} is 0. Therefore, the local stress intensity factors k_I and k_{II} are

$$k_I = \frac{1}{4}\left(3\cos\frac{\alpha}{2} + \cos\frac{3\alpha}{2}\right)K_I \tag{6.6.34}$$

$$k_{II} = -\frac{1}{4}\left(\sin\frac{\alpha}{2} + \sin\frac{3\alpha}{2}\right)K_I \tag{6.6.35}$$

The effective local stress intensity factor is assumed to be

$$k_{eq} = \sqrt{k_I^2 + k_{II}^2} \tag{6.6.36}$$

Assume that the local stress intensity factors remain constants during the crack propagation and the Paris law is applicable to characterize the fatigue crack growth. In this case, the material constants C and m are chosen for martensitic steels as $C = 1.36 \times 10^{-7}$ and $m = 2.25$. The value of C is obtained with the units of MPa $\sqrt{\text{m}}$ for ΔK and mm/cycle for da/dN.

1. Determine the range of local stress intensity factors for the kinked crack.
2. Determine the number of cycles when the kinked crack grows through the thickness and the cup specimen is separated.

Solution. Based on Equation 6.6.31,

$$\Delta K_I = \Delta K_{I,\ max} - \Delta K_{I,\ min} = \frac{P_{max} - P_{min}}{2\pi}\left(\frac{3}{t^3}\right)^{1/2} \left[\frac{2\left(\frac{D}{d}\right)^2 \ln\left(\frac{D}{d}\right)}{\left(\frac{D}{d}\right)^2 - 1} - 1\right] = 5.62\ \text{MPa}\sqrt{\text{m}} \tag{6.6.37}$$

For the cup specimen under pure opening loading conditions, the Mode II stress intensity factor of the original crack is 0. The range of local stress intensity factors of the kinked crack can be calculated by Equations 6.6.34 and 6.6.35 with the kink angle $\alpha = 90°$ as

$$\Delta k_{\mathrm{I}} = 1.99\,\mathrm{MPa}\sqrt{\mathrm{m}}$$
$$\Delta k_{\mathrm{II}} = -1.99\,\mathrm{MPa}\sqrt{\mathrm{m}}$$

The range of effective local stress intensity factors for the kinked crack can be calculated by using Equation 6.6.36 as

$$\Delta k_{eq} = \sqrt{\Delta k_{\mathrm{I}}^2 + \Delta k_{\mathrm{II}}^2} = 2.81\,\mathrm{MPa}\sqrt{\mathrm{m}}$$

Based on the Paris law in Equation 6.6.25, the number of cycles N_f for the cup specimen as the kinked crack propagates through the thickness can be obtained as

$$\int_0^{N_f} N = \int_0^{t/\sin\alpha} \frac{da}{C(\Delta k_{eq})^m} \qquad (6.6.38)$$

Because the local stress intensity factors are assumed to be constant during crack propagation, Equation 6.6.35 can be integrated as

$$N_f = \frac{t}{C(\Delta k_{eq})^m \sin\alpha} \qquad (6.6.39)$$

Therefore, the number of cycles N_f to failure is calculated as 719,236.

REFERENCES

American Society for Testing and Materials (ASTM), *Annual Book of ASTM Standards*, ASTM E399, ASTM, West Conshohocken, PA, 1997.

Barsom, J. M., and Rolfe, S. T., *Fracture and Fatigue Control in Structures: Applications of Fracture Mechanics*, 2nd ed., Prentice Hall, Upper Saddle River, NJ, 1987.

Ben-Aoun, Z. E. A., and Pan, J., Effects of nonsingular stresses on plane-stress near-tip fields for pressure-sensitive materials and applications to transformation toughened ceramics, *International Journal of Fracture*, Vol. 77, 1996, pp. 223–241.

Bibly, B. A., Cardew, G. E., and Howard, I. C., Stress intensity factors at the tips of kinked and forked cracks. In *Fourth International Conference on Fracture*, Pergamon Press, New York, Vol. 3A, 1978, pp. 197–200. University of Waterloo, Ontario, June 19–24, 1977.

Cotterell, B., and Rice, J. R., Slightly curved or kinked cracks, *International Journal of Fracture*, Vol. 16, 1980, pp. 155–169.

Dieter, G. E., *Mechanical Metallurgy*, SI Metric ed., McGraw-Hill, New York, 1988.

Dowling, N. E., Fatigue at notches and the local strain and fracture mechanics approaches, In *Fracture Mechanics*, C. W. Smith (Ed.), ASTM STP 677, American Society for Testing and Materials, Philadelphia, PA, 1979, pp. 247–273.

Dowling, N. E., *Mechanical Behavior of Materials*, Prentice Hall, Upper Saddle River, NJ, 1999.

Forman, R. G., Keary, V. E., and Engle, R. M., Numerical analysis of crack propagation in cyclic-loaded structures, *Journal of Basic Engineering*, Vol. 89, 1967, pp. 459–464.

Griffith, A. A., The phenomena of rupture and flow in solids, *Philosophical Transaction of the Royal Society of London*, Vol. A221, 1921, pp. 163–197.

Hutchinson, J. W., Singular behavior at the end of a tensile crack in a hardening material, *Journal of Mechanics and Physics of Solids*, Vol. 16, 1968, pp. 13–31.

Irwin, G. R., Analysis of stresses and strains near the end of a crack transversing a plate, *Journal of Applied Mechanics*, Vol. 24, 1957, pp. 361–364.

Newman, J. C., Jr., and Raju, I. S., Stress intensity factor equations for cracks in three-dimensional finite bodies subjected to tension and bending loads. In *Computational Methods in the Mechanics of Fracture*, S. N. Atluri (Ed.), Elsevier, New York, 1986.

Paris, P. C., and Erdogan F., A critical analysis of crack propagation laws, *Journal of Basic Engineering*, Vol. 85, 1963, pp. 528–534.

Paris, P. C., Gomez M. P., and Anderson W. P., A rational analytic theory of fatigue, *The Trend in Engineering*, Vol. 13, 1961, pp. 9–14.

Pook, L. P., Fracture mechanics analysis of the fatigue behaviour of spot welds, *International Journal of Fracture*, Vol. 11, 1975, pp. 173–176.

Raju, I. S., and Newman, J. C., Jr., Stress intensity factors for circumferential surface cracks in pipes and rods under tension and bending loads, In *Fracture Mechanics: Seventeenth Symposium*, J. H. Underwood et al. (Eds.), ASTM STP 905, American Society for Testing and Materials, West Conshohocken, PA, 1986.

Rice, J. R., Limitations to the small scale yielding approximation for crack tip plasticity, *Journal of the Mechanics and Physics of Solids*, Vol. 22, 1974, pp. 17–26.

Rice, J. R., A path independent integral and the approximate analysis of strain concentration by notches and cracks, *Journal of Applied Mechanics*, Vol. 35, 1968, pp. 379–386.

Rice, J. R., and Rosengren, G. F., Plane strain deformation near a crack in a power-law hardening materials, *Journal of Mechanics and Physics of Solids*, Vol. 16, 1968, pp. 1–12.

Sham, T. -L, The determination of the elastic T-Term using higher order weight functions, *International Journal of Fracture*, Vol. 48, 1991, pp. 81–102.

Socie, D. F., Dowling, N. E., and Kurath, P., Fatigue life estimation of notched members. In *Fracture Mechanics: Fifteenth Symposium*, R. J. Sanford (Ed.), ASTM STP 833, American Society for Testing and Materials, Philadelphia, PA, 1984, pp. 284–299.

Tada, H., Paris, P., and Irwin, G., *The Stress Analysis of Cracks Handbook*, ASME, New York, 2000.

Walker, K., The effects of stress ratio during crack propagation and fatigue for 2024-T3 and 7075-T6 aluminum. In *Effects of Environment and Complex Load History for Fatigue Life*, STP 462, American Society for Testing and Materials, Philadelphia, PA, 1970, pp. 1–14

Wang, D. -A., Lin, S. -H., and Pan, J, Stress intensity factors for spot welds and associated kinked cracks in cup specimens, submitted for publication in *International Journal of Fatigue*, 2004.

7

FATIGUE OF SPOT WELDS

MARK E. BARKEY
THE UNIVERSITY OF ALABAMA
SHICHENG ZHANG
DAIMLERCHRYLSER AG

7.1 INTRODUCTION

The fatigue-resistant design of spot-welded connections epitomizes the confluence of a number of diverse issues that durability engineers must deal with during the automotive design and analysis process. Some of these issues and their relation to spot-weld fatigue are as follows:

Manufacturing: Properly formed spot welds are the result of a combination of the appropriate current, pressure, and hold time for a particular sheet thickness and material property combination. Residual stresses are normally inevitable.
Metallurgy: Base sheet metal properties will change in the weld nugget and heat affected zone (HAZ) of the spot weld.
Probability and Statistics: Electrode wear through the manufacturing cycle, the nonuniform nature of the joining surface, and missed or incomplete welds create variability at both the local spot-weld level and the structural level.
Stress Analysis: Large-scale automotive structural modeling of spot welds typically relies on finite element analysis techniques based on crudely simplified spot-weld models, and single-weld analysis typically relies on mechanics of materials, elasticity, and fracture mechanics approaches in which detailed models are often employed.
Fatigue Analysis: Fatigue critical weld locations must be determined and then appropriate damage parameters and damage accumulation techniques must be chosen to model spot-weld fatigue behavior.

Automotive Design: The number and placement of spot-weld lines within the load path must be determined, and the weld sequence must be specified to reduce assembly stresses.

This chapter focuses on particular spot-weld analysis techniques and the relation of basic specimen test results to automotive design. However, it should be understood that many issues, including those listed previously, directly affect the fatigue performance of spot-welded structures. Therefore, where appropriate, advice should be sought from welding engineers, designers, statisticians, and others. In addition, the durability engineer will often work with vehicle crash analysis engineers to also consider the ultimate strength and dynamic behavior of spot welds.

7.2 ELECTRICAL RESISTANCE SPOT WELDING

Electrical resistance spot welds are formed by bringing electrodes in contact with sheet metal. Electrical current flows though the electrodes and encounters high resistance at the interface, or faying surface, between the sheets. This resistance creates a large amount of heat, which locally melts the sheet material. The current flowing through the electrodes is then stopped, but the electrodes remain in place as the weld nugget forms from the molten material (Figure 7.1). This molten material is sometimes ejected from

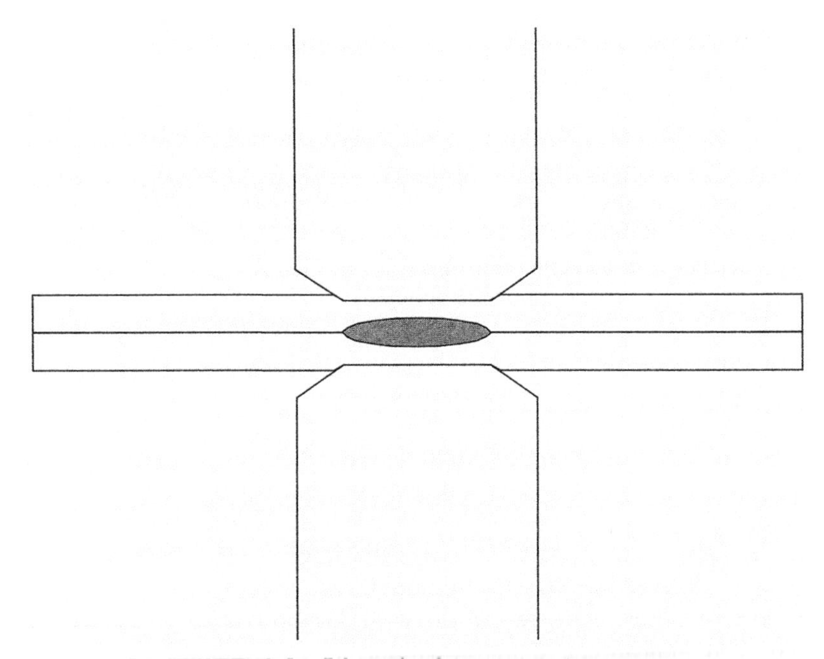

FIGURE 7.1 Schematic of the spot-welding process.

between the sheets and is referred to as expulsion, sometimes resulting in a visible display of sparks. The formation of an acceptable weld depends on the proper combination of current, electrode tip force, and the timing of these quantities, called the weld schedule. The weld schedule can be programmed with modern spot-welder controllers, which allows for nominally repeated welding schedules. A pedestal-type spot welder is shown in Figure 7.2.

Although the welding parameters of a certain weld schedule may be nominally the same, slight changes in the response of the welder and the sheet material can produce slightly different spot weld nugget diameters and fusion zone characteristics. Additionally, although spot welds are often schematically represented as perfect circles, weld expulsion and joint characteristics can produce spot welds that are nonuniform around the circumference. A consequence of this variability in manufacturing is that fatigue test results of spot welds will usually have more variability than fatigue test results of polished laboratory fatigue specimens typically used to determine fatigue properties of materials. An example of the variability of the button size with welding parameters is listed in Table 7.1 and shown schematically in Figure 7.3.

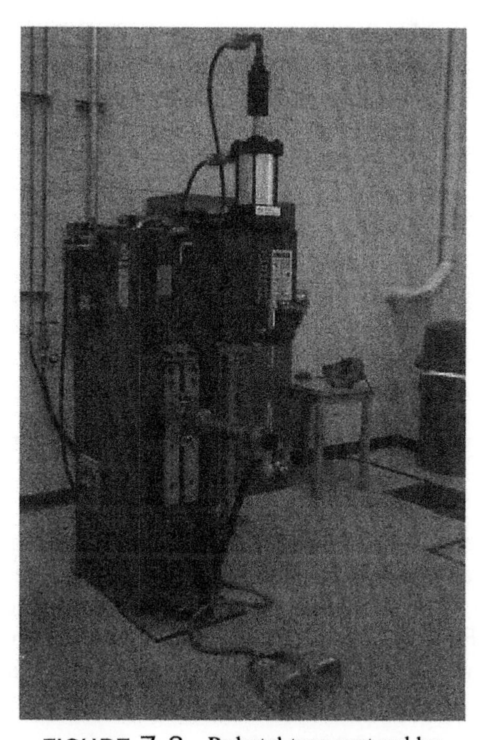

FIGURE 7.2 Pedestal-type spot welder.

TABLE 7.1 Spot-Weld Button Size Variability

Sample	% Current	Current (kA)	Button Size (mm)
A-1	82	11.60	8.57
A-2	82	11.60	8.10
A-3	82	11.60	8.17
A-4	80	11.00	6.83
A-5	80	11.10	6.93
A-6	80	11.10	6.67
A-7	78	10.60	5.45
A-8	78	10.60	5.43
A-9	78	10.60	6.13
A-10	76	10.10	4.77
A-11	76	10.10	4.87
A-12	76	10.10	4.57
A-13	74	9.75	3.43
A-14	74	9.72	3.53
A-15	74	9.74	3.20

Note: Material—galvanized steel.

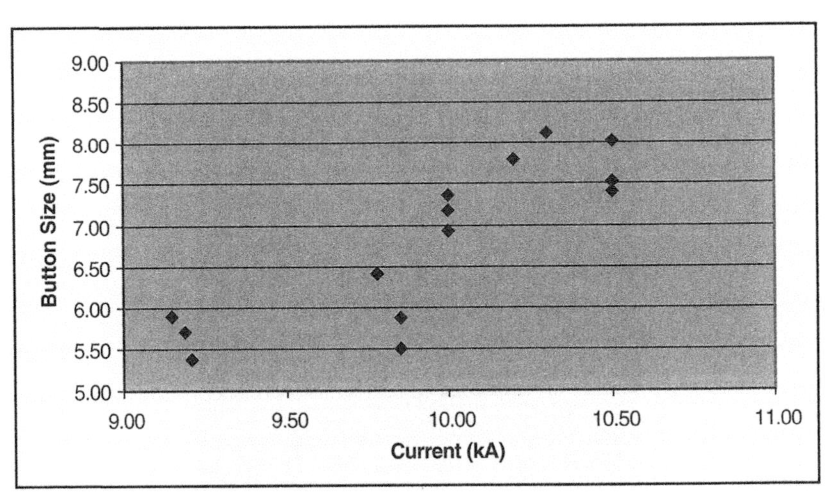

FIGURE 7.3 Weld button diameter variability.

The button diameter is roughly the same as the diameter of the weld nugget and is usually determined by peeling one of the sheets back over the spot weld, leaving a hole in one sheet and a button of attached material to the other sheet, as in Figure 7.4. If the button is elliptical in diameter, the average of the major and minor button diameters is usually reported.

FIGURE 7.4 Spot-weld button.

7.3 SPECIMEN TESTING

Fatigue analysis of spot-welded joints is often based on experimental results from test specimens, or coupons, that contain a single spot weld (Figure 7.5). Various standard and nonstandard specimens have been used, but the most common specimen type is the tensile-shear specimen (Figure 7.6).

The tensile-shear specimen consists of two overlapping strips of sheet metal joined by a single electrical resistance spot weld. Tensile load is applied to the ends of the sheet, nominally placing the weld surface between the sheets in shear. Tensile-shear specimens are used for both strength testing and fatigue testing of spot welds.

Fatigue testing is often conducted in the load-controlled, constant-amplitude loading mode, and, because of the possibility of buckling of the sheet, with R-ratios greater than zero. Data from fatigue tests of these specimens are often presented in the form of load–life data, in which the abscissa is either load amplitude or maximum load. Experimental data from tensile-shear specimens of a high-strength galvannealed sheet steel with two different nominal nugget diameters are shown in Figure 7.7.

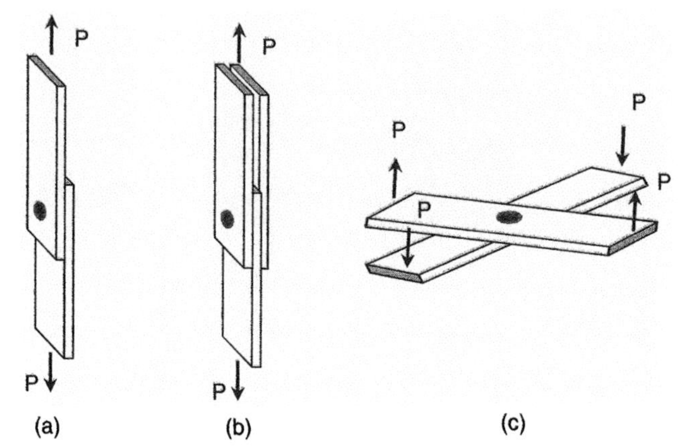

FIGURE 7.5 Examples of single spot-welded specimens: (a) tensile shear, (b) double shear, (c) cross-tension.

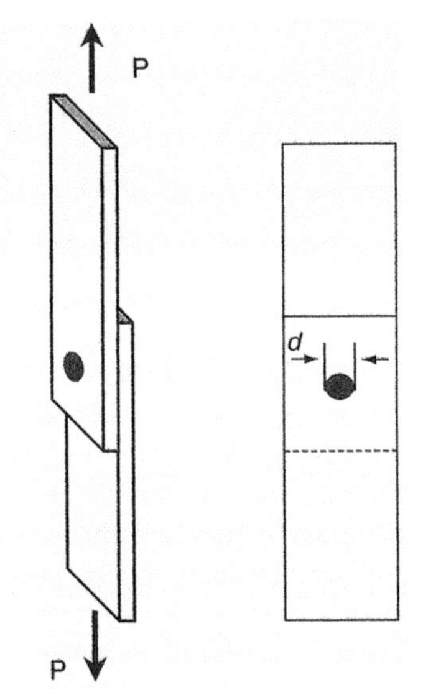

FIGURE 7.6 Tensile-shear specimen.

Interpretation of tensile-shear data and application of the data to automotive structural analysis must be done carefully, as the deformation characteristics of the tensile-shear specimens do not reflect those of spot welds in typical automotive configurations; that is, the deformation of an automotive spot weld is constrained by that of the surrounding welds and structure. In addition, the definition of failure for the laboratory test may not reflect an appropriate definition of failure for a weld in an automotive structure. Common specimen failure definitions are separation of the specimen into two pieces, percentage of load drop from desired constant-amplitude levels, or a specified elongation limit on the specimen.

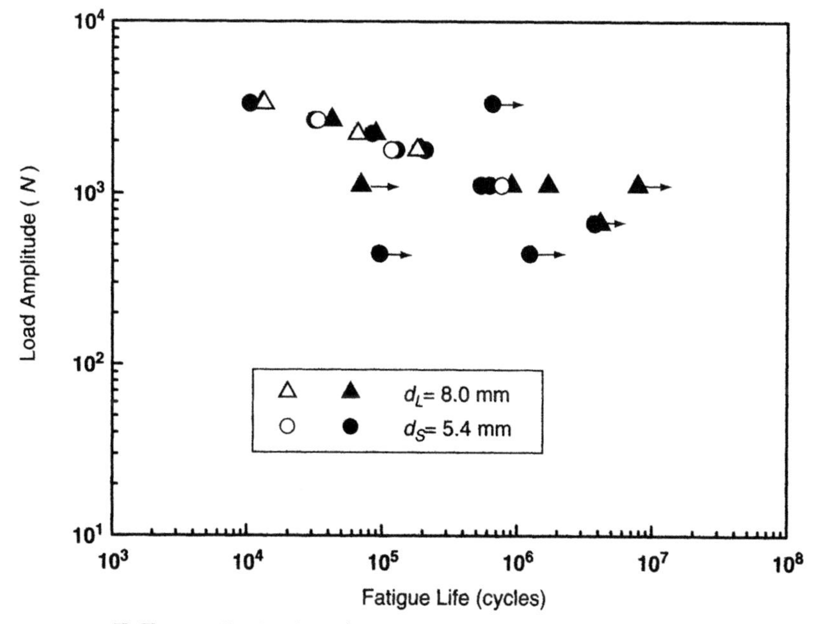

FIGURE 7.7 Applied load amplitude versus fatigue life for two nugget diameters.

For multiaxial spot-weld testing, two basic types of multiaxial spot-welded specimens have been recently proposed by Hahn (Gieske and Hahn, 1994; Hahn et al., 2000) and Lee (Lee et al., 1998; Barkey and Kang, 1999; Barkey and Han, 2001; Barkey et al., 2001), by which the loading condition at a spot weld can be investigated with the same specimen by changing the loading direction.

Whereas fatigue properties of spot welds are most commonly evaluated by using single-weld specimens, other specimens with multiple welds, such as box-beams, have been proposed for more direct application to automotive structures. However, at present, no particular standard multiply welded specimen has been developed.

7.4 FATIGUE LIFE CALCULATION TECHNIQUES

If enough experimental data are available, a multivariable load–life type of approach can be employed, as in Kang and Barkey (1999). However, it can advantageous to employ a fatigue damage parameter for spot welds. In damage parameter approaches, an analytical model of the joint is developed to determine how the stress, deformation, or stress intensity depends on the applied load. These quantities are then related to a fatigue damage parameter and calibrated by specimen tests.

To apply these damage parameters to automotive structures, the loads on the spot welds are typically calculated by unit elastic finite element analysis (Figure 7.8), and time histories of the damage parameters are determined by the linear application of structural loading histories.

The result of the unit load analysis is a set of influence factors, c_{ij}^k, where i refers to the spot-weld number from 1 to m, j refers to the unit load number from 1 to n, and k refers to the force or moment component (1, 2, 3 for forces; 4, 5, 6 for moments). Local spot-weld force histories are determined from a superposition of loading histories, $L_j(t)$, that correspond to the unit load applications. The local spot-weld force histories for spot-weld number i are computed from

$$F_i^k(t) = \sum_{j=1}^{n} c_{ij}^k L_j(t), \tag{7.4.1}$$

where

$$F_i^k = \begin{Bmatrix} F_x \\ F_y \\ F_z \\ M_x \\ M_y \\ M_z \end{Bmatrix}_i \tag{7.4.2}$$

Typically, a spot-weld damage parameter time history such as an equivalent force, structural stress, or stress intensity is determined from the local force components, because direct computation of local spot-weld stresses is not practical for large-scale structural finite element analysis. For variable-amplitude loading, accumulated fatigue damage is then determined from a rainflow cycle count of the damage parameter history, a damage

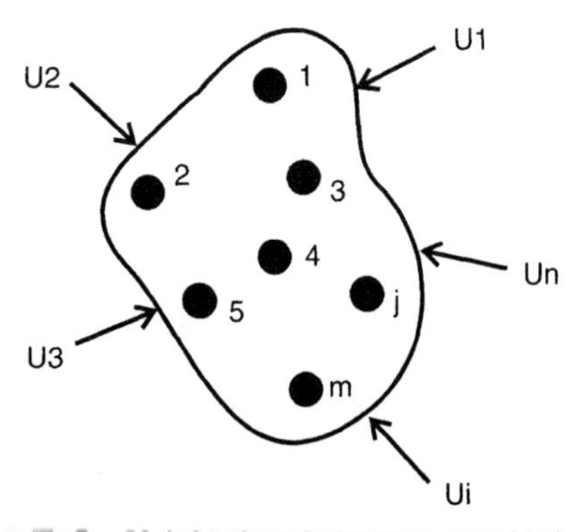

FIGURE 7.8 Unit load analysis for spot-welded structure.

parameter–life curve, and summation of the calculated damage for each identified cycle.

Several common finite element techniques for determining loads are applied to spot welds. Often, rigid elements or stiff beams are used to represent the spot weld between sheets of material that are modeled with shell elements. More sophisticated techniques for representing spot welds are being proposed by some researchers. Examples of common damage parameters are discussed later, and a more detailed explanation of using finite element analysis to determine spot-weld loads and stresses is presented in Sections 7.5.2 and 7.5.3.

7.4.1 BASIC LOAD–LIFE ANALYSIS

The most direct and least sophisticated approach for determining fatigue life is a basic load–life analysis approach. In this approach, load is the basic damage parameter; for example, the shear force in the plane of the nugget is determined from

$$V = (F_x^2 + F_y^2)^{1/2} \tag{7.4.3}$$

The shear force load history of a spot weld is rainflow counted, and the resulting fatigue damage based on each load level is summed. The drawback of this approach is that basic test data must exist for each geometric and material parameter of the structural spot welds. Geometrical parameters include sheet thickness and spot-weld nugget diameter. Material parameters include not only the base metal properties but also the HAZ properties, which could depend on the welding process parameters. However, if data are available, this approach can lead to useful results without detailed stress analysis assumptions.

7.4.2 STRUCTURAL STRESS APPROACHES

The objective of structural stress approaches is to characterize some critical aspect of the stress state at the crack initiation location of the spot-welded joint and incorporate this stress state into a damage parameter that depends on the nugget forces, moments, and geometry of the connection, such as the nugget diameter d and sheet thickness t.

$$\sigma_{ij} = f(F_x, F_y, F_z, M_x, M_y, M_z, d, t, \ldots) \tag{7.4.4}$$

These parameters are usually expressed in terms of structural (or nominal) stresses that are linearly related to nugget loads. These approaches sometimes also include factors to account for loading mode or geometry. The feature that distinguishes these approaches from each other is the specific details of how the stress is determined from the loads and geometry of the joint.

Because the general nature of each structural stress approach is similar, only one representative approach is discussed in detail.

7.4.2.1 Rupp's Structural Stress Approach[1]

Rupp et al. (1995) noted that very detailed stress analyses from finite element calculations are generally not practical or even useful for most automotive applications, because of the large expense of extremely detailed analyses and general lack of appropriateness of commonly available plastic deformation models. Hence, they reasoned that because exact stress calculations at spot welds are not effective, a nominal local structural stress directly related to the loads carried by the spot-welded joint and correlated with the fatigue life would be of more use in the engineering and design of structures. These local structural stresses were calculated based on the nugget cross-sectional forces and moments by using the beam, sheet, and plate theory, using the loads determined from a finite element model of the structure in which the spot-welded connections are represented with stiff beam elements. For cases of nonproportional loading, in which the critical location for fatigue crack initiation was not known in advance, stresses were computed at several stations around the circumference of the nugget, and the station with the most calculated damage was determined to be the critical location. The fatigue life calculated at this location is life of the welded joint.

For this approach, two types of failure modes of the spot-welded joints were considered: cracking in the sheet metal or cracking through the weld nugget. They determined the cracking mode with a generally accepted rule of thumb: when the diameter of the weld is more than $3.5\sqrt{t}$, where t is the sheet thickness (mm), the crack will be in the sheet. Otherwise, the crack will be through the weld nugget.

In either case, a central idea of Rupp's approach was to calibrate the structural stress parameter with the fatigue life results of spot-welded structures and not just with single spot-welded coupons. For St1403 (300HV), St1403 (200HV), StW24, and St1203, they have shown that the mean-stress-corrected fatigue life results of spot welds in various structures reduced scatter quite well in a log–log plot of the structural stress parameter versus fatigue life.

7.4.2.2 Cracking in the Sheet Metal

For the case of cracking in the sheet metal, the formulae of circular plate with central loading are applied to calculate local structural stresses. For

[1] Portions of this section, including Figures 7.10–7.17, reprinted from Kang et al. (2000), with permission.

deriving these formulae, the spot-welded connection was treated as a circular plate with a central, rigid circular kernel, and the outer edges of the plate were treated as fixed, as shown schematically in Figure 7.9.

The solution of the radial stresses for this plate problem is presented in Roark's formulas for stress and strain (Cook, 1989). The maximum radial stress ($\sigma_{r,\ max}$) resulting from lateral forces was determined as follows:

$$\sigma_{r,\ max} = \frac{F_{x,y}}{\pi d t} \tag{7.4.5}$$

where t is sheet thickness, d is nugget diameter, and $F_{x,\ y}$ is a lateral force in the X or Y direction, as shown in Figure 7.10. The radial stress σ_r due to normal force F_z on the weld nugget is given by

$$\sigma_r = \frac{\kappa_1 F_z}{t^2} \tag{7.4.6}$$

where κ_1 is 1.744, a parameter that depends on the ratio of the nugget radius and specimen span, and the maximum radial stress ($\sigma_{r,\ max}$) due to applied moment occurs at the edge of the nugget and is

$$\sigma_{r,\ max} = \frac{\kappa_2 M_{x,y}}{d t^2} \tag{7.4.7}$$

where κ_2 is 1.872, another parameter that depends on the ratio of the nugget radius and specimen span.

The parameters κ_1 and κ_2 are determined for an assumed ratio of $\frac{radius}{span} = 0.1$.

The equivalent stresses for the damage parameter are calculated by the appropriate combination and superposition of the local radial structural stress. For nonproportional loading, the equivalent stresses can be calculated as a function of the angle θ around the circumference of the spot weld as follows:

$$\begin{aligned}\sigma_{eq1}(\theta) = & -\sigma_{max}(F_x)\cos\theta - \sigma_{max}(F_y)\sin\theta + \sigma(F_z) \\ & + \sigma_{max}(M_x)\sin\theta - \sigma_{max}(M_y)\cos\theta\end{aligned} \tag{7.4.8}$$

where

$$\sigma_{max}(F_x) = \frac{F_x}{\pi d t}$$

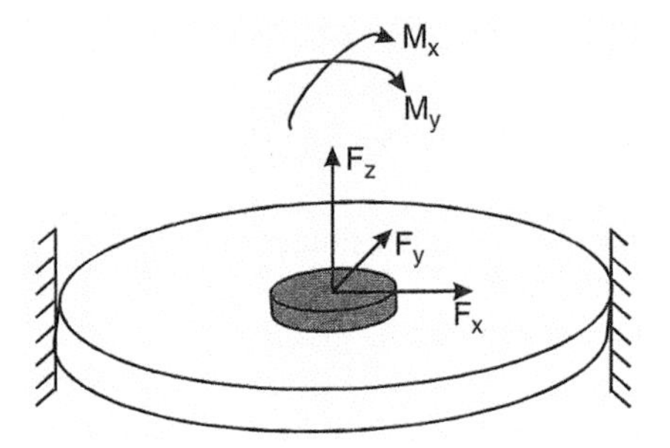

(a) Circular plate model for sheet metals.

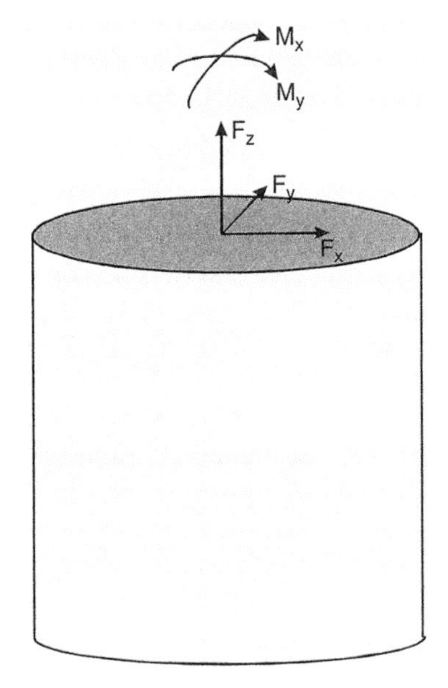

(b) Beam model for nugget subjected to tension, bending, and shear.

FIGURE 7.9 Models of spot-welded connection for stress calculations. From Kang et al. (2000), with permission.

$$\sigma_{max}(F_y) = \frac{F_y}{\pi dt}$$

$$\sigma(F_z) = \kappa\left(\frac{1.744F_z}{t^2}\right) \text{ for } F_z > 0 \quad (7.4.9a–g)$$

$$\sigma(F_z) = 0 \quad \text{for } F_z \leq 0$$

$$\sigma_{max}(M_x) = \kappa\left(\frac{1.872M_x}{dt^2}\right)$$

$$\sigma_{max}(M_y) = \kappa\left(\frac{1.872M_y}{dt^2}\right)$$

$$\kappa = 0.6\sqrt{t}$$

θ is the angle measured from a reference axis in the plane of the spot-weld nugget.

7.4.2.3 Cracking Through Weld Nugget

For the nugget failure mode, structural stresses for cracking through the nugget were calculated based on the formulae of a beam subjected to tension, bending, and shear loading, as shown in Figure 7.9(b). This failure mode can occur when a spot weld is used to connect relatively thick sheets. In this case, the spot-weld nugget is modeled as a circular cross-section of a beam subjected to tension, bending, and shear loading. The normal stress (σ_n), bending stress (σ_b), and the maximum shear stress (τ_{max}) are given by mechanics of materials formulae for these stresses:

$$\sigma_n = \frac{4F_z}{\pi d^2}$$

$$\sigma_b = \frac{32M_{x,y}}{\pi d^3} \quad (7.4.10a–c)$$

$$\tau_{max} = \frac{16F_{x,y}}{3\pi d^2}$$

where F_z is normal force, $F_{x,y}$ is a shear force in the X or Y direction, and d is the nugget diameter.

The nominal nugget stresses in this case can be calculated by superpositioning these formulae, and a failure orientation can be determined by using a stress-based critical plane approach. The resolved tensile stress on the critical plane is taken as the damage parameter. As a function of angle θ along the circumference of the spot-weld, these stresses are

$$\tau(\theta) = \tau_{max}(F_x)\cos\theta + \tau_{max}(F_y)\cos\theta$$
$$\sigma(\theta) = \sigma(F_z) + \sigma_{max}(M_x)\sin\theta - \sigma_{max}(M_y)\cos\theta \quad (7.4.11)$$

where

$$\tau_{max}(F_x) = \frac{16F_x}{3\pi d^2}$$

$$\tau_{max}(F_y) = \frac{16F_y}{3\pi d^2}$$

$$\sigma(F_z) = \frac{4F_z}{\pi d^2} \quad \text{for } F_z > 0 \qquad (7.4.12a\text{–}f)$$

$$\sigma(F_z) = 0 \text{ for } F_z \leq 0$$

$$\sigma_{max}(M_x) = \frac{32M_x}{\pi d^3}$$

$$\sigma_{max}(M_y) = \frac{32M_y}{\pi d^3}$$

7.4.2.4 Mean Stress Corrections for Fatigue Calculations

Stress histories of plate or nugget stresses detailed previously are used to calculate the fatigue life of spot-welded components. Rainflow cycle counting of these histories can be used to determine the equivalent stress amplitude and mean stress associated with each cycle. If the structure is subjected to proportional loading, a single crack initiation site near each nugget can be readily determined. If the structure is subjected to nonproportional loading, then the many potential sites for crack initiation around the circumference of each weld nugget must be examined.

In either proportional or nonproportional loading, corrections may be made for mean stress sensitivity. Rupp et al. (1995) proposed modifying the equivalent stress based on a Goodman-type mean stress correction procedure, determined from the equivalent stress amplitude at $R = 0$:

$$S_0 = \frac{S + MS_m}{M + 1} \qquad (7.4.13)$$

where S is the stress amplitude at each cycle, M the mean stress sensitivity, and S_m the mean stress at each cycle. Then, the total fatigue life is correlated with the calculated maximum equivalent mean-stress-corrected stress amplitude, S_0.

Example. The approach by Rupp and coworkers was developed to use for typical spot-welded structures that contain multiple welds. However, the approach will be demonstrated for fatigue life data collected from combined tension and shear specimens (Kang et al., 2000).

The data include 140 fatigue test results of high-strength steel specimens. The thickness of the specimens was 1.6 mm and the nominal spot-weld diameters of the specimens were 5.4 and 8.0 mm. The typical shape and dimensions of the specimen are shown in Figure 7.10. Combined loads were applied by using a special test fixture developed by the DaimlerChrysler spot-weld design and evaluation committee. This fixture allowed the application of several combinations of tension and shear loading by varying the plane of the spot-welded nugget with respect to the applied loading axis. The three loading directions, 30°, 50°, and 90°, were employed

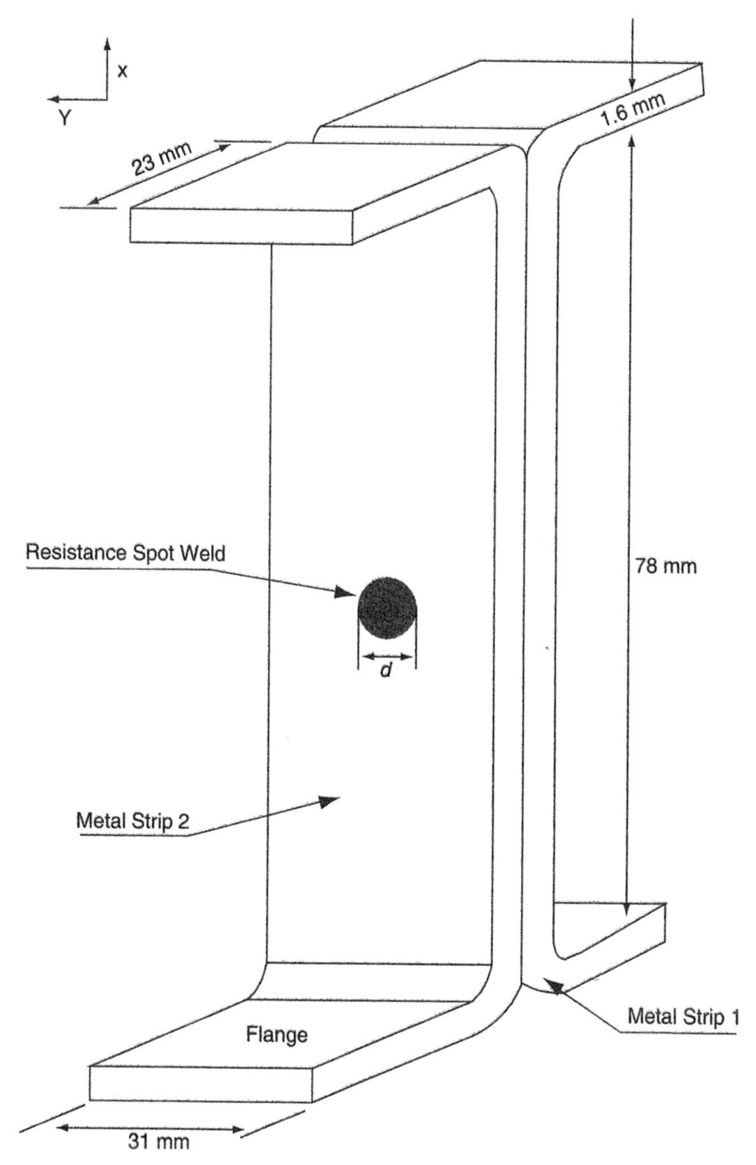

FIGURE 7.10 A typical specimen subjected to the combined tension and shear loads ($d = 8.0$ or 5.4 mm). From Kang et al. (2000), with permission.

to apply the multiaxial loads on the weld nugget, as shown in Figure 7.11. Three mean loads of 1110, 2220, and 3340 N (250, 500, and 750 lb) were applied for each testing angle, and load ratios were varied from 0 to 0.76. The results of these fatigue tests are presented in the load–life plots in Figures 7.12–7.14.

First, the failure mode was determined by using the rule of thumb ($3.5\sqrt{t}$). All specimens were anticipated to fail in the sheet metal because all the nugget diameters are greater than $3.5\sqrt{t}$. Therefore, the plate theory equations were applied to calculate local structural stresses at the center plane of the sheet metal. The next step would typically be to account for the mean stress sensitivity of the material and structure. In this particular case, the test data exhibited very little mean stress sensitivity, and therefore the mean stress sensitivity factor was set at zero. This results in the total fatigue life (N_t) versus maximum equivalent stress amplitude plot shown in Figure 7.15.

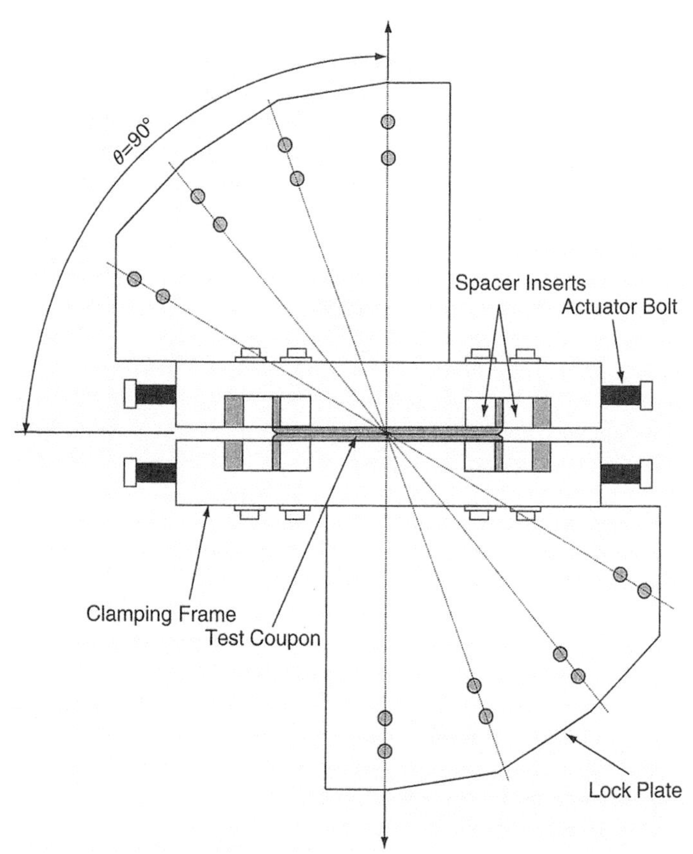

FIGURE 7.11 The test fixture to apply the combined tension and shear loads on the spot welded specimens. From Kang et al. (2000), with permission.

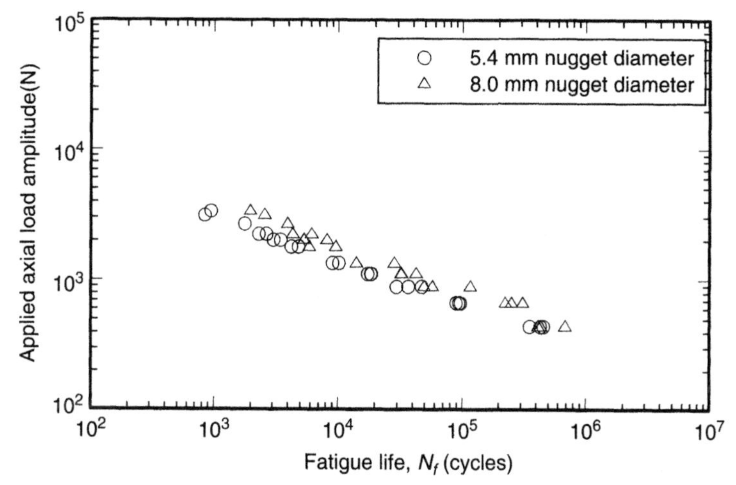

FIGURE 7.12 Experimental results of applied axial load amplitude versus fatigue life for 30° loading direction. From Kang et al. (2000), with permission.

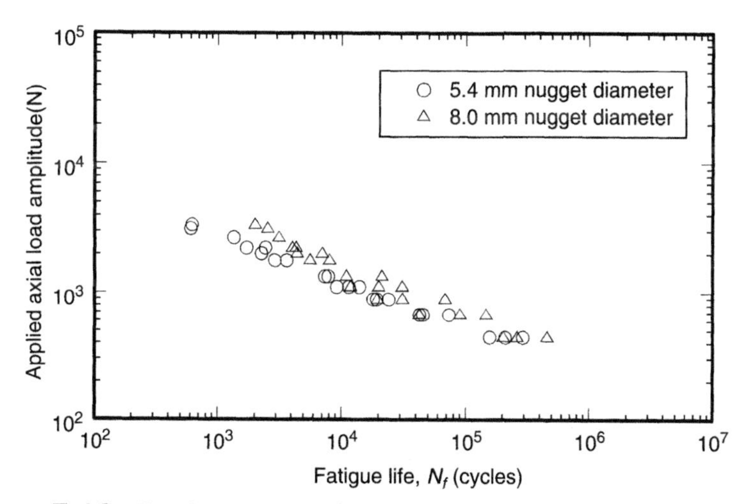

FIGURE 7.13 Experimental results of applied axial load amplitude versus fatigue life for 50° loading direction. From Kang et al. (2000), with permission.

The equation of the power law relation from Figure 7.15 becomes the governing equation to calculate fatigue life. The correlation coefficient (cc) between maximum equivalent stress amplitude and experimental fatigue life was −0.91. Because this comparison is for constant-amplitude data, a comparison can be readily made, and the calculated versus measured fatigue life for multiaxial fatigue test data is presented in Figure 7.16. The dotted line in the figure represents a perfect correlation between the measurements and the calculations, and the solid lines represent a variation by a factor of three from a perfect correlation. As shown in figure, the approach by Rupp and coworkers

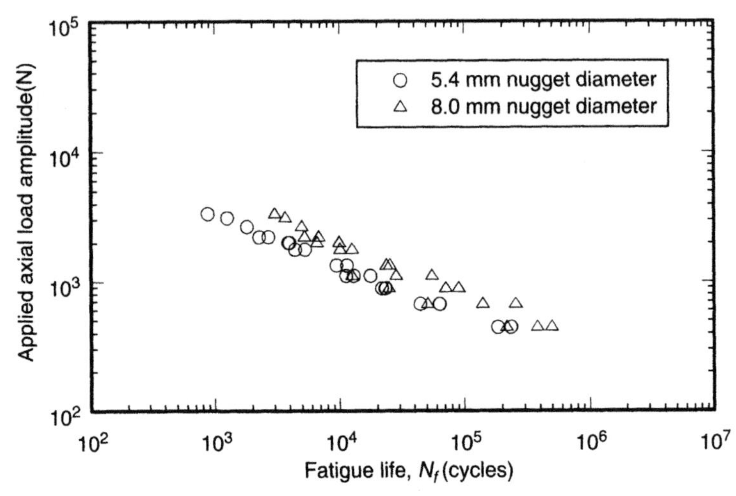

FIGURE 7.14 Experimental results of applied axial load amplitude versus fatigue life for 90° loading direction. From Kang et al. (2000), with permission.

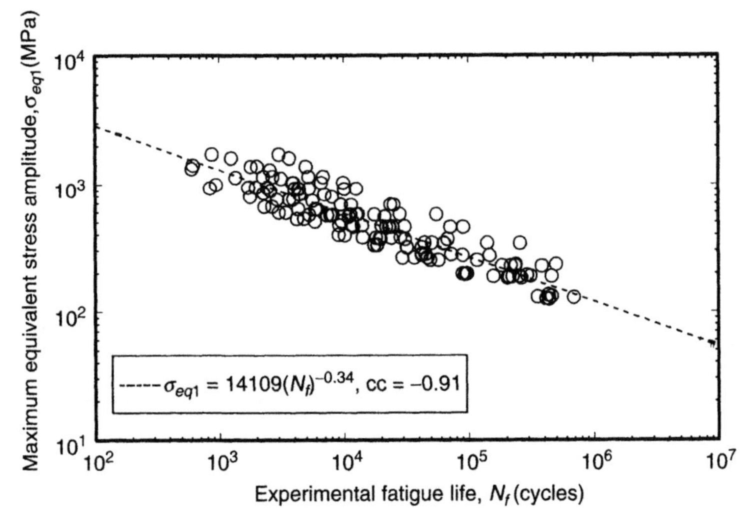

FIGURE 7.15 Experimental life versus Rupp and coworkers' maximum equivalent stress amplitude for the specimens subjected to multiaxial loads. From Kang et al. (2000), with permission.

resulted in a reasonable correlation with measured fatigue life for the multiaxial results.

7.4.2.5 Summary

Rupp and coworkers calculated local structural stresses by using the plate, sheet, and beam theory based on the cross-sectional forces and moments.

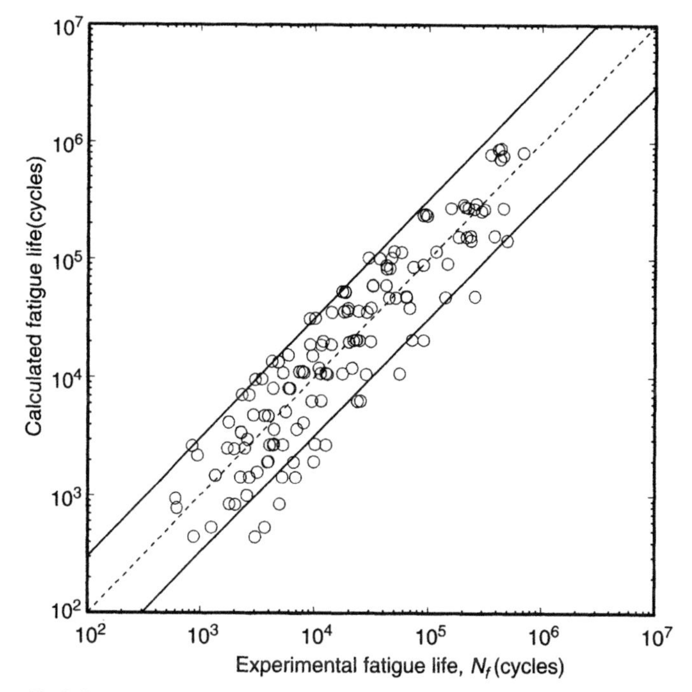

FIGURE 7.16 Experimental life versus calculated life using Rupp and coworkers' method for multiaxial load test data. From Kang et al. (2000), with permission.

This method is suitable for application to large finite element models, because mesh refinement is not necessary for spot welds.

This approach relies on a rule of thumb to determine failure mode that determines the applying theory to calculate local structural stresses. This method requires the following information:

1. The estimated failure mode
2. The forces and moments around nugget
3. Local structural stresses as given by the equations presented
4. Mean stress sensitivity
5. A relation between maximum equivalent stress amplitude and total fatigue life, such as a power law relation, obtained from a fit of experimental data.

7.4.3 STRESS INTENSITY APPROACH

7.4.3.1 Survey of the Approach

The fatigue failure of spot welds is normally considered from two aspects: failure driving force and failure resistance. The former is a load to be calculated and the latter a material property to be experimentally determined.

The driving force of failure is mainly given by external loads and structural geometry, whereas the failure resistance is mainly dictated by the internal composition and microstructure of the material. By comparing the two, one makes safety assessments or establishes failure criteria. The comparison needs a common parameter, which can also be considered as a damage parameter. Structural stress, notch stress, stress intensity factors or other fracture mechanics parameters are common damage parameters. The approaches based on structural stresses have been discussed in Section 7.4.2, and the approach based on stress intensity factors is discussed here.

The stress intensity approach for spot welds was first proposed by Pook (1975a, b), in which the fatigue strength of tensile-shear spot welds is assessed in terms of the stress intensity factor at the spot weld. The work by Yuuki et al. (1986) first showed the feasibility of the stress intensity factor in correlating the fatigue strength of spot welds across different specimens. The essential point of the approach was the determination of stress intensity factors at spot welds. The finite element method is widely used to determine stress intensity factors at spot welds. A three-dimensional finite solid element model was given by Cooper and Smith (1986) and Smith and Cooper (1988), and a similar but substantially refined model was developed by Swellam et al. (1992). A three-dimensional coupled finite/boundary element analysis of spot welds was given by Yuuki and Ohira (1989). Radaj (1989) developed another three-dimensional finite element model in which the spot weld was modeled by solid elements and the sheets by shell elements, with an appropriate coupling of the two kinds of elements. This method was substantially improved by Zhang (1997, 1999a–c, 2001, 2003). All these models relied on refined finite element meshes of and near spot welds.

Theoretically, stress intensity factors at a spot weld subjected to general loads can be directly calculated by the finite element method, but it often demands three-dimensional solid elements and a high degree of mesh refinement near the spot weld. Special expertise is also needed for extracting stress intensity factors from the stresses and displacements in the vicinity of the nugget edge. Two kinds of solutions to stress intensity factors at spot welds are introduced here, without using solid elements: (1) the structural-stress-based solution, in which the stress intensity factors are determined by stresses in the shell elements around the spot weld, and (2) the force-based solution, in which the stress intensity factors are estimated by the interface forces and moments in the beam element that simulates the spot weld.

7.4.3.2 Structural-Stress-Based Method

Some classical work in fracture mechanics as compiled in Tada et al. (1985, pp. 16.2, 16.5, and 29.9) showed that stress intensity factors at a crack between two sheets under edge loads can be expressed alternatively by structural stresses (plate theory stresses) at the crack tip times square root of sheet thickness. The work of Pook (1979, Equations 11, 48, and 94) resulted in the same finding and actually states stress intensity factors in the following form:

$$K_{\mathrm{I}} = k_{\mathrm{I}}\sigma_b\sqrt{t} \tag{7.4.14}$$

$$K_{\mathrm{II}} = k_{\mathrm{II}}\sigma\sqrt{t} \tag{7.4.15}$$

$$K_{\mathrm{III}} = k_{\mathrm{III}}\tau\sqrt{t} \tag{7.4.16}$$

where σ_{b}, σ, and τ are structural stresses and t the sheet thickness. Symmetrical and nonsymmetrical loads are differentiated from each other and analytic values of $k_{\mathrm{I}} = 1/\sqrt{3}$, $k_{\mathrm{II}} = 1/2$, and $k_{\mathrm{III}} = \sqrt{2}$ are given for three special cases. Pook concluded in his work that local crack tip stresses obtained from simple beam/plate theory can be used directly to calculate stress intensity factors. This was the pioneering work to determine stress intensity factors based on structural stresses. Note that sheet thickness plays the same role as crack length in conventional fracture mechanics.

The structural-stress-based method proposed by Pook (1979) was further developed by Radaj (1989) and Zhang (1997, 1999a–c, 2001) for spot welds. Spot-welded lap joints are considered in three groups in terms of sheet thickness and material combinations: equal thickness and identical material, unequal thickness and identical material, and unequal thickness and dissimilar materials. The recent results based on J-integral analysis and elementary plate theory in the later references (Zhang [1997, 1999a–c, 2001]) provide a number of formulae for stress intensity factors expressed by structural stresses. For example, the formulae for the first group of spot welds are as follows (Zhang, 1999b):

$$K_{\mathrm{I}} = \frac{1}{6}\left[\frac{\sqrt{3}}{2}(\sigma_{ui} - \sigma_{uo} + \sigma_{li} - \sigma_{lo}) + 5\sqrt{2}(\tau_{qu} - \tau_{ql})\right]\sqrt{t} \tag{7.4.17}$$

$$K_{\mathrm{II}} = \left[\frac{1}{4}(\sigma_{ui} - \sigma_{li}) + \frac{2}{3\sqrt{5}}(\tau_{qu} + \tau_{ql})\right]\sqrt{t} \tag{7.4.18}$$

$$K_{\mathrm{III}} = \frac{\sqrt{2}}{2}(\tau_{ui} - \tau_{li})\sqrt{t} \tag{7.4.19}$$

The structural stresses (Figure 7.17) in these equations include normal stresses σ_{ui}, σ_{uo}, σ_{li}, σ_{lo}; circumferential shear stresses τ_{ui} and τ_{li}; and transverse shear stresses τ_{qu} and τ_{ql} on the verge of the spot weld. Note that these structural stresses are plate theory stresses without stress singularity. They represent external loads generally. The stress singularity lost is recovered by Equations 7.4.17–7.4.19. The transverse shear stresses τ_{qu} and τ_{ql} are in average values through the sheet thickness and the other stresses are linearly distributed through the sheet thickness.

An appropriate finite element model is needed to extract plate theory stresses at spot welds. A special spoke pattern as shown in principle in Figure 7.18

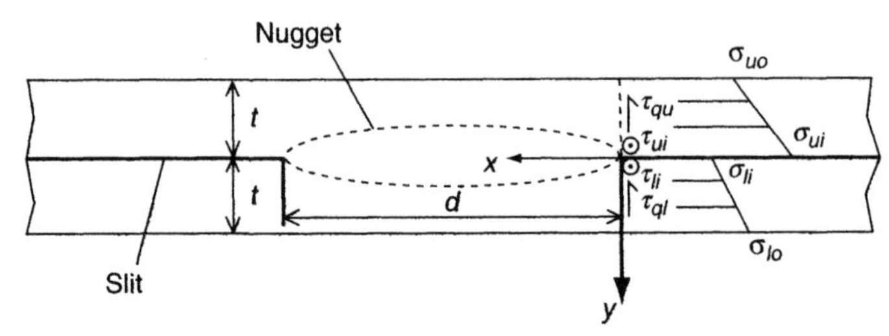

FIGURE 7.17 Structural stresses (plate theory stresses) around a spot weld according to Zhang (2001).

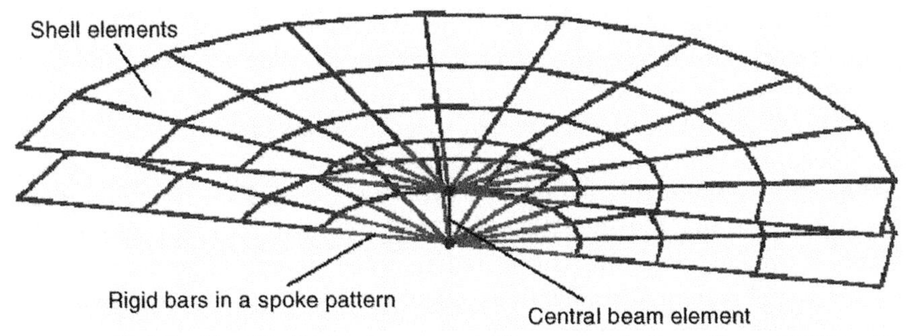

FIGURE 7.18 Simplified finite element model for spot welds. From Zhang (1999b), with permission.

is considered as a simplified model without using solid elements. The central beam in the spoke pattern is actually a cylindrical elastic beam element with a diameter as that of the nugget. The elasticity parameters such as Young's modulus and Poisson's ratio of the beam element may be set equal to those of the base material. The stress results are actually insensitive to these elasticity parameters. The pattern itself consists of rigid bar elements transferring or connecting all six translational and rotational degrees of freedom between the nodes of beam element (master nodes) and those (slave nodes) of the shell elements. The diameter of the pattern is also equal to the nugget diameter. To avoid overconstraints, the rigid bars may be set to transfer only three translational and one rotational (in the radial direction) degrees of freedom (1235 according to the MSC/NASTRAN convention), with the other two rotational degrees of freedom being set free. The effect of overconstraint is negligible for most engineering applications. Adequate mesh refinements should be introduced around the spoke pattern in order to obtain the structural stresses accurately.

Attention should be paid to the coordinate system (Figure 7.17) that must be employed consistently. The origin of the coordinate system is moving along the periphery of the nugget with the positive direction of the abscissa always pointing at the nugget center, as shown in Figure 7.17. The positive direction of the ordinate runs downward through the lower sheet. The positive direction of the shear stresses is also shown in the figure. The sign convention for the normal stresses is as usual, i.e., tensile stress is positive and compressive stress negative. The sign of stress intensity factors is referred to the same coordinate system. The approach given previously, i.e., Equations 7.4.17–7.4.19, plus the simplified finite element model may be applied straightforwardly in finite element simulations for determining stress intensity factors at spot welds. Stresses at the first ring of nodes on the spot-weld verge in the spoke pattern should be used. Alternatively, the stresses from the first ring of elements in the vicinity of the spoke pattern may also be used if heavy mesh refinements are introduced. The output stresses from finite element simulations are normally based on other coordinate systems and must be carefully converted to the current system.

7.4.3.3 Interface-Force-Based Method

Mesh refinements around spot welds are required in finite element analysis for obtaining reliable structural stresses around the spot welds. For automotive structures with a large number of spot welds, one cannot afford to spend too many elements for a single spot weld. A common practice is then to model spot welds with beam elements, which connect two sheets of shell elements without mesh refinements. The crude structural stresses around the spot weld are then not applicable for the structural-stress-based method. In this case, the interface forces (F_x, F_y, and F_z) and moments (M_x, M_y, and M_z) in the beam element are applicable, i.e., the forces and moments transferred by the spot weld. A methodology for estimating the structural stress based on the interface forces and moments was developed by Rupp et al. (1994), Sheppard (1993), and Maddox (1992). A similar estimation was given by Zhang (1997) additionally for the notch stress and stress intensity factors at spot welds. The stress intensity factors at a spot weld were estimated by Zhang (1997, 1999a) as follows:

$$K_{\mathrm{I}} = \frac{\sqrt{3}F}{2\pi d\sqrt{t}} + \frac{2\sqrt{3}M}{\pi d t\sqrt{t}} + \frac{5\sqrt{2}F_z}{3\pi d\sqrt{t}} \tag{7.4.20}$$

$$K_{\mathrm{II}} = \frac{2F}{\pi d\sqrt{t}} \tag{7.4.21}$$

$$K_{\mathrm{III}} = \frac{\sqrt{2}F}{\pi d\sqrt{t}} + \frac{2\sqrt{2}M_z}{\pi d^2\sqrt{t}} \tag{7.4.22}$$

with $F = \sqrt{F_x^2 + F_y^2}$, $M = \sqrt{M_x^2 + M_y^2}$ in the interface of the joint and F_z, M_z along or about the z-axis, which is perpendicular to the interface. The interface forces and moments are also termed as joint face forces and moments or cross-sectional forces and moments. The interface forces and moments are obtained from the output of the beam elements, but they should be referred to the interface of the joint. The stress intensity factors given by the previous equations are maximum values on the spot-weld verge, with K_{I} and K_{II} at the leading vertex of the spot weld in line with the principal loading direction and K_{III} at the side vertex which is at 90° from the leading direction. Equations 7.4.20–7.4.22 are primarily valid for spot welds between sheets of identical material of equal thickness and otherwise t is to be replaced by the smaller sheet thickness as a crude approximation.

7.4.3.4 Examples of Application

The tensile-shear specimen with a single spot weld (Figure 7.19) is well analyzed and widely used in spot-weld testing. The stress intensity factors for the specimen are available from different sources, inclusive of the detailed three-dimensional finite element results. The approach given previously, i.e., the structural-stress-based solutions plus the simplified finite element model, is applied to the specimen. The stress intensity factors at the spot weld with $d = 5\,\mathrm{mm}$ and $t = 1\,\mathrm{mm}$ under tensile-shear force of $F = 1\,\mathrm{kN}$ are predicted by the approach and compared in Table 7.2 with results from other sources. The width of specimen differs to some extent among the

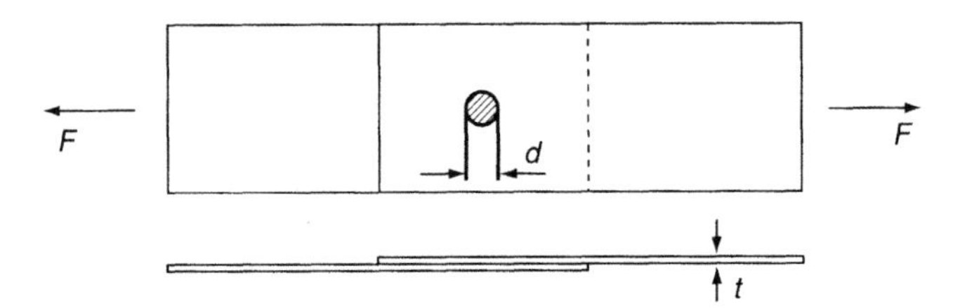

FIGURE 7.19 Tensile-shear specimen.

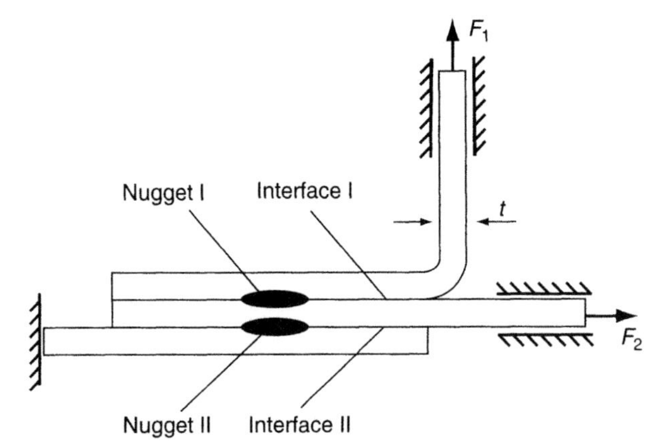

FIGURE 7.20 Multisheet spot welds with diameter $d = 5\,\mathrm{mm}$ of both nuggets, sheet thickness $t = 1\,\mathrm{mm}$, notch root radius $\rho = 0.2\,\mathrm{mm}$, and peel (F_1) eccentricity $e = 13.5\,\mathrm{mm}$ under loads of $F_1 = F_2 = 100\,\mathrm{N}$. From Zhang (2001), reprinted with kind permission of Kluwer Academic Publishers.

TABLE 7.2 Stress Intensity Factors K_I and K_{II} at the Leading Vertex and K_{III} at the Side Vertex of the Weld Spot, all in $N/mm^{3/2}$, Under Tensile-Shear Force $F = 1$ kN with Nugget Diameter $d = 5$ mm and Sheet Thickness $t = 1$ mm

Authors	K_I	K_{II}	K_{III}
Pook (1975b), analytic approximation	163.3	199.8	—
Radaj (1989) FEM	60.0	147.8	65.4
Smith and Cooper (1986), FEM	63.2	132.4	99.0
Swellam et al. (1992), FEM	64.2	134.6	—
Yuuki and Ohira (1989), FEM + BEM	55.5	123.2	90.3
Zhang (1997), analytic approximation	55.1	127.3	90.0
Zhang (1999b), FEM	77.9	130.0	86.9

different sources, but this does not matter because it is significantly larger than the nugget diameter. In the current finite element analysis, a local refinement of the mesh with smallest shell elements of about 0.25 mm (the nugget periphery is divided by 64 elements and it is much finer than the mesh shown in Figure 7.18) is introduced around the spot weld and the structural stresses are evaluated from the first ring of the shell elements adjacent to the nugget edge. The table shows that the stress intensity factors predicted by the current approach are quite close to predictions from other sources, especially for the leading stress intensity factor of K_{II} and the circumferential K_{III}. The K_I value predicted by the current model seems too high. This is attributed to the rigid support in the radial direction. The effect of the deviation in K_I is limited in this case in terms of the equivalent stress intensity factor that should be really considered in the strength assessment.

As another example of application, the stress-based solutions are applied to a three-sheet spot-welded joint (Figure 7.20). The joint with $d = 5$ mm of both nuggets, $t = 1$ mm of all the sheets, and $e = 13.5$ mm of load eccentricity is subjected to shear and peel forces of $F_1 = F_2 = 100$ N. The notch root radius at the nugget edge is assumed as $\rho = 0.2$ mm. The distributions of the stress intensity factors around the two spot welds are shown in Figure 7.21. As expected, the opening effect is dominant at interface I whereas there is a closure effect at interface II as a result of the reaction to the peel force F_1.

7.4.3.5 Summary

Stress intensity factors at a spot weld are expressed by the structural stresses around the spot weld or estimated by the interface forces and moments in the spot weld. The determination of the stress intensity factors at spot welds is as simplified as determining the structural stresses or the interface forces and moments. Force-based solutions are less accurate than stress-based ones.

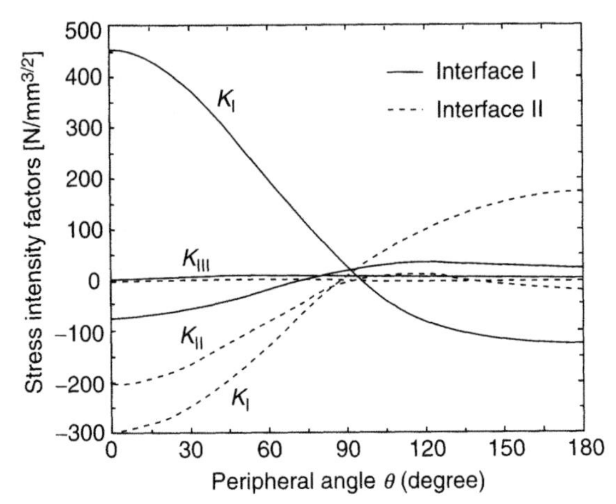

FIGURE 7.21 Stress intensity factors around the spot welds at the two interfaces I and II. From Zhang (2001), reprinted with kind permission of Kluwer Academic Publishers.

The limitations of these results are stressed here. First, they are linear solutions and therefore bounded to elastic and small-deformation behavior of the material. The results should be applicable to brittle fracture, high-cycle fatigue, and high-speed impact failure of spot welds in which plasticity is contained by a large elastic stress field. For low-cycle fatigue and ultimate failure of spot welds in which plasticity or large deformation may prevail, the results are hardly applicable, although they will not necessarily always fail. Second, material heterogeneity, residual stress, welding imperfection, and other welding-related factors have not been considered here.

REFERENCES

Barkey, M. E., Kang, H., and Lee, Y., Failure modes of single resistance spot welded joints subjected to combined fatigue loading, *International Journal of Materials and Product Technology*, Vol. 16, No. 6/7, 2001, pp. 510–526.

Barkey, M. E. and Han, J., Fatigue analysis of spot welds subjected to a variable amplitude loading history, SAE Technical Paper 2001-01-0435, 2001.

Barkey, M. E. and Kang, H., Testing of spot welded coupons in combined tension and shear, *Experimental Techniques*, Vol. 23, No. 5, 1999, pp. 20–22.

Cook, R. D., *Roark's Formulas for Stress & Strain*, 6th ed., McGraw-Hill, New York, 1989.

Cooper, J. F. and Smith, R. A., Initial fatigue crack growth at spot welds. In *Fatigue of Engineering Materials and Structures*, Institution of Mechanical Engineers, London, 1986.

Gieske, D. and Hahn, O., Neue Einelement-probe zum Prüfen von Punktschweiß verbindungen unter kombinierten Belastungen, *Schweißen und Schneiden*, Vol. 46, 1994, pp. 9–12.

Hahn, O., Dölle, N., Jendrny, J., Koyro, M., Meschut, G., and Thesing, T., Prüfung und Berechnung geklebter Blech-Profil-Verbindungen aus Aluminium, *Schweißen und Schneiden*, Vol. 52, 2000, pp. 266–271.

Kang, H. and Barkey, M. E., Fatigue life estimation of resistance spot-welded joints using an interpolation/extrapolation technique, *International Journal of Fatigue*, Vol. 21, 1999, pp. 769–777.

Kang, H., Barkey, M. E., and Lee, Y., Evaluation of multiaxial spot weld fatigue parameters for proportional loading, *International Journal of Fatigue*, Vol. 22, 2000, pp. 691–702.

Lee, Y., Wehner, T., Lu, M., Morrissett, T., and Pakalnins, E., Ultimate strength of resistance spot welds subjected to combined tension and shear, *Journal of Testing and Evaluation*, Vol. 26, 1998, pp. 213–219.

Maddox, S. J., Fatigue design of welded structures. In *Engineering Design in Welded Constructions*, Pergamon Press, Oxford, 1992, pp. 31–56.

Pook, L. P., Approximate stress intensity factors for spot and similar welds, NEL Report No. 588, 1975a.

Pook, L. P., Fracture mechanics analysis of the fatigue behavior of spot welds, *International Journal of Fracture*, Vol. 11, 1975b, pp. 173–176.

Pook, L. P., Approximate stress intensity factors obtained from simple plate bending theory. *Engineering Fracture Mechanics* 12, 1979, pp. 505–522.

Radaj, D., Stress singularity, notch stress and structural stress at spot welded joints, *Engineering Fracture Mechanics*, Vol. 34, 1989, pp. 495–506.

Rupp, A., Grubisic, V., and Buxbaum, O., Ermittlung ertragbarer Beanspruchungen am Schweißpunkt auf Basis der übertragenen Schnittgrößen, *FAT Schriftenreihe*, Vol. 111, 1994.

Rupp, A., Storzel, K., and Grubisic, V., Computer aided dimensioning of spot welded automotive structures, SAE Technical Report No. 950711, 1995.

Sheppard, S. D., Estimation of fatigue propagation life in resistance spot welds, ASTM STP 1211, 1993, pp. 169–185.

Smith, R. A. and Cooper, J. F., Theoretical predictions of the fatigue life of shear spot welds. In *Fatigue of Welded Constructions*, The Welding Institute, Abington, Cambridge, 1988.

Swellam, M. H., Ahmad, M. F., Dodds, R. H., and Lawrence, F. V., The stress intensity factors of tensile-shear spot welds, *Computing Systems in Engineering*, Vol. 3, 1992, pp. 487–500.

Tada H., Paris P., and Irwin, G., *The Stress Analysis of Cracks Handbook*, Paris Productions Inc and Del Research Corp, 1985, pp. 16.2, 16.5, 29.9.

Yuuki, R. and Ohira, T., Development of the method to evaluate the fatigue life of spot-welded structures by fracture mechanics, IIW Doc. III-928-89, 1989.

Yuuki, R., Ohira, T., Nakatsukasa, H., and Yi, W., Fracture mechanics analysis of the fatigue strength of various spot welded joints. In *Symposium on Resistance Welding and Related Processes*, Osaka, 1986.

Zhang, S., Stress intensities at spot welds, *International Journal of Fracture*, Vol. 88, 1997, pp. 167–185.

Zhang, S., Approximate stress intensity factors and notch stresses for common spot-welded specimens, *Welding Journal*, Vol. 78, 1999a, pp.173-s–179-s.

Zhang, S., Recovery of notch stress and stress intensity factors in finite element modeling of spot welds. In *NAFEMS World Congress on Effective Engineering Analysis*, Newport, Rhode Island, 1999b, pp. 1103–1114.

Zhang, S., Stress intensities derived from stresses around a spot weld, *International Journal of Fracture*, Vol. 99, 1999c, pp. 239–257.

Zhang, S., Fracture mechanics solutions to spot welds, *International Journal of Fracture*, Vol. 112, 2001, pp.247–274.

Zhang, S., Stress intensity factors for spot welds joining sheets of unequal thickness, *International Journal of Fracture*, Vol. 122, 2003, pp. L119–L124.

8

Development of Accelerated Life Test Criteria

Yung-Li Lee
DaimlerChrysler
Mark E. Barkey
University of Alabama

8.1 Introduction

Product validation tests are essential at later design stages of product development. In the automotive industry, fatigue testing in a laboratory is an accelerated test that is specifically designed to replicate fatigue damage and failure modes from proving grounds (PG) testing. Detailed damage analysis is needed to correlate the accelerated test to PG testing. Therefore, accurate representation of PG loading is essential for laboratory durability test development.

PG loading can be measured by driving an instrumented vehicle over the PG with various test drivers. The vehicle is equipped with transducers for component loading histories and sensors for other important vehicle parameters, such as temperature, speed, and displacement. A typical test schedule that was derived on a target customer (usually the 95th percentile customer usage) comprises numerous repeated cycles, usually varying from 800 to 1000. Each cycle then combines several mixed events (e.g., rough road, high-speed laps, city route, and country route). Data acquisition is usually performed on one test cycle. As illustrated in Figure 8.1, the one-cycle loading measurement for various drivers should be extrapolated to the expected complete loading history. The statistical techniques for cycle

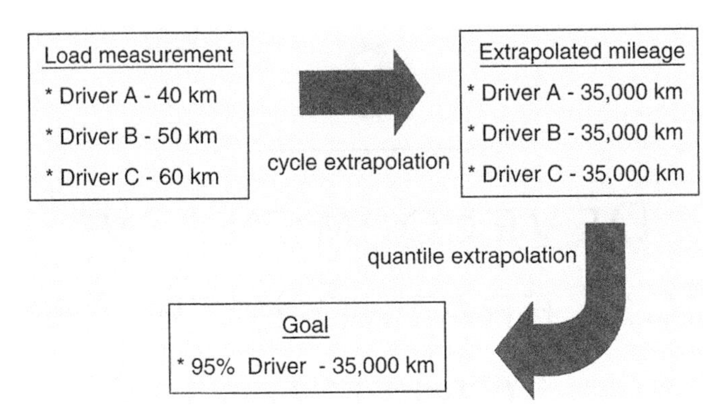

FIGURE 8.1 Schematic illustration of cycle extrapolation and quantile extrapolation.

extrapolation (Dreβler et al., 1996; Roth, 1998; Socie, 2001; Nagode et al., 2001; Socie and Pompetzki, 2003) are briefly described here.

It is believed that vehicle usage, operational conditions (such as weather and road roughness), and driver variability can affect the loading profile. Weather can influence traction force of the tires. For example, wet weather induces a smaller coefficient of friction between a tire and the ground than dry weather does. Roughness of a dirt road varies from time to time and can be washed out abruptly after rains. A study (Socie and Park, 1997) has shown that even for professional drivers, the driving patterns are not repeatable and can significantly alter the measured loading profile. In general, the weather and road roughness conditions have less significant impact to powertrain components than to the chassis and body components. To design components with a very low probability of fatigue failure, it is essential to quantify the severe damage-loading spectrum with a very low probability of occurrence (usually less than or equal to 5%) and the component fatigue resistance with a very high probability of survival. In this chapter, the quantile extrapolation techniques (Dreβler et al., 1996; Roth, 1998; Socie and Pompetzki, 2003) in the rainflow cycle domain are reviewed.

This chapter demonstrates how to develop an accelerated fatigue test criterion to account for test driver variability. An equivalent damage concept is used to correlate the accelerated test to PG loading, meaning that the damage value due to a laboratory test is equivalent to that from a PG loading. Based on the specific percentile damaging loading profile, product validation tests for durability are developed accordingly. This chapter specifically describes the development procedures of two types of product validation tests: one for powertrain system durability testing (dynamometer testing) and the other for general mechanical component fatigue testing (component life testing).

8.2 DEVELOPMENT OF DYNAMOMETER TESTING

A dynamometer, referred to as a differential or a transmission dynamometer, has been widely used in the automotive industry to validate durability and mechanical integrity of a gear set. For example, a typical differential dynamometer shown in Figure 8.2 consists of a driving shaft, a differential gear set, and two driven shafts. With a control module, a driving shaft will develop horsepower to turn the differential gears and the driven shafts from which the dynamometer motors absorb power and generate electricity. The electricity is then used to provide power back to the driving shaft. The control module defines the torque level and speed of the driving shaft with a specific running time. For example, according to a particular vehicle

FIGURE 8.2 A rear differential dynamometer test setup.

application, the bogey (a term borrowed from golf, whose meaning here is the minimum life requirement or target life) for a differential dynamometer requires x-hour driving shaft running time at y rpm with a z% of wheel slip torque.

Dynamometer testing is usually conducted to correlate PG testing in terms of fatigue damage to the associated gear set. Gear-tooth analysis indicates that the local gear-tooth bending moment and the number of cycles are linearly related to the driving shaft torque and revolutions. Thus, it is fair to assess the gear-set fatigue damage based on information of the driving shaft torque and revolutions (the so-called rotating moment histograms) experienced from PG and dynamometer testing. Figure 8.3 schematically demonstrates the use of rotating moment histograms for a driving shaft to develop either the transmission or the differential dynamometer schedule.

Fatigue testing is conducted to establish the fatigue resistance (T–N) curve (shown in Figure 8.4) for fatigue life estimates of a dynamometer system with a driving shaft under various torque levels. The linear damage rule as described in Chapter 2 is then used to calculate the amount of damage accumulated in the driving shaft using the T–N curve and the PG rotating moment histograms. It should be noted that other failure mechanisms, such as clutch slippage, thermal variation, and wear, are not a part of the analysis because of the lack of appropriate mathematical models and material data. Therefore, PG testing is always required for final validation.

At the PG, durability data on driveline components of a vehicle were first recorded on a powertrain endurance (PTE) test schedule at PG and then reduced to rotating moment histograms for driveline shafts. The PTE test is

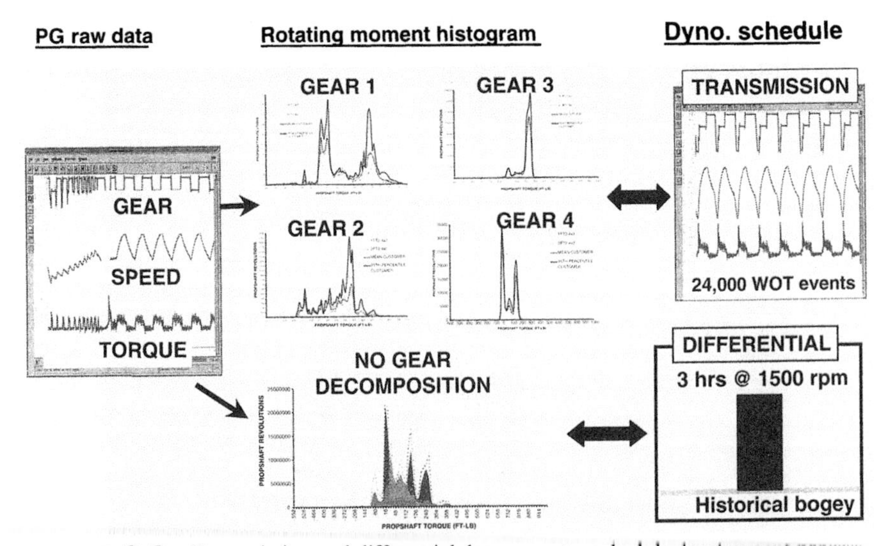

FIGURE 8.3 Transmission and differential dynamometer schedule development process.

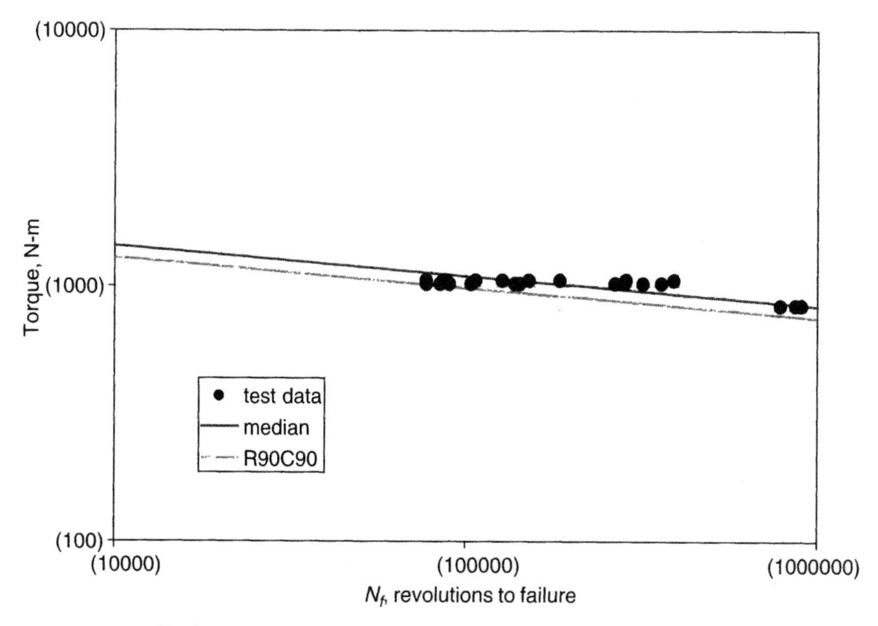

FIGURE 8.4 Fatigue test data of a differential gear set with gear ratio of 3.92.

specially designed to test the durability of the engine, transmission, torque converter, axle shafts, bearings, and seals. Depending on a vehicle application, a PTE test can comprise 700 test cycles in 35,000 PG miles, equivalent to 100,000 customer miles with a 95th percentile customer severity that has a low probability of occurrence of 5%.

The following sections detail the analysis procedures for driver variability in the PG load profile required for the dynamometer schedule development. A differential dynamometer test schedule for a rear wheel drive vehicle is illustrated for demonstration.

8.2.1 ROTATING MOMENT HISTOGRAMS

The torque versus revolution histogram (also named the rotating moment histogram, RMH) is an application of range-based counting methods to rotating components such as shafts. It has been widely used in gear-train designs and analyses. To construct the RMH, simultaneous measurements of both torque and rotational speed of a driving shaft are needed. As shown in Figure 8.5, the number of revolutions (n_i) is counted at a fixed torque level (T_i), given the discrete time intervals (Δt_i, $i = 1, m$) in rotational speed time history ($rpm(t)$). The number of revolutions at T_i can be expressed as follows:

$$n_i = \sum_{i=1}^{m} \int_{\Delta t_i} rpm(t)dt \tag{8.2.1}$$

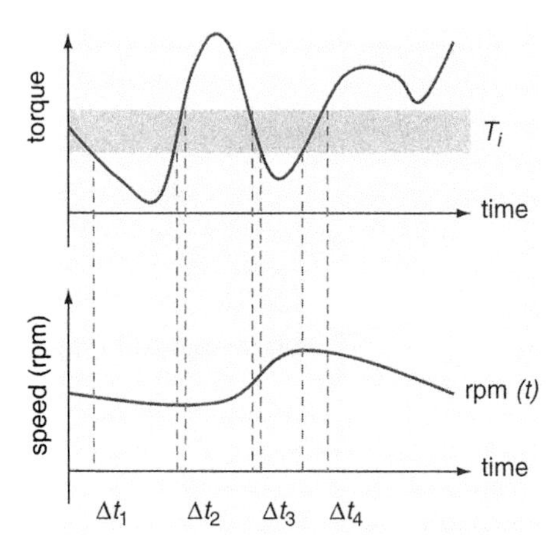

FIGURE 8.5 Generation of torque versus number of revolutions of a rotating component.

To make statistical analysis possible, the RMH from each driver must have the same bin limits, therefore resulting in identical bin size and torque values at each bin. One can specify the bin limits by taking the absolute maximum torque as the upper limit and the absolute minimum torque as the lower bin limit.

8.2.2 RMH CYCLE EXTRAPOLATION

The purpose of the RMH cycle extrapolation is to estimate the RMH for a much longer time period based on one short-term load measurement. The method for including loading variability in longer times is to shift the RMH upward with an extrapolated factor for higher numbers of cycles and then to extrapolate the loading spectrum to the two extremes for higher torque estimates. For example, Figure 8.6 demonstrates the RMHs for one lap and 700 laps of the PTE schedule for the 95th percentile driver profiles. Given one lap of the RMH, the loading histogram for 700 laps is extrapolated by a factor of 700. Then, the extrapolation technique for higher torques can be approximately done by graphically extending the data backward and intercepting with the torque axis at one revolution. A check needs to be performed to determine whether the estimated maximum torque value meets or exceeds the physical limit of the driveline system, which can be governed either by the wheel slip torque or by the maximum engine output torque. In Figure 8.6, the y-axis is a log scale of the number of propshaft revolutions calculated and the x-axis is for torque values. Any point on the histograms represents a certain number of revolutions found at the corresponding torque level. In this example, the positive torque and the negative torque represent the driving shaft torque due to the test vehicle under acceleration and coasting, respectively. This is the type of information used for a dynamometer test schedule development.

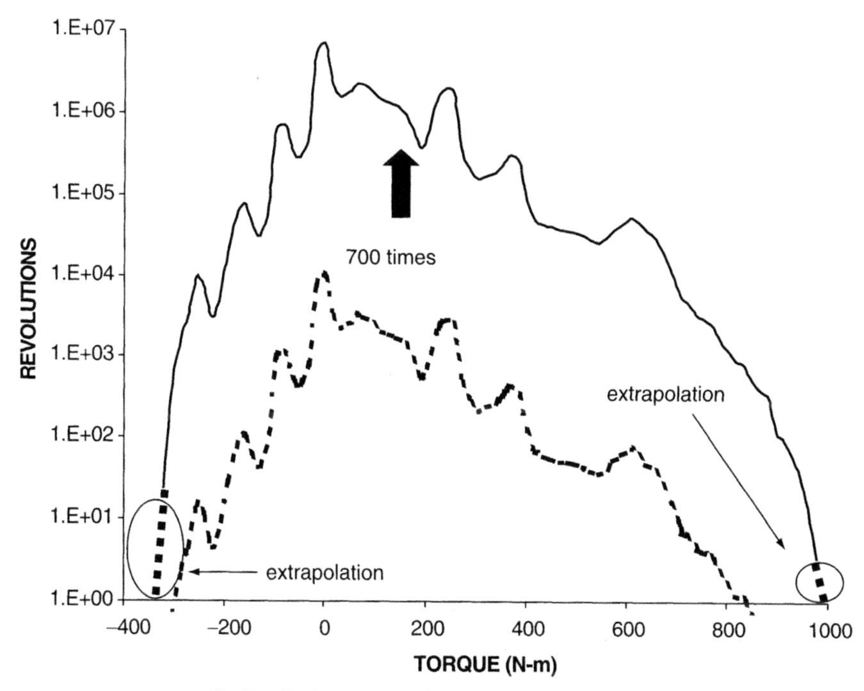

FIGURE 8.6 Cycle extrapolation of rotating moment histograms.

8.2.3 RMH QUANTILE EXTRAPOLATION

The purpose of the RMH quantile extrapolation is to predict the RMH of the single most damaging driver that would exist in a much larger set of data, based on a set of several time history measurements due to driver variability. It is assumed that the damage generated by a loading history correlates with the total number of revolutions in that history. This implies that the most damaging RMH contains the most revolutions.

When the sample size is six or more, the median and the 95th percentile histograms can be generated with an assumption that the revolutions in each torque bin follow a certain statistical distribution. However, in the case where the sample size is limited to less than five, a special statistical method is required to calculate the median and the 95th percentile driver profile with reasonable accuracy. Therefore, a process is developed to make statistical analysis possible. First, the total number of revolutions for all drivers is calculated by adding the number of revolutions in each bin. Second, the average number of revolutions is calculated. Next, the number of revolutions for each driver in each bin is normalized (i.e., average revolutions divided by total revolutions). The average customer revolution profile is determined by the following equation:

TABLE 8.1 Factors for the Standard Deviation for a Range

n_s	CF
2	0.886
3	0.591
4	0.486
5	0.430

$$\bar{x} = \frac{1}{n}\sum_{1}^{n} x_i \tag{8.2.2}$$

Next is the task of determining a 95th percentile customer revolution profile. To do so, per the study by Lipson and Sheh (1973), the standard deviation (σ) from a given data range (range) and sample size (n_s) can be estimated as follows:

$$\sigma = \text{range} \times \text{CF} \tag{8.2.3}$$

where CF is the conversion factor, given for various values of sample size (Table 8.1). The range is the difference between the largest and smallest revolutions in each torque bin. The 95th percentile customer revolution profile ($x_{95\%}$) is then derived by Equation 8.2.4:

$$x_{95\%} = \bar{x} + 1.645 \times \sigma \tag{8.2.4}$$

in which the constant of 1.645 is the 95% value of the standard normal distribution.

Figure 8.7 shows the RMHs for a driving shaft of a vehicle under eight different PG drivers over one cycle of the PTE schedule. The various dots represent the effect of PG drivers to the response of the driving shaft and the two solid lines are the median and the 95th percentile driver PG loading profiles. For each torque level, significant scatter in the eight measured revolutions is observed. This driver variability study indicates that each driver has different driving patterns.

8.2.4 DAMAGE ANALYSIS

Among the many mathematical models to describe progressive and accumulated damage, a linear damage rule has been widely accepted by practicing engineers because of its simplicity, despite its shortcomings of unpredictability and the exclusion of the load-sequence effect. The linear damage rule assumes damage (life-used up) is additive and defines failure to occur when

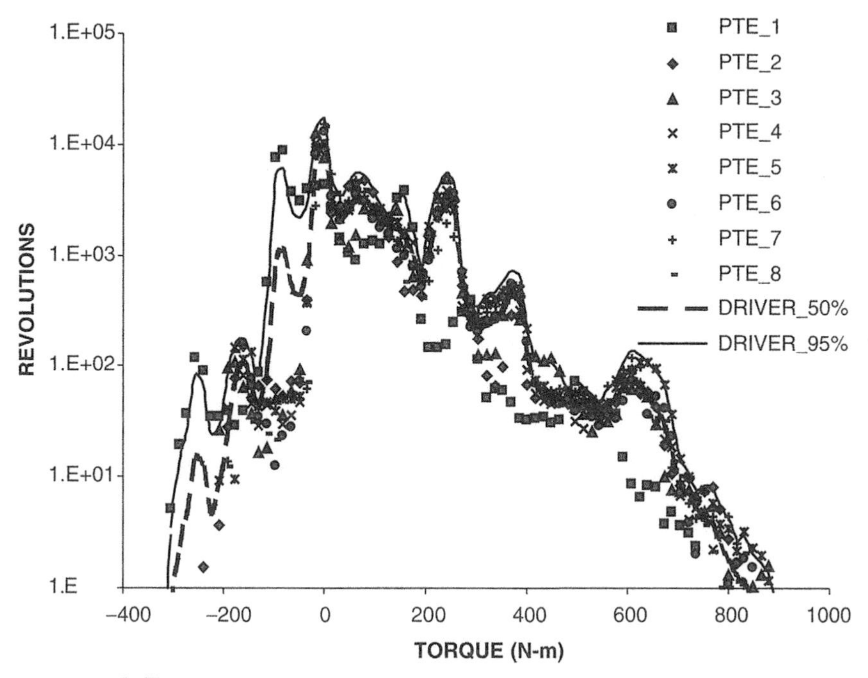

FIGURE 8.7 Quantile extrapolation of rotating moment histogram for one PTE test cycle.

$$D = \sum d_i = \sum \frac{n_i}{N_i} \geq 1.0 \tag{8.2.5}$$

The summation notion represents the accumulation of each individual damage (d_i), which is defined by the ratio of n_i to N_i, where n_i is the counted number of revolutions at a driving shaft torque level T_i from PG testing and N_i is the fatigue life in revolutions to failure at the same torque level determined from a torque versus life (T–N) curve.

For a gear-train system with a rotating driving shaft subjected to a variable torque history with various speeds in revolutions per minute (rpm), the range-pair counting algorithm is used to identify the number of revolutions (n_i) at each torque level with reference to the RMH. Fatigue testing with a differential dynamometer as shown in Figure 8.2 is required to generate the baseline fatigue resistance curve of the system. The T–N curve for the rear differential dynamometer in Figure 8.4 can be expressed as follows:

$$T_i = T_f'(N_i)^b \tag{8.2.6}$$

where T_i is a torque level, $T_f'(=2500\text{N}\cdot\text{m})$ is the fatigue strength coefficient, and b $(=-0.1)$ is the fatigue strength exponent. With number of revolutions (n_i) counted at a torque level (T_i), the fatigue life (N_i) and damage (d_i) for each bin can be easily calculated from Equations 8.2.5 and 8.2.6.

8.2.5 DEVELOPMENT OF A DYNAMOMETER TEST SCHEDULE

To reproduce the same damage value from PG, a differential dynamometer test requires the driving shaft running at a critical torque level for a certain amount of revolutions. The critical torque level is chosen as the torque associated with the highest damage value calculated from the PG loading profile. The total revolutions required can be reached by running the driving shaft at a specific speed for a certain time duration. Either the most frequent speed occurrence or the speed corresponding to the critical torque level is assigned as the test speed.

Damage distributions for the median and the 95th percentile PG drivers are illustrated in Figure 8.8, in which negative torque is assumed to be positive torque for a conservative reason, and the damage values shown in the positive torque range include the contribution from the negative torque range. The highest damage value to both driver profiles occurs at the torque level of 650 N·m and the accumulated damage values of the two drivers at that level are 0.52 and 0.95, respectively.

If the driving shaft to the rear differential dynamometer were subjected to a critical torque level of 650 N·m, the expected life to failure would be 710,000 revolutions estimated from Equation 8.2.6, the *T–N* curve with a damage of 1.0. Therefore, to replicate the same damage values from the median and the 95th percentile PG drivers (i.e., 0.52 and 0.95), the corresponding lives to failure are 370,000 and 670,000 revolutions, respectively. Given that 1500 rpm of the driving shaft speed is of interest, the running time required is 4.1 h for 370,000 revolutions to the median PG driver and 7.4 h for 670,000 revolutions to the 95th percentile PG driver.

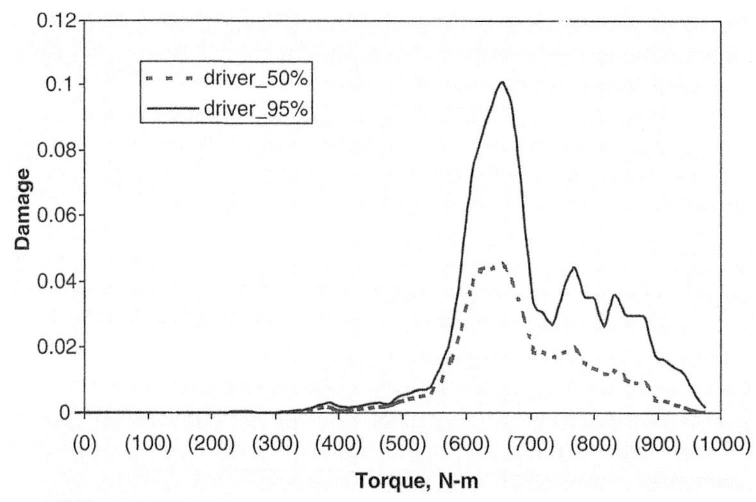

FIGURE 8.8 Damage density profile of a driving shaft in a complete PTC test.

Example 8.2.1. A transmission dynamometer has been widely used in the automotive industry to validate durability and mechanical integrity of a gear set. A transmission dynamometer consists of a transmission input shaft, a transmission gear set, and a transmission output shaft. With a control module, the input shaft will develop horsepower to turn the transmission gears and the output shaft, from which the dynamometer motors absorb power and generate electricity. The control module defines the torque level and speed of the input shaft, with a specific turning time, which will produce an equivalent damage level from a PG schedule. Figure 8.9 shows the schematic of a transmission dynamometer test step-up.

Development of a transmission dynamometer schedule often requires information of the transmission input shaft torque and revolutions, which are often determined by propshaft torque and speed from a PG data acquisition. The number of revolutions experienced by the propshaft is counted at a fixed torque level given the discrete time intervals in the speed time history. Finally, the input shaft torque and revolutions can approximately be related as follows:

$$T_{\mathrm{I}} = \frac{T_P}{\mathrm{GR}} \tag{8.2.7}$$

$$\mathrm{REV_I} = \mathrm{REV}_P \times \mathrm{GR} \tag{8.2.8}$$

where T_I and T_P are the torque levels for input shaft and propshaft; $\mathrm{REV_I}$ and REV_P are the revolutions of the input shaft and propshafts, and GR is the transmission gear ratio. It should be noted that transmission gear efficiency (nearly 0.95) is ignored in this calculation.

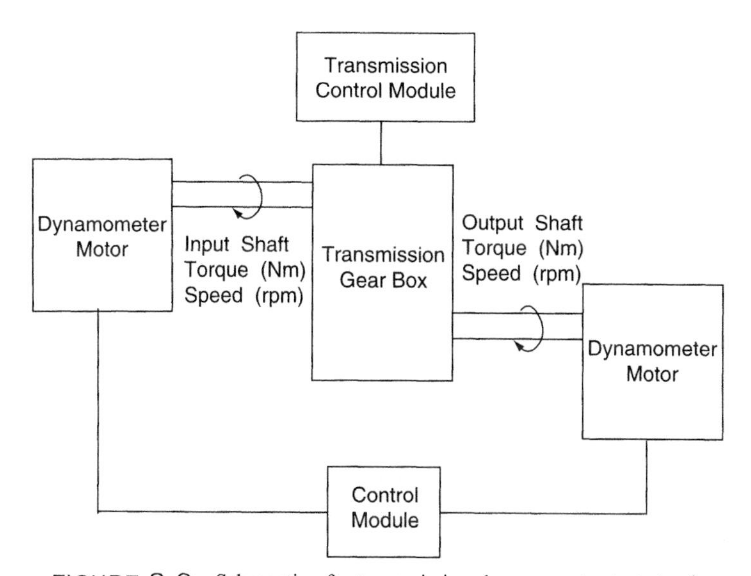

FIGURE 8.9 Schematic of a transmission dynamometer test stand.

Determine a transmission dynamometer schedule for an input shaft torque level of 165 N·m running at 3000 rpm in a first-gear position. The dynamometer schedule should produce the same damage level from a target PG schedule. It is given that the slope (b) of the torque versus revolution (T–N) curve is −0.1 from previous transmission dynamometer testing. The transmission equipped in a rear wheel drive vehicle has a transmission gear ratio of 4.0 in the first gear. The propshaft torque and speed data for this vehicle were acquired based on a PG PTE schedule. The driveline data in first-gear position were further reduced and are presented in Table 8.2 in which the positive (+) torque and the negative (−) torque represent the transmission input shaft torque due to the test vehicle under acceleration and coasting, respectively.

Solution. The dynamometer lab test should be preformed at an input shaft torque level of 165 N·m and an input shaft speed of 3000 rpm. Because

TABLE 8.2 Propshaft Torque and Revolutions at First Gear in the PG Schedule

Bin No.	T_p(N·m)	REV_P
1	−500	2788
2	−450	3448
3	−400	6589
4	−350	7949
5	−300	8934
6	−250	11189
7	−150	14250
8	−100	43115
9	−50	456087
10	50	415545
11	100	640668
12	150	484805
13	200	243964
14	250	180651
15	300	163882
16	350	133320
17	400	101181
18	450	85591
19	500	73170
20	550	51583
21	600	50111
22	650	36108
23	700	15537
24	750	7060
25	800	2148
26	850	1306
27	900	206

the PG torque and revolution information is measure from the propshaft, the transmission output, the data must be converted to determine the PG transmission input torque and revolutions. To perform this conversion, the transmission first-gear ratio and the previous equations are used. For conservative reasons, the coast (negative) torques will be converted to positive torques for this analysis. The converted torque level and revolutions for each bin of the transmission input shaft are tabulated in Table 8.3.

To determine the number of revolutions (REV_{eq}) at a specific torque level (T_{target}) inducing the same amount of damage as to a number of revolutions (REV_I) at a different torque level (T_I), the equation of equivalent damage revolution is introduced as follows:

$$REV_{eq} = REV_I \left(\frac{T_I}{T_{target}} \right)^{-1/b} \tag{8.2.9}$$

TABLE 8.3 Propshaft Torque and Revolutions at First Gear in the PG Schedule

Bin No.	T_P(N·m)	REV_P	T_I(N·m)	REV_I	REV_{eq} at T_{target}	
1	−500	2788	125	11152	694	
2	−450	3448	113	13792	313	
3	−400	6589	100	26356	176	
4	−350	7949	88	31796	59	
5	−300	8934	75	35739	13	
6	−250	11189	63	44756	3	
7	−150	14250	38	57000	0	
8	−100	43115	25	172460	0	
9	−50	456087	13	1824348	0	
10	50	415545	13	1662180	0	
11	100	640668	25	2562672	0	
12	150	484805	38	1939220	0	
13	200	243964	50	975856	6	
14	250	180651	63	722604	48	
15	300	163882	75	655528	247	
16	350	133320	88	533280	993	
17	400	101181	100	404724	2706	
18	450	85591	113	342364	7770	
19	500	73170	125	292680	18224	
20	550	51583	138	206332	34556	
21	600	50111	150	200444	77280	
22	650	36108	163	144432	127850	
23	700	15537	175	62148	111936	
24	750	7060	188	28240	104137	
25	800	2148	200	8592	58824	
26	850	1306	213	5224	67137	
27	900	206	225	824	18319	
					631291	TOTAL

where $b = -0.1$ is the slope of the fatigue resistance curve and $T_{target} = 165$ N·m is the input shaft controlled torque in the transmission dynamometer testing.

Application of Equation 8.2.9 leads to the equivalent damage revolutions for each pair of (T_I and REV_I), as shown in Table 8.3. It is also found that the torque level at 163 N·m causes more damage to the system than others. Therefore, the critical torque value (165 N·m) to the dynamometer system is a reasonable choice. Knowing the total equivalent damage revolutions for the input shaft (631,291 revolutions) and the input shaft rotational speed (3000 rpm), the time to produce the equivalent damage to the PG data set is 3.5 h (=631,291/3000/60).

8.3 DEVELOPMENT OF MECHANICAL COMPONENT LIFE TESTING

Component life testing is commonly designed to validate fatigue strength of a component based on a target customer usage or a PG schedule. The knowledge of the loads acting on a component is crucial for such a loading spectrum generation. Normally, strain gages are employed to correlate the relationship between strains and loads. The placement of strain gages in a smooth section in which the gross material behaves elastically is by far preferred. If possible, real-time load measurement should be conducted for an actual load time history, but because of the costs of long-term measurement and safety measures, the measurement period is usually not long enough to be used directly in a test. The acquired short-term load time data have to be extrapolated in rainflow cycle matrices instead of in the time domain. This is called the cycle extrapolation process. The scatter of loading depends on operating conditions and customer usage. It is often not feasible to quantify the scatter of loading (quantile extrapolation) in the time domain, and therefore this often has been done in the rainflow cycle domain.

After the rainflow cycle counting procedure, the measured loading spectrum can be available for development of component life test criteria. There are two widely accepted methods for component life tests: constant amplitude and block load cycle. The constant-amplitude fatigue test is the fastest, simplest, most straightforward, least expensive, and most frequently used laboratory fatigue test. Over the years, much vital information has been created via this type of testing. However, the constant-amplitude test has the limitation that it does not take into account the effect of loading sequence on the fatigue life of a component. Occasional overloads may cause plasticity at the high area of stress concentration, resulting in stress redistribution and a residual stress that will affect overall fatigue performance of the part. The block load cycle test (or so-called programmed fatigue test) is preferable over the constant-amplitude fatigue test, because it resembles the actual service or PG loading spectrum, which is broken into a series of short constant-amplitude loading tests. The test set-up requires six to eight load blocks (steps) in a program and at least 10 repetitions of the program to test the part. Both tests are valid only when the following conditions are met:

(1) The failure location and mode in the service/PG loading and on the test must be same.
(2) The damage value associated with the test criterion should be identical to that calculated on the service life or the PG schedule.

The following section discusses the cycle extrapolation technique from a short-term measurement to a longer time span, the quantile extrapolation technique from an average loading spectrum to an extreme loading, and their applications to the component life test development.

8.3.1 RAINFLOW CYCLE EXTRAPOLATION

The purpose of the rainflow cycle extrapolation is to predict the rainflow histogram for a much longer time period based on a short-term load measurement. In general, fatigue analysis is performed on a recorded load history, assuming that the possibility of overload occurrence in a longer time period is ignored. This is an incorrect postulate because the load measurement may be different because a driver can maneuver a vehicle over a repeatable event each time in a different manner. That means, for example, that a driver will not hit the pothole every time with the same speed and the same steering angle.

One traditional technique (Grubisic, 1994) is to construct a cumulative frequency distribution (cumulative exceedance histogram) from the rainflow cycle counting matrix based on a short-term measurement and then to extrapolate the cycles and load levels to the target life. In this approach, the rainflow cycle counting matrix must be reduced to a simple frequency histogram that contains the number of cycles at each specific load range or amplitude. This can be accomplished by first converting the number of cycles counted for each offset load at any given load range to an equivalent number of damage cycles with a zero mean load range. Second, the cumulative exceedance cycles are generated at each load amplitude level in such a manner that all the cycles for the load levels are accumulated, which exceed and equal the specific load amplitude. The load amplitude versus the cumulative cycle curve is termed the *cumulative exceedance diagram*, which is a simple way to represent an envelope of the service/PG loading spectrum. In Figure 8.10, the long-term loading spectrum can be generated by shifting this loading spectrum rightward to higher cycles by an extrapolation factor and by fitting

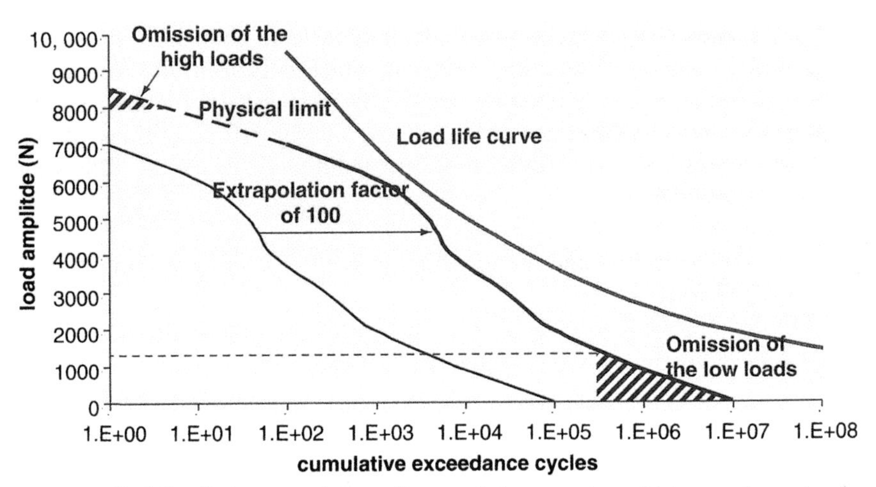

FIGURE 8.10 Cycle extrapolation of a cumulative exceedance histogram for a chassis component under PG loading.

and projecting the data curve to the upper left corner for higher load estimates. This technique fails to take into account the effect of mean load variability.

A life test criterion can be developed after the extrapolated PG loading spectrum is developed. There are two ways to accelerate the test very easily. One is that the small stress cycles that are below the fatigue limit of the material can be left out. As a rule of thumb (Heuler and Seeger, 1986), about half of the fatigue limit for smooth coupons can be the allowable filter level of omission of small cycles for notched specimens. Or, if the fatigue limit is unknown, 15% of the maximum load is chosen as the cut-off load level for omission of nondamaging cycles, because this load is generally below half of the fatigue limit. The other is that the test frequency may be increased as long as there is no resonance within the system or otherwise does not change the damage mechanism of the component, such as what may occur with non-metallic components. Based on the historical cycle extrapolation technique, the following example illustrates the procedures to develop a block load cycle test from a PG loading spectrum.

Example 8.3.1. One axial load cell was designed to measure PG loads (in newtons) on one vehicle suspension component (a tie rod). The calibrated tie rod was instrumented to a vehicle running at the PG with one cycle of a PG endurance schedule. It is known that one complete PG schedule consists of 100 PG cycles. The acquired axial loads on the tie rod per PG cycle were then rainflow-counted, and the rainflow matrix with a range-mean format is listed in Table 8.4. Four component fatigue tests at two different load levels were carried out to determine the approximate fatigue properties of the tie rod. The load-life equation is

TABLE 8.4 Rainflow Cycle Counting Matrix: One Cycle of Tie Rod PG Loads

		Mean (N)							
		−5000	−3000	−1000	1000	3000	5000	7000	9000
Range (N)	2000	10	158	250	500	410	300	30	9
	4000		7	28	60	50	36	2	
	6000			17	35	23	7		
	8000			1	10	5			
	10000				9	4			
	12000				5	1			
	14000				1				

$$P_a = P_f'(2N_f)^b$$

where P_f', the fatigue strength coefficient = 20,000 N and $b = -0.14$. Develop a block load cycle test bogey. Note that the test setup requires six to eight load blocks (steps) in a program and at least 10 repetitions of the program to test the part.

Solution. One needs to perform the damage calculation per bin, using Morrow's mean stress correction formula in which negative mean loads are treated as positive mean loads to be conservative because the polarity of the load is not associated with sign of the stress on the critical location. Results of the damage value for every bin are tabulated in Table 8.5. The next step is to sum all the damage values per row ($d_{i,\,\mathrm{row}}$) and calculate the equivalent damage cycles ($n_{i,\,R=-1}$) for each load amplitude P_a with $R = -1$. Then the cumulative exceedance cycles ($n_{i,\,R=-1,\,\mathrm{cumulative}}$) and load amplitude are calculated in Table 8.6 and plotted on a semilog scale in Figure 8.10.

The tie rod loads in one PG cycle were reduced to a cumulative exceedance histogram form as shown in Figure 8.10, in which the ordinate represents load amplitude and the abscissa the number of cycles accumulated at or greater than a specific magnitude of load amplitude. This histogram needs to extend to a complete PG schedule by translating the histogram toward the right for higher cycles with a multiplying factor of 100 and extrapolating the PG load toward lower cycles for larger field loads. This procedure is shown in Figure 8.10.

One is required to examine that the physical limit of the tie rod governs the maximum extrapolated field load. The physical limit can be derived from the large but infrequent load amplitudes from accidental or exceptional situations, which may occur at most a dozen times in the life time for the vehicle. In this example, the maximum extrapolated load for the tie rod is assumed to be limited to 8000 N, above which all the load amplitudes should be truncated. It is also needed to reduce the test time by eliminating the number of low-amplitude loads that are considered not damaging. The

TABLE 8.5 Associated Damage Values: One Cycle of Tie Rod PG Loads

		Mean (N)							
		−5000	−3000	−1000	1000	3000	5000	7000	9000
Range (N)	2000	0.000000	0.000001	0.000000	0.000001	0.000001	0.00002	0.000001	0.000001
	4000	0.000003	0.000006	0.000012	0.000023	0.000040	0.000006		
	6000			0.000064	0.000132	0.000191	0.000142		
	8000			0.000029	0.000293	0.000325			
	10,000				0.001300	0.001297			
	12,000				0.002656	0.001176			
	14,000				0.001598				

TABLE 8.6 Tabulated Cumulative Exceedance Cycles: One Cycle of Tie Rod PG Loads

P_a(N)	$d_{i,\text{row}}$	$N_{f,R=-1}$ (cycles)	$n_{i,R=-1}$ (cycles)	$n_{i,R=-1,\text{cumulative}}$ (cycles)
1000	0.000007	981841539	6610	7515
2000	0.000091	6947477	633	905
3000	0.000529	3837477	203	272
4000	0.000647	49160	32	69
5000	0.002579	9986	26	37
6000	0.003832	2715	10	11
7000	0.001598	903	1	1

magnitude of load amplitude (1200 N), 15% of the maximum load amplitude (8000 N) is the cut-off value for omission of nondamaging cycles. This reduces the test time from 10^7 cycles to 400,000 cycles (a factor of 25).

To simulate the complete PG loads in the laboratory, the cumulative exceedance histogram is reduced to a step load histogram (Figure 8.11). The step load histogram is developed based on the fact that the total damage done to the part under testing conditions would be the same as the damage in the PG schedule. In this example, the histogram is divided into six load steps and ten repetitions, as this appears to be sufficient to duplicate the PG loads.

Recently, state-of-the-art cycle extrapolation techniques (Dreßler et al., 1996; Roth, 1998; Socie, 2001; Nagode et al., 2001; Socie and Pompetzki, 2003) have been developed to account for mean load variability. An illustration of this approach is shown in Figure 8.12, which shows two rainflow matrices represented by a from–to format, instead of the one matrix in the range–mean form. The number of cycles in each bin is depicted by a different color or a larger size. The left and right figures are the cycle counting results from a short-term measurement and a long-term extrapolation, respectively. The left rainflow matrix is also equivalent to a two-dimensional probability density function that can be obtained by dividing the number of

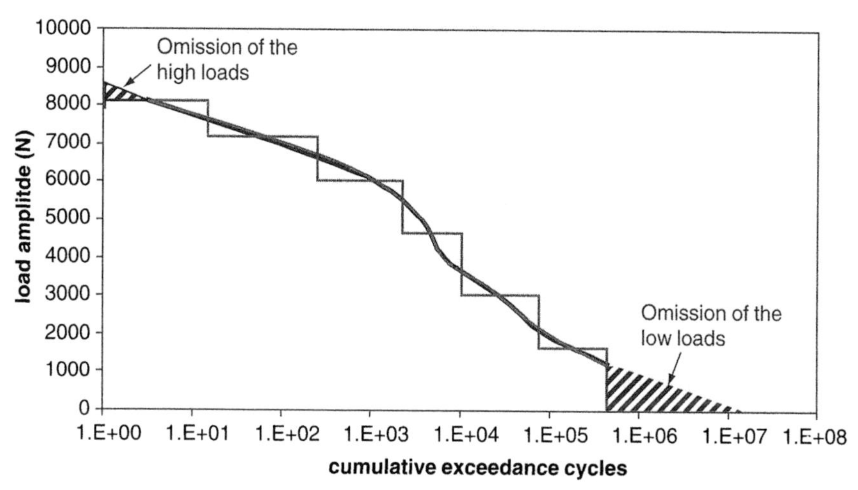

FIGURE 8.11 Development of a step load histogram.

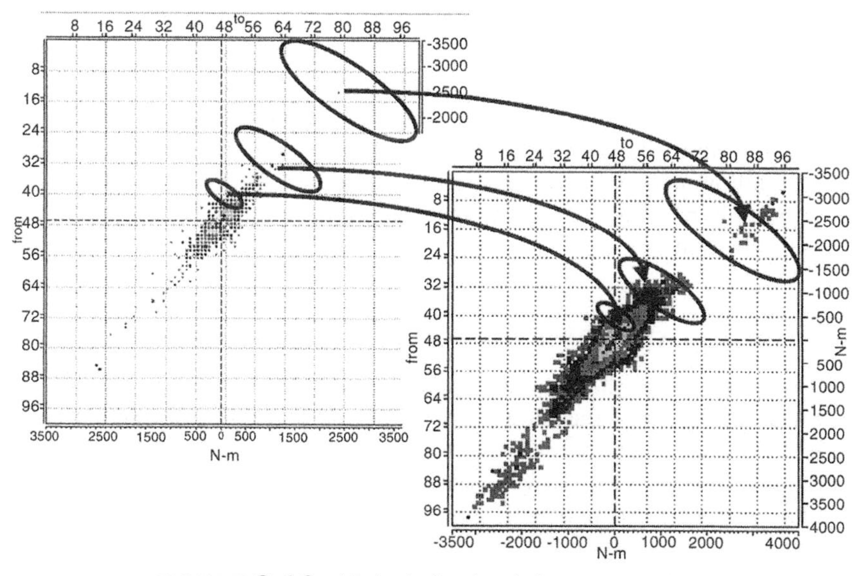

FIGURE 8.12 Method of cycle rainflow extrapolation.

cycles in each bin of the matrix by the total number of cycles. To any number of total cycles, a new rainflow matrix can be constructed by randomly inserting the cycles in each bin based on their probability of occurrence at each bin and around its neighborhood bins, described as the elliptic bandwidth in Figure 8.12.

An adaptive variable bandwidth in Figure 8.12 is selected to account for variability of the centered cycles in the boundary. The cycles at extreme loading vary much more than low load cycles. That means that considerable variability, indicted as a bigger bandwidth, is expected at extreme load cycles and little variability, as indicated by a smaller bandwidth, is used at a lower load region.

For extrapolation purposes, a nonparametric density estimator converts the discrete probability density distribution into a continuous probability density function. Kernel estimators (Dreβler et al., 1996; Roth, 1998) provide a convenient way to estimate the probability density. This method is considered as fitting an assumed probability distribution to a local area of the rainflow matrix. The size of the local area is determined by the bandwidth of the estimator. Consider the rainflow matrix to be the representative of $x - y$ data points. The probability density at any point $(f(x, y))$ can be obtained by summing the contribution from all of the kernels that lie over that point.

$$f(x, y) = \frac{1}{n}\sum_{i=1}^{n}\left[\frac{1}{(h\lambda_i)^2}K\left(\frac{x - X_i}{h\lambda_i}, \frac{y - Y_i}{h\lambda_i}\right)\right] \tag{8.3.1}$$

where n is the total number of cycles in the rainflow matrix, K is the kernel estimator corresponding to points X_i and Y_i, h and λ_i are the kernel and adaptive bandwidths.

Once the adaptive bandwidths have been determined, the overall density histogram can easily be obtained from Equation 8.3.1. Finally, the extrapolated rainflow matrix to any desired total number of cycles is constructed by randomly placing cycles in the matrix with the appropriate probability density function.

8.3.2 RAINFLOW CYCLE QUANTILE EXTRAPOLATION

The purpose of the rainflow quantile extrapolation is to predict the rainflow histogram of the single most damaging driver that would exist in a much larger set of data, based on a set of several time histories due to driver variability. In the past, it was assumed that the damage generated by a loading history correlated with the total number of cycles in that history, which implies that the most damaging rainflow matrix contains the most cycles. For example, several load-time histories (say, quantity m) have been converted into m rainflow histograms with the same number of bins and sizes. Then, based on the m measured cycles in each bin and the appropriate statistical distribution, one may construct the x-percentile damaging rainflow histogram by determining the x-percentile of the number of cycles in each bin. This approach totally excludes the dependency of the number of cycles in different bins of a rainflow matrix.

To include the dependency of the cycle entries in the bins, the quantile extrapolation techniques have been published elsewhere (Dreβler et al., 1996; Roth, 1998; Socie and Pompetzki, 2003). A brief discussion is given here for illustration. To optimize the concept of the multidimensional statistical analysis, numbers of bins in a rainflow matrix have to be divided into a series of clusters (regions) with similar damage characteristics. In general, three regions are chosen for simplicity. The damage in each bin was calculated by an assigned *S–N* curve and was later clustered into the three defined regions. Figure 8.13 shows two figures: the left one with a from–to rainflow matrix and the right one corresponding to three similar damage clusters. Each cluster is represented by the total damage values of all the bins within. As a result, for each rainflow matrix, three total damage values for the three similar clusters are presented.

In the case of eight rainflow histograms corresponding to eight different drivers, there will be an 8×3 damage matrix formed e.g., $(D_1(1), D_1(2), D_1(3))$, $(D_2(1), D_2(2), D_2(3))$, ..., $(D_8(1), D_8(2), D_8(3))$. One can calculate the average vector and a covariance matrix and finally determine the x percentile damage vector $(D_{xp}(1), D_{xp}(2), D_{xp}(3))$ based on the assumption that the measured damage matrix follows a multidimensional normal distribution. A new rainflow matrix will be calculated by superimposing all the measured rainflow histograms and by extrapolating the resulting histogram to a desired total number of cycles. The new damage vector $(D_{new}(1), D_{new}(2), D_{new}(3))$ for the newly generated matrix is determined as before. Finally, one will generate the x-percentile damaging rainflow histogram by scaling the new matrix by a factor of the x-percentile damage vector to the new damage vector in each cluster, i.e, $D_{xp}(i)/D_{new}(i)$, where $i = 1, 2$, and 3.

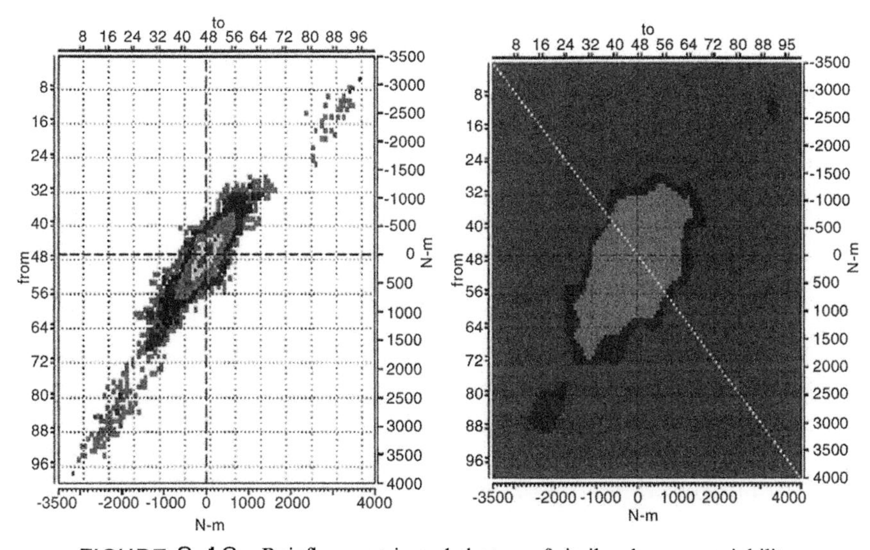

FIGURE 8.13 Rainflow matrix and clusters of similar damage variability.

8.3.3 DEVELOPMENT OF A BLOCK LOAD CYCLE TEST

A block load cycle test schedule for a driveshaft based on the 95th percentile PG driver in the PTE testing is illustrated in this section. A test vehicle was instrumented with the strain-gauged driveshaft for torque measurement. For driver variability, eight PG drivers were assigned to maneuver the test vehicle over the same test track for only one PTE cycle. It is given that a complete PTE schedule consisted of 700 test cycles.

After data acquisition, the raw data in the time domain were cleaned up by eliminating anomaly signals. To predict a complete PTE schedule with 700 cycles based on one cycle of a short-term measurement, the torque time data for the eight different drivers should first be converted to rainflow cycle counting histograms and then be extrapolated to the long-term histograms by a factor of 700. It is noted that the physical torque limit for the driveshaft (= 800 ft·lb) governed by wheel slip was assigned in the cycle extrapolation analysis. The quantile extrapolation technique was used to identify the target histogram for the 95th percentile driver profile, based on the eight extrapolated rainflow-counting histograms.

The three-dimensional target histogram was finally condensed to a two-dimensional histogram. This was accomplished by firstly converting the number of cycles counted for each offset load at any given load range to an equivalent damage cycles with a zero mean load range. Second, cumulative exceedance cycles were generated for each load amplitude level in such a manner that all the cycles for the load levels that exceed and equal to the specific load amplitude are accumulated. Figure 8.14 demonstrates the cumulative exceedance diagrams for the driveshaft response due to eight different PG drivers and the 95th percentile driver profile.

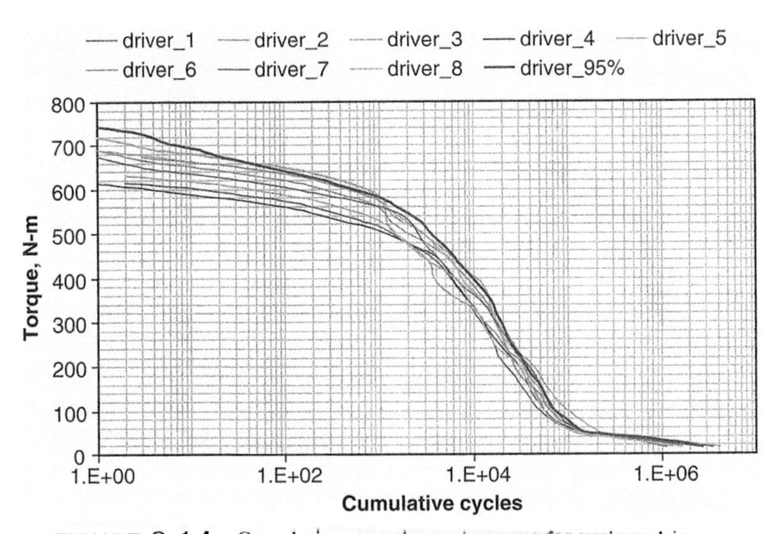

FIGURE 8.14 Cumulative exceedance diagrams for various drivers.

Development of a block cycle test bogey for the driveshaft needs the PTE cumulative exceedance histogram with the 95th percentile driver severity. As shown in Figure 8.15, six block loads were chosen to resemble the PG loading history. Given ten repetitions of the block load program, the block load levels, the corresponding cycles, and the cycles per repetition are summarized in Table 8.7. Figure 8.16 schematically shows the block cycle test program.

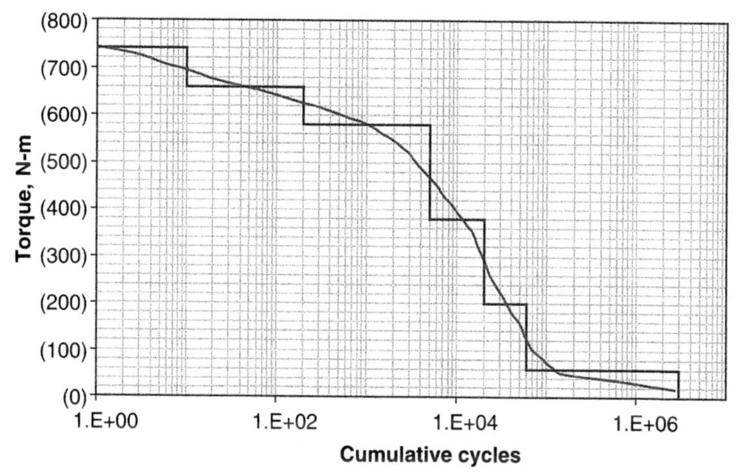

FIGURE 8.15 Development of block cycle load testing.

TABLE 8.7 Summary of Block Load Cycle Testing

Block Load Amplitude (N·m)	Cycles	Cycles Per Repetition
740	10	1
660	190	19
580	4800	480
380	15000	1500
200	40000	4000
60	3000000	300000

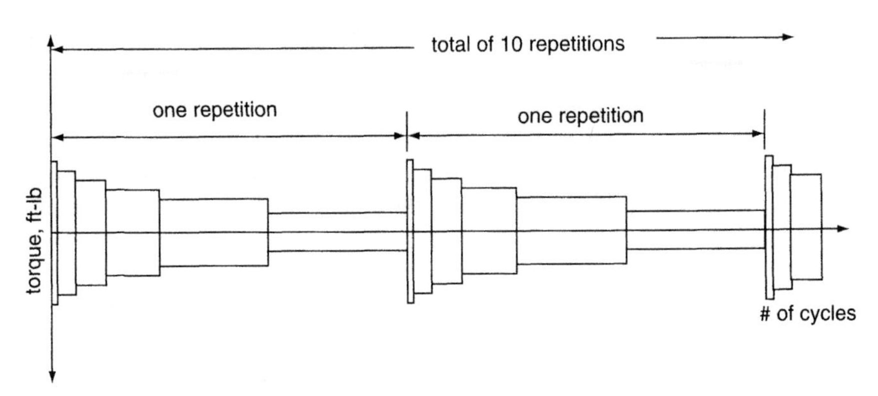

FIGURE 8.16 Block cycle load test schedule.

REFERENCES

Dreβler, K., Grunder, B., Hack, M., and Kottgen, V. B., Extrapolation of rainflow matrices, SAE No. 960569, 1996.

Grubisic, V., Determination of load spectra for design and testing, *International Journal of Vehicle Design*, Vol. 15, Nos. 1/2, 1994, pp. 8–26.

Heuler, P., and Seeger, T., A criterion for omission of variable amplitude loading histories, *International Journal of Fatigue*, Vol. 8, No. 4, 1986, pp. 225–230.

Lipson, C., and Sheh, N. J., *Statistical Design and Analysis of Engineering Experiments*, McGraw-Hill, New York, 1973.

Nagode, M., Klemenc, J., and Fajdiga, M., Parametric modeling and scatter prediction of rainflow matrices, *International Journal of Fatigue*, Vol. 23, 2001, pp. 525–532.

Roth, J. S., Statistical modeling of rainflow histograms, Report No. 182 / UILU ENG-98-4017, M.S. Thesis, University of Illinois at Urbana-Champaign, IL, 1998.

Socie, D. F., and Park, K., Analytical descriptions of service loading suitable for fatigue analysis. In *Proceedings of the Tenth International Conference on Vehicle Structural Mechanics and CAE*, SAE P308, 1997, pp. 203–206.

Socie, D. F., Modeling expected service usage from short-term loading measurements, *International Journal of Materials and Product Technology*, Vol. 16, Nos. 4/5, 2001, pp. 295–303.

Socie, D. F., and Pompetzki, M. A., Modeling variability in service loading spectra. In *Probabilistic Aspects of Life Predictions*, ASTM STP 1450, ASTM International, West Conshohocken, PA, 2003.

9

RELIABILITY DEMONSTRATION TESTING

MING-WEI LU
DAIMLERCHRYSLER

9.1 INTRODUCTION

Prevention is the strategy of employing appropriate precautionary actions during the product development cycle to eliminate nonconformance in products delivered to customers. The ultimate goal is to design and manufacture products that will meet the needs and expectations of customers. When adequate preventive actions are not taken, there may not be sufficient time to fix nonconformance found late in the testing program, and the goal of delivering a defect-free product may not be met. The primary role of reliability testing is to measure the reliability of a component or system released to the conformance to customer requirements. To ensure that customer expectations are fully addressed, product development specifications must require that products provide reliable performance over the entire customer ownership period. Verification of design conformance to customer-driven specifications is a key step within a prevention methodology development program. Reliability testing is the only practical means to verify that products meet life and functional specifications.

Assessing the reliability of products during the product development cycle is an important task in satisfying customers. Severity of usage is the prime factor affecting field reliability. The engineer must be knowledgeable of usage severity and incorporate this factor into the product design. Laboratory testing is a critical step in the development of a product. It allows the design engineer to evaluate design reliability early in the development phase. The lab test is performed under controlled conditions, often in an accelerated manner. Several methods can be used to establish a correlation between the laboratory and field. These methods are, for example, life data comparison (lab test life data to field test life data), analytical models (mathematical model based on the part's physical conditions), and duty cycle simulation (duplication of the duty cycle imposed by the field test schedule such as proving ground test for vehicle). Once the required correlation is established between the laboratory and field, the engineer can perform lab reliability demonstration testing. Establishing a meaningful lab test criterion at which reliability quantification is made is an important task for the engineer in order to provide a reliable product. This test criterion (i.e., required test time under a specific loading condition) is called a test bogey. The reliability is defined as the probability that a component/system will function at customer expectation levels at a test bogey under specified environmental and duty cycle conditions. One of the situations encountered frequently in testing involves a tradeoff between the number of samples for a test and testing time.

There are two unique test phases within a reliability development process: reliability development testing and reliability demonstration testing. Reliability development testing is performed to identify problems and evaluate the effectiveness of corrective redesign actions. The primary purpose is to improve reliability performance rather than measure the reliability. The most important actions are to log all failures generated during testing and to ensure that corrective actions are taken. This verifies through testing that failure-prone areas of the design are fixed. Reliability quantification during this development phase is intent on measuring improvement. The purpose of a reliability demonstration test is to determine whether a product meets or exceeds the established minimum reliability requirement.

No two parts produced are exactly the same. Given a small sample out of a population, how may the information derived from the sample be used to obtain valid conclusions about the population? This kind of problem can be treated with the aid of statistical tools. In engineering tests, it is generally assumed that the test sample was chosen at random. The objective of statistics is to make inferences about a population based on information contained in a sample. Two types of estimations are in common use: (1) point estimation and (2) interval estimation. The first type of estimate is called a point estimate because a single number represents the estimate. The second type is called an interval estimate by using the data in the sample to calculate two points that are intended to enclose the true value of the parameter estimated.

It should be recognized that when population parameters are estimated on the basis of finite samples, errors of estimation are unavoidable. The significance of such errors is not reflected in the point estimation. A point estimate of a parameter is not very meaningful without some measure of the possible error in the estimate. The point estimation is a single value. This estimated value rarely coincides with the true value of the parameter being estimated. An interval is needed that is expected to cover the true value of the parameter with some specified odds with a prescribed confidence level. This interval is called the confidence interval. Thus, a 90% confidence interval for a given parameter implies that in the long sequence of replications of an experiment, the computed limits of the interval will include the true value of the parameter about 90 times in 100. This fraction 90% is called the confidence level. The confidence interval can be one sided or two sided. It is true that higher the degree of confidence, longer the resulting interval. In reliability life testing with a demonstration target, for example, *R95C90* requirement, it means that the 90% confidence lower limit of a product's reliability estimate at a measurement point should be greater than 95%. The reliability and confidence are related when one attempts to project reliability determined from a sample to the population.

Appropriate statistical tools can assist engineers in dealing with variability in product characteristics, such as life and performance, and thus obtain a valid conclusion about populations. Insufficient testing increases the risk of introducing an unsatisfactory product whereas excessive testing is costly, time consuming, and a misuse of engineering resources. This chapter presents approaches for (1) determining the quantity of test samples required and the test length that must be tested to meet the reliability program objectives (or reliability demonstration target requirement), and (2) reliability predictions.

9.2 BINOMIAL TEST METHOD

The binomial test method is a "success/failure," "go/no go," or "acceptable/not acceptable" type of analysis. A system or component is submitted to a minimum test or performance bogey. If a test sample makes it to the bogey, it is a success; if it does not, it is a failure. Such tests are typical in verifying minimum reliability levels for new products prior to production release. The following are some limitations of using the binomial method: (1) requires multiple test samples, (2) no test failures are allowed, and (3) failure modes and variability are not disclosed.

For "success/failure" type situations, the binomial distribution applies. The relationship between reliability, confidence level, sample size, and number of failures is given by the following equation:

$$\sum_{x=0}^{r} \binom{n}{x} R^{n-x}(1-R)^{x} \leq 1 - C \tag{9.2.1}$$

where n is the sample size, r the number of failures in the sample size, R the reliability (probability of success), and C the confidence level.

For special cases in which no failures occur ($r = 0$), Equation 9.2.1 becomes:

$$R^{n} \leq 1 - C \tag{9.2.2}$$

The lower limit of R with C confidence level can also be given by (Kapur and Lamberson, 1977):

$$\frac{y}{[y + (n - y + 1)\, F_{1-C,\, 2(n-y+1),\, 2y}]} \tag{9.2.3}$$

where y is the number of successes in the sample of size n and $F_{1-C,\, 2(n-y+1),\, 2y}$ is a value with the upper tail area equal to $(1 - C)$ from the F distribution with $2(n - y + 1)$ and $2y$ degrees of freedom.

Example 9.1. What is the minimum number of samples required with no failures in order to verify a reliability of 90% at a 90% confidence level?

Solution. The use of Equation 9.2.2 with $R = 0.9$ and $C = 0.9$ gives $n = 22$. Thus, 22 samples tested without failures will demonstrate 90% reliability at a 90% confidence level.

Example 9.2. A company purchased 14 new machines and found that 2 of the 14 failed to operate satisfactorily. What is the reliability of the new machine at 90% confidence?

Solution. We can use either Equation 9.2.1 or Equation 9.2.3 to calculate the reliability. Here, by Equation 9.2.3 with $n = 14$, $r = 2$, $y = 12$, $C = 0.9$, and $F_{1-C,\, 2(n-y+1),\, 2y} = F_{0.1,\, 6.24} = 2.04$ (Appendix 9.3), the lower limit of R is 66.2%. Hence, the reliability is 66.2% at 90% confidence.

9.3 BINOMIAL TEST METHOD FOR A FINITE POPULATION

In many situations, we want to accept or reject a population with a small or finite population size. In this section, we describe the binomial test method (Section 9.2) for the reliability demonstration test based on the samples from a finite population.

A method for determining the sample size from a given finite population size is described next. The following notations are used in this method.

N = Population size
X = Number of defectives in a population
n = Sample size selected from a population of size N
k = Number of defectives from a sample of size n
E = Event with $\{k$ defectives out of n samples$\}$
$q = X/N$ percent of defective items in a population

The "90% reliability with 90% confidence" type requirement will be used to accept or reject a population. Testing is performed until 90% confidence and 90% reliability is achieved. For use of this type of requirement, it is convenient to replace X (the number of defectives) by the variable $q = X/N$, the fraction of defective items in the population.

For products that are highly critical, it may be useful to describe a criterion of at least, say, 90% confidence that there are zero defectives in the population. Such criteria can only possibly be met if the sample found zero defectives ($k = 0$). A stringent criterion of this "zero defective" type would require almost a 100% sample.

Let K be the number of defectives in a sample of size n. Hence (by hypergeometric distribution), the probability of $K = k$ is

$$P(K = k) = \frac{\binom{X}{k}\binom{N-X}{n-k}}{\binom{N}{n}} \qquad k = a, a+1, \ldots, b-1, b \qquad (9.3.1)$$

where $a = \max\{0, n - N + X\}$ and $b = \min\{X, n\}$

The relationship between population size (N), sample size (n), number of defectives (k) in the sample of size n, confidence level (C), and percent defective (X_o/N) is given by (Lu and Rudy, 2001)

$$\frac{\sum_{i=0}^{k}\binom{X_0}{i}\binom{N-X_0}{n-i}}{\binom{N}{n}} \leq 1 - C \qquad (9.3.2)$$

where X_o is the highest X value satisfying Equation 9.3.2.

For the special case in which no defectives occurred ($k = 0$), Equation 9.3.2 becomes

$$\frac{\binom{N-X_0}{n}}{\binom{N}{n}} \leq 1 - C \qquad (9.3.3)$$

As N is high, let $q = X/N$ be the percent of defective items in the population. Then, from Equation 9.3.1,

$$P(K = k) = \binom{n}{k} q^k (1-q)^{n-k} \qquad (9.3.4)$$

That is, when the sampling fraction n/N is small enough, binomial probabilities can be used as an approximation to the hypergeometric probabilities in Equation 9.3.1. Hence, one is back to the usual binomial distribution. The relationship between sample size (n), number of defectives (k) in the sample of size n, confidence level (C), and percent defective (q) is given by

$$\sum_{i=0}^{k}\binom{n}{i} q^i (1-q)^{n-i} \leq 1 - C \qquad (9.3.5)$$

TABLE 9.1 Minimum Sample Size Requirements with No Failures by Lot Sizes and Quality Targets

90% Reliability 90% Confidence		95% Reliability 90% Confidence		90% Reliability 95% Confidence		95% Reliability 95% Confidence	
N	n	N	n	N	n	N	n
20	14	20	18	20	16	20	19
40	17	40	27	40	21	40	31
60	19	60	32	60	23	60	38
80	20	80	35	80	24	80	42
100	21	100	37	100	25	100	45
230	22	120	38	150	26	150	48
		140	39	270	27	200	51
		160	40	290	28	250	52
		200	41	300	29	300	54
		260	42			400	55
		340	43			500	56
		460	45			600	59

For the special case in which $k = 0$,

$$(1 - q)^n \leq 1 - C \tag{9.3.6}$$

From Equation 9.3.3, the minimum sample size (n) required with no failures by different lot size (N) and quality targets (reliability 90%, 95% with confidence 90%, 95%) are given in Table 9.1.

Example 9.3. (1) For a population size $N = 40$ with a 95% reliability with 90% confidence target requirement, what is the minimum sample size that without failures will demonstrate 95% reliability with 90% confidence target? (2) For a lot size of 284, with the number of defectives $k = 1$ in a sample of size of $n = 30$, what is the percent defective of the lot with $C = 90\%$ confidence?

Solution. (1) By Equation 9.3.3, $n = 27$; that is, $n = 27$ samples tested without failures will demonstrate 95% reliability with 90% confidence target for a population size of 40.
(2) By Equation 9.3.2, the lot percent defective is 12%.

9.4 WEIBULL ANALYSIS METHOD (TEST TO FAILURE)

The Weibull distribution is a useful theoretical model of life data that has been successfully applied in engineering practices. It is a general distribution that can be made (by adjustment of the distribution parameters) to model a wide range of different life distributions. For component testing, it is recommended to run the test to failure. This is advantageous because the component's life distribution and failure modes can be obtained.

The cumulative distribution function for the three-parameter Weibull of the random variable T is given by (Kapur and Lamberson, 1977)

$$F(t) = 1 - \exp\{-[(t-\delta)/(\theta-\delta)]^{\beta}\},\ t > \delta \tag{9.4.1}$$

where θ is the characteristic life (scale parameter), β the Weibull slope (shape parameter), and δ the minimum life (location parameter). The reliability at t is defined as

$$R(t) = 1 - F(t) = \exp\{-[(t-\delta)/(\theta-\delta)]^{\beta}\} \tag{9.4.2}$$

The mean life of the Weibull distribution is

$$\mu = \delta + (\theta - \delta)\Gamma(1 + 1/\beta) \tag{9.4.3}$$

where $\Gamma(x)$ is the gamma function. B_q life is the life by which q% of the items or less is expected to fail. For example, for a B_{10} life of 1000 cycles, the percent failure must be 10% or less at 1000 cycles. This means that at 1000 cycles, the reliability is at least 90%. The B_q life is given by

$$B_q = \delta + (\theta - \delta)\ \{\ln[(100/(100-q)]\}^{1/\beta} \tag{9.4.4}$$

The two-parameter Weibull has a minimum life of zero ($\delta = 0$). The density function $f(t)$ and the distribution function $F(t)$ are given by

$$f(t) = \beta/\theta(t/\theta)^{\beta-1}\exp[-(t/\theta)^{\beta}],\ t > 0 \tag{9.4.5}$$

$$F(t) = 1 - \exp[-(t/\theta)^{\beta}],\ t > 0 \tag{9.4.6}$$

$$B_q = \theta\{\ln[100/(100-q)]\}^{1/\beta} \tag{9.4.7}$$

The rate at which failures occur in a certain time interval is called the failure rate during that interval. The hazard rate function $h(t)$ is defined as the limit of the failure rate as the interval approaches zero. Thus the hazard rate function is the instantaneous failure rate. It is defined as $h(t) = f(t)/R(t)$.

$$h(t) = f(t)/R(t) = \beta t^{\beta-1}/\theta^{\beta} \tag{9.4.8}$$

When $\beta = 1$, h(i) is a constant. When $\beta > 1$ $h(t)$ is strictly increasing. When $\beta < 1$ $h(t)$ is strictly decreasing. The mean life is

$$\mu = \theta\Gamma(1 + 1/\beta) \tag{9.4.9}$$

and the variance is

$$\sigma^2 = \theta^2[\Gamma(1 + 2/\beta) - \Gamma^2(1 + 1/\beta)] \tag{9.4.10}$$

Two basic parameter estimation methods (rank regression method and maximum likelihood estimator, MLE, method) are used in most commercially available software packages. The rank regression method is a graphical plotting technique by using linear regression along with median rank plotting (Kapur and Lamberson, 1977). The MLE method is a statistical procedure for parameter estimation by maximizing the likelihood function (Mann et al., 1974). The MLE method is recommended because the maximum likelihood estimates often have desirable statistical properties.

A graphical plot provides a representation of the data that is easily understood by the engineer. It is extremely useful in detecting outliers, in estimating the minimum life (location parameter), and in helping with the decision as to whether observed failure times are from a Weibull distribution.

Example 9.4. Nine lower control arms were tested to failure on a jounce to rebound test fixture. The cycles to failure were 35,000; 38,000; 46,500; 48,000; 50,000; 57,000; 58,000; 68,000; and 71,000.

1. Determine the Weibull slope, the characteristic life, and the B_{10} life of the lower control arm.
2. Does the lower control arm meet the 90% reliability at 30,000 cycles with 90% confidence reliability requirement?

Solution. 1. The Weibull plot is shown in Figure 9.1 by using the MLE method and the likelihood ratio method for 90% confidence lower bound (WINSmith, 2002). From Figure 9.1, we have

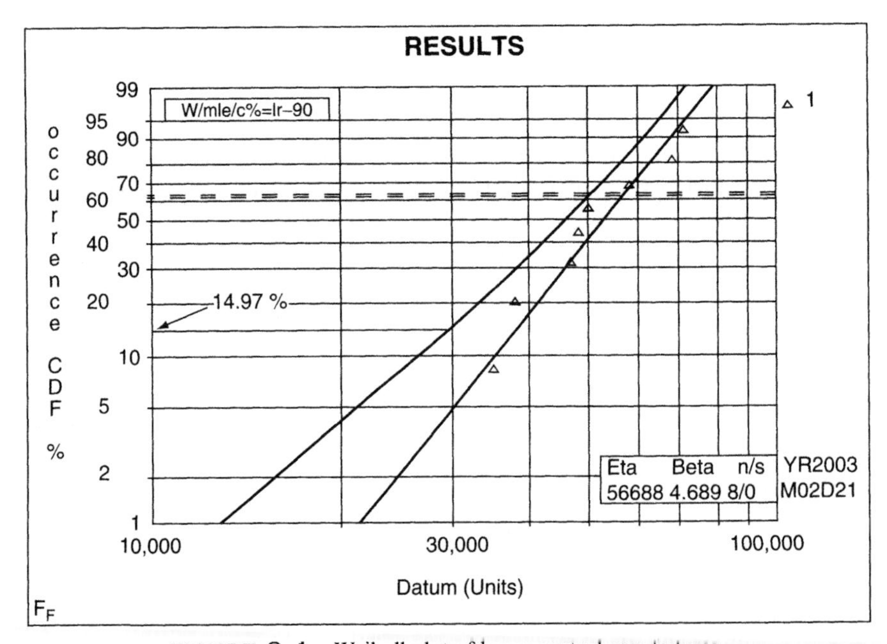

FIGURE 9.1 Weibull plot of lower control arm failures.

Weibull slope = 4.689
Characteristic life = 56,688 cycles
B_{10} life = 26,328 cycles

2. Using the 90% lower bound line, the percent failure is 14.97% (or reliability = 85.03%) at 30,000 cycles. Therefore, the lower control arm does not meet the 90% reliability at 30,000 cycles with 90% confidence reliability requirement.

9.5 EXTENDED TEST METHOD

One of the situations encountered frequently in testing involves a tradeoff between sample size and testing time. If the test item is expensive, the number of test items can be reduced by extending the time of testing on fewer items. Extended testing is a method to reduce sample size by testing the samples to a time that is higher than the test bogey requirement. Two limitations of using the extended test method are: (1) requires knowledge of Weibull slope and (2) extended test time and no failures are allowed. In this section, a relationship among sample size, test time, confidence level, and reliability is developed.

The approach is based on the Weibull distribution with parameters β and θ and Equation 9.2.2. It is assumed that there are no failures in the samples while running the test for a time and the Weibull slope is also given. The assumed Weibull slope is based on prior knowledge from similar component testing data or engineering experience. The following notations are used in this method:

t_1 = Extended testing time
n_1 = Number of items running without failure to t_1
t_2 = Test bogey
n_2 = Number of items running without failure to t_2
R = Reliability
C = Confidence level
β = Weibull slope

The reliability at t_2 with a confidence C is

$$R_2 = (1 - C)^{1/n_2} = \exp[-(t_2/\theta)^{\beta}] \tag{9.5.1}$$

Similarly, the reliability at t_1 with a confidence C is

$$R_1 = (1 - C)^{1/n_1} = \exp[-(t_1/\theta)^{\beta}] \tag{9.5.2}$$

From Equations 9.5.1 and 9.5.2,

$$\frac{n_2}{n_1} = \left(\frac{t_1}{t_2}\right)^\beta \tag{9.5.3}$$

Also, from Equation 9.2.2,

$$R^{n_2} = 1 - C \tag{9.5.4}$$

where t_1/t_2 is the ratio of test length to test bogey.

Combining Equations 9.5.3 and 9.5.4, the relationship between sample size (n_1), extended testing time (t_1), confidence level (C), and reliability (R) can be found in Equation 9.5.5, assuming a Weibull slope (β) and a given test bogey (t_2).

$$R^{(n_1)(t_1/t_2)\beta} = 1 - C \tag{9.5.5}$$

Example 9.5. A bearing engineer has developed a new bearing. The new bearing must have a B_{10} life of 1000 h. The company has set a requirement of a 90% confidence level for all decisions. Only 10 bearings are available for testing. How long must 10 bearings run without failure to meet the $R90C90$ reliability requirement (90% reliability at 1000 hours with 90% confidence)? The Weibull slope is assumed to be 1.5.

Solution. By using Equation 9.5.5 with $n_1 = 10$, $t_2 = 1000$, $b = 1.5$, $R = 0.9$, and $C = 0.9$,

$$t_1 = \{\frac{\ln(1-C)}{(n_1)\ln(R)}\}^{1/b} = 1000\{\frac{\ln(1-0.9)}{(10)\ln(0.9)}\}^{1/1.5} = 1684\,\text{h}$$

Hence, 10 bearings must run 1684 h to meet the $R90C90$ at a 1000-h requirement.

9.6 WEIBULL ANALYSIS OF RELIABILITY DATA WITH FEW OR NO FAILURES

In fitting a Weibull distribution to reliability data, one might have few or no failures. However, there must be at least two failures to calculate estimates for Weibull parameters. This section presents more accurate estimates of reliability and confidence limits that apply to few or no failures with an assumed shape parameter value of β. The reliability function at time t from a two-parameter Weibull distribution is given by

$$R(t) = \exp[-(t/\theta)^\beta],\ t > 0 \tag{9.6.1}$$

For a given sample of size n, suppose that the $r \geq 0$ failure times and the $(n-r)$ times of nonfailures are $t_1, t_2, \ldots, t_n$. Assuming that β is given, the corresponding lower C% confidence limit for the true θ is given by (Nelson, 1985):

$$\theta_C = \{2T/\chi^2_{(C;\,2r+2)}\}^{1/\beta} \tag{9.6.2}$$

where $T = t_1^\beta + t_2^\beta + \ldots + t_n^\beta$, and $\chi^2_{(C;\,2r+2)}$ is the Cth percentile of the chi-square distribution with $(2r + 2)$ degrees of freedom. Hence, the reliability at t cycles with $C90$ is given by

$$R(t) = 1 - \exp[-(t/\theta_C)^\beta] \tag{9.6.3}$$

Example 9.6. For vehicle liftgate life testing, a total of 12 tests were conducted with no failures (six suspended at 15,000 cycles and six suspended at 30,000 cycles). The test bogey is 15,000 cycles. What is the reliability value at 15,000 cycles with $C90$?

Solution. Assuming that $\beta = 1.5$, from Equation 9.6.2, the corresponding lower 90% confidence limit for the true θ is given by

$$\theta_C = 69,514 \text{ cycles}$$

where $n = 12$, $r = 0$, $C = 90$, $T = t_1^\beta + t_2^\beta + \cdots + t_{12}^\beta = 42,199,618$, and $\chi^2_{(C;2r+2)} = \chi^2_{(90;\,2)} = 4.605$ (Appendix 9.2).

Hence, the reliability at 15,000 cycles with $C90$ is given by

$$R(15,000) = \exp[-(15,000/69,514)^{1.5}] = 90.5\%$$

9.7 BIAS SAMPLING APPROACH

The bias sampling approach offers a means to achieve design verification from a reduced number of samples. The reduction is achieved by appropriate selection of test specimens prior to the actual test. The samples selected for testing must be from that portion of the population that can confidently be expected to yield results in the lower 50th percentile. In effect, it is worse-case testing, by using a subpopulation of samples bias to the worst case rather than a single worst-case sample. For related information, see Allmen and Lu (1994).

Analysis of the test results assumes data will be distributed as if a random sample were tested, except that all data points will fall below the midpoint. With this assumption, equally confident predictions can be achieved by using around half the sample size of a random sample. The approach is most applicable when applied to components containing a single predominant variable effecting test results, e.g., BHN versus fatigue life. When two predominate variables exist, the sample savings are somewhat reduced by a need to add some conservatism to the analysis and prediction. The following process illustrates the methodology employed for application of the bias sampling approach:

1. *Identify design factors affecting test results.* Application of the bias sampling approach necessitates the knowledge of the various design factors and how variability in these factors affect test results. Such knowledge is gained through prior observations, experiments, and the engineer's intuition.
2. *Determine relationship between design factors and results.* Although variability exists in all design factors, certain factors have a dominant effect on test results whereas others have a negligible effect. Application of the bias sampling approach requires identification of the predominant factors and the establishment of their effect on test results.
3. *Devise the appropriate means for selection of the test samples.* Many components will contain only a single predominant factor (Factor A) affecting test results. When this is the situation, test specimens from a random sample lot should be selected such that all Factor A characteristics lie within the minimal to nominal range. When possible, this scheme should be employed by isolating a random sample and then sorting out the desired samples from the total samples. Often, it is not possible to sort from a large sample when sample availability is restricted. In this case, samples should be procured with the critical Factor A specified within the desired range, i.e., minimal to nominal. The objective then is to obtain a sample with variability in the lower specified range. When two predominant factors exist, additional attention must be given to sample selection or specification. In addition, there should be an understanding of the interaction between the two factors and treated appropriately in the sampling process.
4. *Conduct tests and analyze test results by the Weibull analysis method.* The Weibull distribution is one of the most versatile and widely used distributions within the fields of strength testing and life testing. The Weibull distribution is represented by a family of curves rather than one curve. Here, test to failure is employed, and the Weibull analysis method is used to analyze the data and predict population performance. For any given n test to failure results, the Weibull analysis is run on a total of $2n$ samples. The additional n samples are handled as suspended items with test results greater than any of the actual tested parts. The reliability- and confidence-bound calculation are also based on $2n$ observations.

Example 9.7. The reliability verification test plan for a new stamped suspension arm requires demonstration of $R90C90$ at 50,000 cycles on a vertical jounce to rebound loading text fixture (for testing in the up and down directions). Four suspension arms were selected and tested to failure on a jounce to rebound test fixture. The four selected samples were confidently worse than average samples from a population, e.g., minimum material properties. The cycles to failure data were 75,000, 95,000, 110,000 and 125,000. Does the part conform to the reliability demonstration target requirement?

Solution. Because of the biased samples, one can run the Weibull analysis, assuming eight samples were tested. Thus, the test data are treated as follows—four failures and four suspended items:

Cycles to Failure	Status
75,000	Failure
95,000	Failure
110,000	Failure
125,000	Failure
125,000	Suspension
125,000	Suspension
125,000	Suspension
125,000	Suspension

Figure 9.2 shows the graphical plot of the results of the Weibull paper with 90% confidence lower bound (WINSmith, 2002). For the typical data illustrated, the reliability demonstrated at 50,000 cycles with 90% confidence is 93.8%, indicating conformance to the objective.

Attribute analysis can be used if all the lower region samples pass the test criteria. When this is the case, it can be assumed that almost twice the actual number of samples tested met the criteria. The test sample size with no failure at a selected reliability level (R) with 90% confidence is shown in Table 9.2.

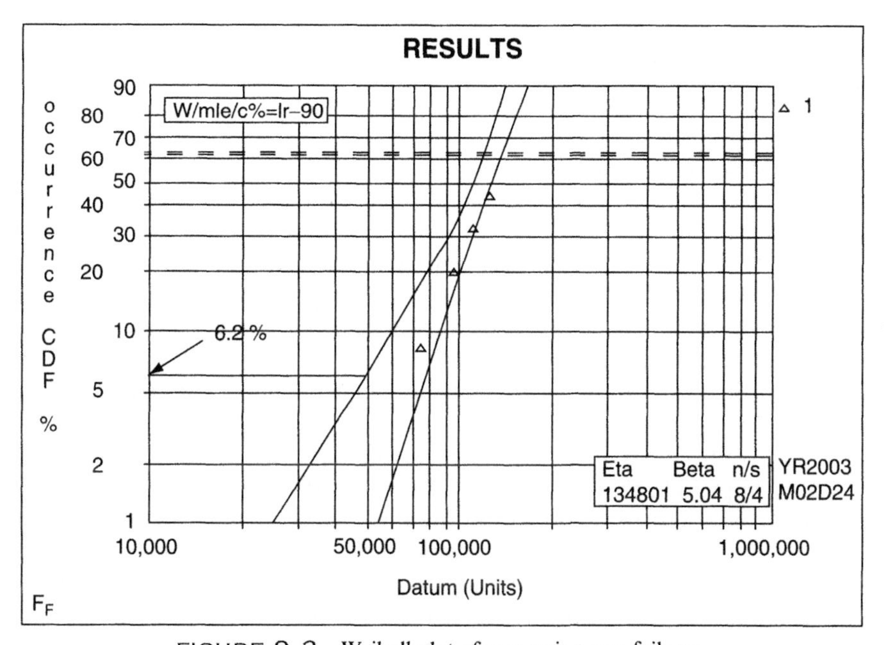

FIGURE 9.2 Weibull plot of suspension arm failures.

TABLE 9.2 Test Sample Size Determination (90% Confidence)

	Reliability (%)			
	80	90	95	99
Bias sampling	5	11	22	114
Binomial (Section 9.2)	11	22	45	230

The appropriate quantification of the results is presented in Section 9.2 with a modified reliability level R', where $R' = 1 - (2)(1 - R)$ for a given R.

Example 9.8. Periodic production samples of engine connecting rods are tested to confirm conformance to reliability specifications. The specification requires a minimum reliability demonstration of R90C90 at the established bogey life. The experience has proved that material hardness is the predominant variable affecting life. A total of 22 parts are selected, and all are then checked for hardness in a noncritical area. The 11 parts with the lowest hardness are selected to provide the bias sample. If all samples successfully reach the bogey (no failures), do the parts meet the $R90C90$ specification?

Solution. From Table 9.2, which shows 11 parts with no failures, the parts are deemed to be in conformance to the $R90C90$ specification.

9.8 BAYESIAN APPROACH

In product development and testing of automotive systems or components, the designer, drawing on a wealth of past experience, strives to meet or exceed given reliability goals. However, the traditional approach to statistical inference does not account for past experience. Thus, a large amount of hard data obtained by testing is necessary to demonstrate a reliability level with a high degree of confidence. In practice, the number of systems or components actually life tested is always very small. One answer to this dilemma is provided by Bayesian statistics, which combines subjective judgment or experience with hard data to provide estimates similar to those obtained from the traditional statistical inference approach. To use the Bayesian approach, the degree of belief in the performance of a new system or component must be quantified before the reliability demonstration test.

Assume that the reliability at t_0 is at least R_o before the reliability demonstration test. If there is no strong information on our prior distribution on system unreliability q, it can be assumed that the prior distribution on system unreliability is uniformly distributed between 0 and $B = (1 - R_o)$.

The density function of q is

$$f(q) = 1/B,\ 0 \le q \le B \tag{9.8.1}$$

Let X be the number of successes in n tests. The probability function is

$$f(x|q) = P[X = x|q] = \binom{n}{x}(1-q)^x q^{n-x}, x = 0, 1, \ldots, n \tag{9.8.2}$$

The posterior distribution of q given x is

$$h(q|x) = f(q)f(x|q)/f(x)$$

$$= (1/B)\left[\binom{n}{x}(1-q)^x q^{n-x}\right] / \left(\int_0^B (1/B)\binom{n}{x}(1-q)^x q^{n-x} dq\right) \tag{9.8.3}$$

When $x = n$ (no failures), then

$$\begin{aligned} h(q|x) &= (1-q)^n / \left(\int_0^B (1-q)^n dq\right) \\ &= (n+1)(1-q)^n/(1-R_o^{n+1}) \end{aligned} \tag{9.8.4}$$

Given n tests without failure. Let R' be such that

$$\int_0^{1-R'} h(q|x)dq = C \tag{9.8.5}$$

From Equation 9.8.4,

$$\int_0^{1-R'} (n+1)(1-q)^n/(1-R_o^{n+1})dq = C \tag{9.8.6}$$

That is,

$$(1-R'^{n+1})/(1-R_o^{n+1}) = C \tag{9.8.7}$$

Hence,

$$R' = [1 - C(1-R_o^{n+1})]^{1/(n+1)} \tag{9.8.8}$$

R' is the reliability at t_0 with C confidence.

Example 9.9. Given 10 tests without failure, what is the reliability value R' at $t_0 = 10,000$ cycles with 90% confidence assuming the reliability at t_0 is at least 70% before the reliability demonstration test?

Solution. From Equation 9.8.8 with $n = 10$, $R_o = 70\%$, and $C = 90\%$, the reliability is 82.3% with 90% confidence.

TABLE 9.3 R' Values with 90% Confidence by $n = 5, 10, 15, 20, 22$ and $R = 70\%, 80\%, 85\%, 90\%$

	R_o (%)			
n	70	80	85	90
5	76.8	83.4	87.2	91.3
10	82.3	85.4	88.2	91.6
15	86.8	87.8	89.4	92.1
20	89.6	90.0	90.7	92.6
22	90.5	90.7	91.2	92.8

Example 9.10. What is the minimum number of tests required without failures for a reliability of 90% at 10,000 cycles with a 90% confidence reliability demonstration requirement assuming the reliability at t_0 is at least 70% before the reliability demonstration test?

Solution. From Equation 9.8.8 with $R_o = 70\%$, $R' = 90\%$, and $C = 90\%$, the minimum number of tests required without failures is 21.

Table 9.3 shows the R' values with 90% confidence by different n and R values.

9.9 STEP-STRESS ACCELERATED TEST METHOD

Most products are designed to operate for a long period of time, and in such cases life testing can be a relatively lengthy procedure. Lengthy tests tend to be expensive, and the results become available too late to be of much use. An accelerated test can be used if the physical law leading to failure is known. For example, in fatigue testing, the life over a range of stresses is known to follow the relationship (Equation 4.1.7)

$$S_a = S_f' N^b \tag{9.9.1}$$

where N is life (number of fatigue cycles), S_a is the stress amplitude, and S_f' and b are material constants.

For a newly developed product, there usually are not many test units available for life testing. It may not be practical to use constant stress accelerated testing. This roadblock must be overcome; the step-stress test is discussed here. An example of a rear suspension aft lateral link is used to validate the SSAT method.

The step-stress test is a test where a product is subjected to successively higher levels of stress. A product is tested initially at the normal stress level for a specified length of time. If it does not fail, it is subjected to a higher stress level for a specified time. The stress on a product is thus increased step by step until it fails. The stress level cannot be raised too high and change the material response from linear to nonlinear or change the failure mode. Usually, all products go through the same specified pattern of stress levels and test times. The parameters can be temperature, voltage, amplitude of vibration, etc. Some related topics are discussed in Lu and Rudy (2003) and Nelson (1990). The cumulative exposure model (Nelson, 1990) is also discussed. This model can predict a product's life quickly. Thus, the cost of conducting a life test can be reduced significantly.

9.9.1 STEP-STRESS ACCELERATED TEST (SSAT) MODEL

In the SSAT model, instead of a time axis, there is a stress–time axis for the probability plot of failure data. As in Section 9.4, the density function $f(t)$ and the cumulative distribution function $F(t)$ of the random variable T with a two-parameter Weibull distribution are given by

$$f(\mathrm{t}) = \beta/\theta(t/\theta)^{\beta-1}\exp[-(t/\theta)^{\beta}], t > 0 \tag{9.9.2}$$

$$F(t) = 1 - \exp[-(t/\theta)^{\beta}],\ t > 0 \tag{9.9.3}$$

where θ is the characteristic life (scale parameter) and β the Weibull slope (shape parameter).

The mathematical formulation is described as follows.

Assumptions:

(1) For any constant stress S, the life distribution is Weibull.
(2) The Weibull shape parameter β is constant.
(3) The Weibull scale parameter $\theta = (\alpha/S_i)^{\gamma}$, where α and γ are positive parameters.

Let S_1 be a normal test stress level for the product. Let $S_1 < S_2 < \cdots < S_m$ denote m levels of stress.

The testing stress S of a SSAT experiment can be defined as

$$\begin{aligned} S &= S_1, \quad 0 \le t < t_1 \\ &= S_2, \quad t_1 \le t < t_2 \\ &\vdots \\ &= \mathrm{S_m},\ t_{m-1} \le t < t_m \end{aligned}$$

The cumulative distribution at a constant stress S_i is defined as

$$F_i(t) = 1 - \exp\{-(t/\theta_i)^{\beta}\},\ i = 1, 2, \ldots, m \tag{9.9.4}$$

where

$$\theta_i = (\alpha/S_i)^{\gamma}$$

Here, α, β, and γ are three positive parameters. Let

$$t_0 = 0 \text{ and } \Delta_i = t_i - t_{i-1} \text{ for } i = 1, 2, \ldots, m$$

The cumulative distribution $F_0(t)$ of an SSAT can be shown as (Nelson, 1990)

$$F_0(t) = 1 - \exp\{-[\Delta(t)]^{\beta}\}, t_{i-1} \leq t < t_i, i = 1, 2, \ldots, m \tag{9.9.5}$$

where $\Delta(t) = \Delta_1/\theta_1 + \Delta_2/\theta_2 + \cdots + (t - t_{i-1})/\theta_i$

9.9.2 ESTIMATIONS OF α, β, AND γ

Given a sample of testing times (n failure and n' suspended) from a population having $F_0(t)$, let $M_i(i = 1, \ldots, n)$ be the corresponding median ranks of n failure times. The least squares estimators (α', β', γ') can be obtained by minimizing Ω,

$$\Omega = \sum (F_0(t) - M_i)^2 \tag{9.9.6}$$

where the summation $\sum$ is over t at n failure times.

Based on the observed failure data, α, β, and γ can be estimated. Hence, the life distribution at a normal test stress can be determined by using

$$\beta = \beta' \text{ and } \theta = (\alpha'/S_1)^{\gamma'}$$

For 90% confidence band, similar to that seen previously, we can use 90% ranks (instead of median rank) in Equation 9.9.6 to estimate α, β, and γ.

Median rank is the method most frequently used in probability plotting, particularly if the data are known not to be normally distributed. When there are suspended data in the test data, median ranks can also be obtained. More information on these ranks can be found in Kapur and Lamberson (1977).

Example 9.11. A rear suspension aft lateral link was selected for SSAT model validation (Lu *et al.* 2003).

Let $S_1 < S_2 < \cdots < S_8$ denote eight levels of stress. $S_1 = 2000$ lb is the normal test stress level for rear suspension aft lateral link and the test bogey is 9184 cycles.

Because one would like to cover at least half of one bogey (9184 cycles) at stress S_1 and have at least a full bogey cycle by the end of stress S_2, the block cycle at each stress S_i is defined as follows: the block cycle at stress S_1 is two thirds of one bogey, and the subsequent block cycles are one half of the previous block cycles. The 400-lb test load step increase is selected for our SSAT experiment.

The testing stress (S) of a SSAT experiment is defined as follows.

		Block Cycles
$S = 2000$	$0 \leq t < 6123$	6123
$= 2400$	$6123 \leq t < 9184$	3061
$= 2800$	$9184 \leq t < 10715$	1531
$= 3200$	$0715 \leq t < 11480$	765
$= 3600$	$11480 \leq t < 11863$	383
$= 4000$	$11863 \leq t < 12054$	191
$= 4400$	$12054 \leq t < 12150$	96
$= 4800$	$12150 \leq t < 12198$	48

The 12 cumulative failure stress cycles along with median ranks (Appendix 9.4) are shown in Table 9.4. What are the estimations of α, β, and γ and the Weibull life distribution at a test load $S_1 = 2000$?

Solution As $\theta = (\alpha/S)^{\gamma}$ and the median life $= \theta[\ln(2)]^{1/\beta}$, from $S_a = S_f' N^b$, we have $\gamma = -1/b$. The calculated b value from the S–N curve is -0.149. The true b value lies most likely within $b \pm 30\%$ of b; that is, b ranges from -0.194 to -0.104. Hence, the γ value ranges from 5.15 to 9.62.

By using the Excel Solver program with constraints $\beta > 1$ and $5 \leq \gamma \leq 10$ from Equation 9.9.6 and median rank and stress cycles in Table 9.4,

$$\alpha' = 6490.98,\ \beta' = 1.584,\ \text{and}\ \gamma' = 10$$

The corresponding minimum value of Ω is 0.011662. Hence, the life distribution at a test load $S_1 = 2000$ is a Weibull distribution with

$$\beta = 1.584$$
$$\theta = (\alpha'/S_1)^{\gamma} = 129,659 \text{ cycles}$$

TABLE 9.4 Cumulative Failure Stress Cycles and Median Rank

Stress Cycles	Median Rank
9,063	0.056
9,319	0.137
10,015	0.218
10,685	0.298
10,905	0.379
10,929	0.460
11,173	0.540
11,247	0.621
11,308	0.702
11,527	0.783
11,576	0.863
11,820	0.944

$$B10 \text{ life} = 31,319 \text{ cycles}$$

$$B50 = 102,876 \text{ cycles}$$

$$\text{Reliability at bogey (9184 cycles)} = 98.50\%$$

9.10 COMPARISON TESTING ANALYSIS METHOD

An extremely effective method to verify the adequacy of a proposed new product (Design B) is to evaluate it against a proven existing or baseline product (Design A). This approach is attractive because it eliminates the need for test correlation, usage severity knowledge, and fixed reliability/confidence level criteria. This section covers the method to compare the B_q life of two designs, using the two-parameter Weibull distribution. The B_q life is given by (Equation 9.4.7)

$$B_q = \theta[\ln(100/(100 - q))]^{1/\beta} \tag{9.10.1}$$

This method assumes that there are three B_q lives (B_q life, 90% lower bound, and 90% upper bound) for Design A and Design B. It also assumes that the B_{10} life distribution is approximately normally distributed. The confidence and the percent improvement with a given confidence level can be determined. This method is best illustrated by Example 9.12 for statistical comparison of B_{10} lives.

Example 9.12. For a given test data (all parts were run to failure).

Design A	Design B
46,927	198,067
76,708	248,359
111,950	290,935
116,764	294,634
168,464	593,354
292,111	794,436

Step 1. Determine Weibull parameters and B_{10} lives.

Based on the maximum likelihood ratio approach, we can have the following results.

	Char. Life	Slope	B_{10} Life (90% lower bound)	B_{10} Life	B_{10} Life (90% upper bound)
Design A	153,559	1.836	19,053	45,090	76,633
Design B	458,934	2.028	69,300	151,286	245,410

Step 2. Compute approximate standard deviations.
For Design A:

$$\sigma_A = (B_{10}\text{ upper} - B_{10}\text{ lower})/(2 \times 1.28) = (76,633 - 19,053)/(2 \times 1.28) = 22,492$$

For Design B:

$$\sigma_B = (B_{10}\text{ upper} - B_{10}\text{ lower})/(2 \times 1.28) = (245,410 - 69,300)/(2 \times 1.28) = 68,793$$

Step 3. Compute confidence (C).
The confidence (C) is defined as

$$C = P\{(B_{10}\text{ life})_B > (B_{10}\text{ life})_A\}$$

Hence,

$$C = \Phi\{(151,286 - 45,090)\}/[(22,492)^2 + (68,793)^2]^{1/2}\} = \Phi(1.47) = 92.9\%$$

where Φ is the cumulative distribution function for the standard normal random variable (Appendix 9.1). We can be 92.9% confident that the life of Design B is longer than that of Design A.

Step 4. Compute improvement.
Let

$$z = [(B_{10}\text{ life})_2 - (B_{10}\text{ life})_1 - (\text{Improvement})]/[(\sigma_A)^2 + (\sigma_B)^2]^{1/2}$$

Here the 90% confidence level is chosen for the improvement factor. Therefore, from the standard normal table, $z = 1.28$ (Appendix 9.1). Let

$$1.28 = (151,286 - 45,090 - (\text{Improvement}))/[(\sigma_A)^2 + (\sigma_B)^2]^{1/2}$$

Then

$$\text{Improvement} = 13,554,\text{ or } 13,554/45,090 = 31\%$$

In other words, we can be 90% confident that Design B has a 31% longer B_{10} life than Design A.

There is no precise statistical test for a significant difference between two samples if the population is assumed to be Weibull. When comparing sample sets with different slopes, one design might be superior in median life (B_{50} life) and the other design superior for B_{10} life. This occurs when Weibull lives cross each other when the two data sets are plotted on Weibull probability paper. Caution must be exercised to ascertain which life (median or B_{10}) is of greater concern. An improved median life will not necessarily improve field reliability or warranty if the B_{10} life is not also improved. Similarly, the caution for a situation where one design might be superior in B_{10} life and the other superior for B_5 life or less (less than 5% failure) and B_5 or less is more concern for safety-related components.

9.11 REPAIRABLE SYSTEM RELIABILITY PREDICTION

This section describes a major system reliability assessment process that utilizes the repairable system reliability method. This technique is used to predict reliability and warranty of the field product. Many systems can be categorized into two basic types: one-time or nonrepairable systems, and reusable or repairable systems. For example, if the system is a vehicle and the water pump fails, then the water pump is replaced and hence the vehicle is repaired. The vehicle is considered a repairable system and the water pump a nonrepairable component. However, if the water pump is serviceable, its failed component replaced (say, the impeller), and the pump placed back in service, then the water pump can be considered a repairable system.

For a repairable system, failures are not necessarily independent or identically distributed. Repairable systems can use stochastic models to assess reliability. The non-homogeneous Poisson process (NHPP) model to assess the repairable system reliability is discussed. For related information, see Ascher and Feingold (1984), Crow (1974), and Lu and Rudy (1994).

Repairable systems require the use of stochastic models to assess reliability. A stochastic model is used to describe random events in a continuum. In our case, the random events are incidents and the continuum is time (or mileage). Given next are the definitions for the NHPP model (Ascher and Feingold, 1984; Crow, 1974). The random variables can be considered incident times.

The NHPP is defined as a nonterminating sequence of nonnegative random variables $X_1, X_2, \ldots,$ such that the number of incidents in any interval of length $(t_2 - t_1)$ has a Poisson distribution with mean

$$\int_{t1}^{t2} u(t)dt \tag{9.11.1}$$

That is, for all $t_2 > t_1 > 0$, and $j = 0, 1, 2, \ldots,$

$$P[N(t_2) - N(t_1) = j] = \frac{e^{-\int_{t1}^{t2} u(t)dt}\left(\int_{t1}^{t2} u(t)dt\right)^j}{j!} \tag{9.11.2}$$

where $N(t)$ is the number of incidents by the end of time t.

It follows that

$$E[N(t_2) - N(t_1)] = \int_{t1}^{t2} u(t)dt \tag{9.11.3}$$

$u(t)$ is called the rate of occurrence of failures (ROCOF) and varies with time rather than being a constant. The $u(t)$ is also called the intensity function (Crow, 1974). The NHPP model is considered for reliability estimation when the process has the following ROCOF (or intensity function):

$$u(t) = \lambda \beta t^{\beta - 1} \tag{9.11.4}$$

where λ and β are parameters estimated from actual test data.

When $\beta = 1$, $u(t)$ is a constant. When $\beta > 1$, $u(t)$ is strictly increasing and the successive interarrival times are stochastically decreasing (characteristic of a wear-out situation). When $\beta < 1$, $u(t)$ is strictly decreasing and the successive interarrival times are stochastically increasing (characteristic of a debugging situation).

The parameters λ and β, of the intensity function, can be determined as follows. Let k be the total number of systems in the study. For the ith system, $i = 1, 2, \ldots, k$, let

S_i = Starting time of the ith system
T_i = Ending (or current) time of the ith system
N_i = Total number of failures experienced by the ith system
x_{ij} = System time of the ith system at the jth occurrence of a failure, $j = 1, 2, 3, \ldots, N_i$

The maximum likelihood estimates of λ and β are values $\hat{\lambda}$ and $\hat{\beta}$ satisfying the equations (Crow, 1974)

$$\hat{\lambda} = \frac{\sum_{i=1}^{k} N_i}{\sum_{i=1}^{k} (T_i^{\hat{\beta}} - S_i^{\hat{\beta}})} \tag{9.11.5}$$

$$\hat{\beta} = \frac{\sum_{i=1}^{k} N_i}{\hat{\lambda} \sum_{i=1}^{k} (T_i^{\hat{\beta}} \ln T_i - S_i^{\hat{\beta}} \ln S_i) - \sum_{i=1}^{k} \sum_{j=1}^{N_i} \ln x_{ij}} \tag{9.11.6}$$

where $0^* \ln 0$ is taken to be 0.

If the $x_{ij}s$ represent all incidents that would be considered warranty conditions, the expected number of incidents' estimation at system age t is then given by

$$\hat{\lambda} t^{\hat{\beta}} \tag{9.11.7}$$

If $x_{ij}s$ represent all incidents that meet the definition of reliability, then the reliability at system age t is given by

$$R(t) = \exp(-\hat{\lambda} t^{\hat{\beta}}) \tag{9.11.8}$$

The reliability of going another d interval without failure at system age t is given by

$$R(t) = \exp[-\hat{\lambda}(t+d)^{\hat{\beta}} - \hat{\lambda} t^{\hat{\beta}}] \tag{9.11.9}$$

Example 9.13 (Lu and Rudy, 1994). A new heating, ventilation, and air conditioning (HEVAC) system was developed and is considering extending its warranty coverage to 50,000 miles or 100,000 miles. The new HEVAC system was put into each of two vehicles for a duration of 100,000 miles of road testing with the following observed incidents:

System No.	Incidents No.	System Miles	Description
1	1	2,300	Fan switch inoperative
	2	15,100	Duct loose, right side I/P
	3	35,500	No air, left side
	4	51,200	Defroster inoperative
	5	60,000	Fan switch inoperative
	6	78,000	Selector switch efforts too high
	7	98,500	Heater core failed
2	1	6,500	Recirculate switch inoperative
	2	13,100	Vent loose
	3	31,100	Drier bottle inoperative
	4	42,000	Fan switch inoperative
	5	49,500	A/C inoperative
	6	61,000	Refrigerant leak
	7	85,000	Condenser failed
	8	94,500	Vent loose

The design life of the HEVAC system is 70,000 miles and the mission length (arbitrary trip length for which a reliability estimate is desired) is 1000 miles. Answer the following questions:

(1) What are the estimations for the two parameters λ and β of the intensity function?

Solution. From Equations 9.11.5 and 9.11.6 with $k = 2$, $S_1 = S_2 = 0$, $T_1 = T_2 = 100,000$, $N_1 = 7$, and $N_2 = 8$, $\hat{\beta} = 0.9169$, and $\hat{\lambda} = 1.9532$ E-4.

(2) What is the reliability and number of expected incidents at the design life of 70,000 miles?

Solution. From Equation 9.11.8, the reliability at 70,000 cycles is 4.48 E-3. From Equation 9.11.7, the expected number of incidents is 5.408.

(3) What is the probability of experiencing no incidents during a 1000-mile trip if the vehicle has 20,000 miles on it at the start of the trip?

Solution. From Equation 9.11.9 with $d = 1000$ miles and $t = 20,000$ miles, the probability of experiencing no incidents during a 1000-mile trip is 92%.

REFERENCES

Allmen, C. R. and Lu, M. W., A reduced sampling approach for reliability verification, *Quality and Reliability Engineering International*, Vol. 10, 1994, pp. 71–77.

Ascher, H. and Feingold, H., *Repairable System Reliability, Modeling, Inference, Misconceptions and Their Causes*, Marcel Dekker, New York, 1984.

Crow, L. H., Reliability analysis for complex, repairable systems, *Reliability and Biometry, Statistical Analysis of Lifelength*, SIAM, Philadelphia, 1974.

Kapur, K. C. and Lamberson L. R., *Reliability in Engineering Design*, Wiley, New York, 1977.

Lipson, C. and Sheh, N. J., *Statistical Design and Analysis of Engineering Experiments*, McGraw-Hill, New York, 1973.

Lu, M. W., Leiman, J. E., Rudy, R. J., and Lee, Y. L., Step-stress accelerated test method: a validation study. In *SAE World Congress*, Detroit, MI, SAE Paper 2003-01-0470, 2003.

Lu, M. W. and Rudy, R. J., Reliability demonstration test for a finite population, *Quality and Reliability Engineering International*, Vol. 17, 2001, pp. 33–38.

Lu, M. W. and Rudy, R. J., Vehicle or system reliability assessment procedure. In *Proceedings of the IASTED International Conference on the Reliability Engineering and its Applications*, Honolulu, August 1994, pp. 37–42.

Mann, N. R., Schafer, R. E., and Singpurwalla, N. D., *Method for Statistical Analysis of Reliability and Life Data*, Wiley, New York, 1974.

Nelson, W., Weibull analysis of reliability data with few or no failures, *Journal of Quality Technology*, Vol. 17, No. 3, 1985, pp. 140–146.

Nelson, W., *Accelerated Testing: Statistical Models, Test Plans, and Data Analysis*, Wiley, New York, 1990.

WINSmith™ Weibull Software, Fulton Findings™, 2002.

APPENDIX 9.1 Standard Normal Distribution

z	0.00	0.01	0.02	0.03	0.04	0.05	0.06	0.07	0.08	0.09
0.0	0.5000	0.5040	0.5080	0.5120	0.5160	0.5199	0.5239	0.5279	0.5319	0.5359
0.1	0.5398	0.5438	0.5478	0.5517	0.5557	0.5596	0.5636	0.5675	0.5714	0.5753
0.2	0.5793	0.5832	0.5871	0.5910	0.5948	0.5987	0.6026	0.6064	0.6103	0.6141
0.3	0.6179	0.6217	0.6255	0.6293	0.6331	0.6368	0.6406	0.6443	0.6480	0.6517
0.4	0.6554	0.6591	0.6628	0.6664	0.6700	0.6736	0.6772	0.6808	0.6844	0.6879
0.5	0.6915	0.6950	0.6985	0.7019	0.7054	0.7088	0.7123	0.7157	0.7190	0.7224
0.6	0.7257	0.7291	0.7324	0.7357	0.7389	0.7422	0.7454	0.7486	0.7517	0.7549
0.7	0.7580	0.7611	0.7642	0.7673	0.7704	0.7734	0.7764	0.7794	0.7823	0.7852
0.8	0.7881	0.7910	0.7939	0.7967	0.7995	0.8023	0.8051	0.8078	0.8106	0.8133
0.9	0.8159	0.8186	0.8212	0.8238	0.8264	0.8289	0.8315	0.8340	0.8365	0.8389
1.0	0.8413	0.8438	0.8461	0.8485	0.8508	0.8531	0.8554	0.8577	0.8599	0.8621
1.1	0.8643	0.8665	0.8686	0.8708	0.8729	0.8749	0.8770	0.8790	0.8810	0.8830
1.2	0.8849	0.8869	0.8888	0.8907	0.8925	0.8944	0.8962	0.8980	0.8997	0.9015
1.3	0.9032	0.9049	0.9066	0.9082	0.9099	0.9115	0.9131	0.9147	0.9162	0.9177
1.4	0.9192	0.9207	0.9222	0.9236	0.9251	0.9265	0.9279	0.9292	0.9306	0.9319
1.5	0.9332	0.9345	0.9357	0.9370	0.9382	0.9394	0.9406	0.9418	0.9429	0.9441
1.6	0.9452	0.9463	0.9474	0.9484	0.9495	0.9505	0.9515	0.9525	0.9535	0.9545
1.7	0.9554	0.9564	0.9573	0.9582	0.9591	0.9599	0.9608	0.9616	0.9625	0.9633
1.8	0.9641	0.9649	0.9656	0.9664	0.9671	0.9678	0.9686	0.9693	0.9699	0.9706
1.9	0.9713	0.9719	0.9726	0.9732	0.9738	0.9744	0.9750	0.9756	0.9761	0.9767
2.0	0.9772	0.9778	0.9783	0.9788	0.9793	0.9798	0.9803	0.9808	0.9812	0.9817
2.1	0.9821	0.9826	0.9830	0.9834	0.9838	0.9842	0.9846	0.9850	0.9854	0.9857
2.2	0.9861	0.9864	0.9868	0.9871	0.9875	0.9878	0.9881	0.9884	0.9887	0.9890
2.3	0.9893	0.9896	0.9898	0.9901	0.9904	0.9906	0.9909	0.9911	0.9913	0.9916
2.4	0.9918	0.9920	0.9922	0.9925	0.9927	0.9929	0.9931	0.9932	0.9934	0.9936
2.5	0.9938	0.9940	0.9941	0.9943	0.9945	0.9946	0.9948	0.9949	0.9951	0.9952
2.6	0.9953	0.9955	0.9956	0.9957	0.9959	0.9960	0.9961	0.9962	0.9963	0.9964
2.7	0.9965	0.9966	0.9967	0.9968	0.9969	0.9970	0.9971	0.9972	0.9973	0.9974
2.8	0.9974	0.9975	0.9976	0.9977	0.9977	0.9978	0.9979	0.9979	0.9980	0.9981
2.9	0.9981	0.9982	0.9982	0.9983	0.9984	0.9984	0.9985	0.9985	0.9986	0.9986
3.0	0.9987	0.9987	0.9987	0.9988	0.9988	0.9989	0.9989	0.9989	0.9990	0.9990
3.1	0.9990	0.9991	0.9991	0.9991	0.9992	0.9992	0.9992	0.9992	0.9993	0.9993
3.2	0.9993	0.9993	0.9994	0.9994	0.9994	0.9994	0.9994	0.9995	0.9995	0.9995
3.3	0.9995	0.9995	0.9995	0.9996	0.9996	0.9996	0.9996	0.9996	0.9996	0.9997
3.4	0.9997	0.9997	0.9997	0.9997	0.9997	0.9997	0.9997	0.9997	0.9997	0.9998

Note: Entries in the table are values of $F(z)$, where $F(z) = P\{Z \leq z\}$, Z is a standard normal random variable.

APPENDIX 9.2 Chi-square Distribution

	p									
n	0.995	0.99	0.98	0.975	0.95	0.9	0.8	0.75	0.7	0.5
1	0.000	0.000	0.001	0.001	0.004	0.016	0.064	0.102	0.148	0.455
2	0.010	0.020	0.040	0.051	0.103	0.211	0.446	0.575	0.713	1.386
3	0.072	0.115	0.185	0.216	0.352	0.584	1.005	1.213	1.424	2.366
4	0.207	0.297	0.429	0.484	0.711	1.064	1.649	1.923	2.195	3.357
5	0.412	0.554	0.752	0.831	1.145	1.610	2.343	2.675	3.000	4.351
6	0.676	0.872	1.134	1.237	1.635	2.204	3.070	3.455	3.828	5.348
7	0.989	1.239	1.564	1.690	2.167	2.833	3.822	4.255	4.671	6.346
8	1.344	1.647	2.032	2.180	2.733	3.490	4.594	5.071	5.527	7.344
9	1.735	2.088	2.532	2.700	3.325	4.168	5.380	5.899	6.393	8.343
10	2.156	2.558	3.059	3.247	3.940	4.865	6.179	6.737	7.267	9.342
11	2.603	3.053	3.609	3.816	4.575	5.578	6.989	7.584	8.148	10.341
12	3.074	3.571	4.178	4.404	5.226	6.304	7.807	8.438	9.034	11.340
13	3.565	4.107	4.765	5.009	5.892	7.041	8.634	9.299	9.926	12.340
14	4.075	4.660	5.368	5.629	6.571	7.790	9.467	10.165	10.821	13.339
15	4.601	5.229	5.985	6.262	7.261	8.547	10.307	11.037	11.721	14.339
16	5.142	5.812	6.614	6.908	7.962	9.312	11.152	11.912	12.624	15.338
17	5.697	6.408	7.255	7.564	8.672	10.085	12.002	12.792	13.531	16.338
18	6.265	7.015	7.906	8.231	9.390	10.865	12.857	13.675	14.440	17.338
19	6.844	7.633	8.567	8.907	10.117	11.651	13.716	14.562	15.352	18.338
20	7.434	8.260	9.237	9.591	10.851	12.443	14.578	15.452	16.266	19.337
21	8.034	8.897	9.915	10.283	11.591	13.240	15.445	16.344	17.182	20.337
22	8.643	9.542	10.600	10.982	12.338	14.041	16.314	17.240	18.101	21.337
23	9.260	10.196	11.293	11.689	13.091	14.848	17.187	18.137	19.021	22.337
24	9.886	10.856	11.992	12.401	13.848	15.659	18.062	19.037	19.943	23.337
25	10.524	11.524	12.697	13.120	14.611	16.473	18.940	19.939	20.867	24.337
26	11.160	12.198	13.409	13.844	15.379	17.292	19.820	20.843	21.792	25.336
27	11.808	12.878	14.125	14.573	16.151	18.114	20.703	21.749	22.719	26.336
28	12.461	13.565	14.847	15.308	16.928	18.939	21.588	22.657	23.647	27.336
29	13.121	14.256	15.574	16.047	17.708	19.768	22.475	23.567	24.577	28.336
30	13.787	14.953	16.306	16.791	18.493	20.599	23.364	24.478	25.508	29.336

Note: Entries in the table are values of χ_p^2, where $p = P\{\chi^2 \geq \chi_p^2\}$, χ^2 has a χ^2 distribution with n degree of freedom.

(*Continued*)

APPENDIX 9.2 Chi-square Distribution (*Continues*)

	p									
n	0.3	0.25	0.2	0.1	0.05	0.025	0.02	0.01	0.005	0.001
1	1.074	1.323	1.642	2.706	3.841	5.024	5.412	6.635	7.879	10.827
2	2.408	2.773	3.219	4.605	5.991	7.378	7.824	9.210	10.597	13.815
3	3.665	4.108	4.642	6.251	7.815	9.348	9.837	11.345	12.838	16.266
4	4.878	5.385	5.989	7.779	9.488	11.143	11.668	13.277	14.860	18.466
5	6.064	6.626	7.289	9.236	11.070	12.832	13.388	15.086	16.750	20.515
6	7.231	7.841	8.558	10.645	12.592	14.449	15.033	16.812	18.548	22.457
7	8.383	9.037	9.803	12.017	14.067	16.013	16.622	18.475	20.278	24.321
8	9.524	10.219	11.030	13.362	15.507	17.535	18.168	20.090	21.955	26.124
9	10.656	11.389	12.242	14.684	16.919	19.023	19.679	21.666	23.589	27.877
10	11.781	12.549	13.442	15.987	18.307	20.483	21.161	23.209	25.188	29.588
11	12.899	13.701	14.631	17.275	19.675	21.920	22.618	24.725	26.757	31.264
12	14.011	14.845	15.812	18.549	21.026	23.337	24.054	26.217	28.300	32.909
13	15.119	15.984	16.985	19.812	22.362	24.736	25.471	27.688	29.819	34.527
14	16.222	17.117	18.151	21.064	23.685	26.119	26.873	29.141	31.319	36.124
15	17.322	18.245	19.311	22.307	24.996	27.488	28.259	30.578	32.801	37.698
16	18.418	19.369	20.465	23.542	26.296	28.845	29.633	32.000	34.267	39.252
17	19.511	20.489	21.615	24.769	27.587	30.191	30.995	33.409	35.718	40.791
18	20.601	21.605	22.760	25.989	28.869	31.526	32.346	34.805	37.156	42.312
19	21.689	22.718	23.900	27.204	30.144	32.852	33.687	36.191	38.582	43.819
20	22.775	23.828	25.038	28.412	31.410	34.170	35.020	37.566	39.997	45.314
21	23.858	24.935	26.171	29.615	32.671	35.479	36.343	38.932	41.401	46.796
22	24.939	26.039	27.301	30.813	33.924	36.781	37.659	40.289	42.796	48.268
23	26.018	27.141	28.429	32.007	35.172	38.076	38.968	41.638	44.181	49.728
24	27.096	28.241	29.553	33.196	36.415	39.364	40.270	42.980	45.558	51.179
25	28.172	29.339	30.675	34.382	37.652	40.646	41.566	44.314	46.928	52.619
26	29.246	30.435	31.795	35.563	38.885	41.923	42.856	45.642	48.290	54.051
27	30.319	31.528	32.912	36.741	40.113	43.195	44.140	46.963	49.645	55.475
28	31.391	32.620	34.027	37.916	41.337	44.461	45.419	48.278	50.994	56.892
29	32.461	33.711	35.139	39.087	42.557	45.722	46.693	49.588	52.335	58.301
30	33.530	34.800	36.250	40.256	43.773	46.979	47.962	50.892	53.672	59.702

Note: Entries in the table are values of χ_p^2, where $p = P\{\chi^2 \geq \chi_p^2\}$, χ^2 has a χ^2 distribution with *n* degree of freedom.

APPENDIX 9.3 *F*-Distribution

		n_1											
		1	2	3	4	5	6	7	8	9	10	11	12
n_2	p												
1	0.90	39.86	49.50	53.59	55.83	57.24	58.20	58.91	59.44	59.86	60.19	60.47	60.71
	0.95	161.4	199.5	215.7	224.6	230.2	234.0	236.8	238.9	240.5	241.9	243.0	243.9
	0.99	4052	4999	5404	5624	5764	5859	5928	5981	6022	6056	6083	6107
2	0.90	8.53	9.00	9.16	9.24	9.29	9.33	9.35	9.37	9.38	9.39	9.40	9.41
	0.95	18.51	19.00	19.16	19.25	19.30	19.33	19.35	19.37	19.38	19.40	19.40	19.41
	0.99	98.50	99.00	99.16	99.25	99.30	99.33	99.36	99.38	99.39	99.40	99.41	99.42
3	0.90	5.54	5.46	5.39	5.34	5.31	5.28	5.27	5.25	5.24	5.23	5.22	5.22
	0.95	10.13	9.55	9.28	9.12	9.01	8.94	8.89	8.85	8.81	8.79	8.76	8.74
	0.99	34.12	30.82	29.46	28.71	28.24	27.91	27.67	27.49	27.34	27.23	27.13	27.05
4	0.90	4.54	4.32	4.19	4.11	4.05	4.01	3.98	3.95	3.94	3.92	3.91	3.90
	0.95	7.71	6.94	6.59	6.39	6.26	6.16	6.09	6.04	6.00	5.96	5.94	5.91
	0.99	21.20	18.00	16.69	15.98	15.52	15.21	14.98	14.80	14.66	14.55	14.45	14.37
5	0.90	4.06	3.78	3.62	3.52	3.45	3.40	3.37	3.34	3.32	3.30	3.28	3.27
	0.95	6.61	5.79	5.41	5.19	5.05	4.95	4.88	4.82	4.77	4.74	4.70	4.68
	0.99	16.26	13.27	12.06	11.39	10.97	10.67	10.46	10.29	10.16	10.05	9.96	9.89
6	0.90	3.78	3.46	3.29	3.18	3.11	3.05	3.01	2.98	2.96	2.94	2.92	2.90
	0.95	5.99	5.14	4.76	4.53	4.39	4.28	4.21	4.15	4.10	4.06	4.03	4.00
	0.99	13.75	10.92	9.78	9.15	8.75	8.47	8.26	8.10	7.98	7.87	7.79	7.72
7	0.90	3.59	3.26	3.07	2.96	2.88	2.83	2.78	2.75	2.72	2.70	2.68	2.67
	0.95	5.59	4.74	4.35	4.12	3.97	3.87	3.79	3.73	3.68	3.64	3.60	3.57
	0.99	12.25	9.55	8.45	7.85	7.46	7.19	6.99	6.84	6.72	6.62	6.54	6.47
8	0.90	3.46	3.11	2.92	2.81	2.73	2.67	2.62	2.59	2.56	2.54	2.52	2.50
	0.95	5.32	4.46	4.07	3.84	3.69	3.58	3.50	3.44	3.39	3.35	3.31	3.28
	0.99	11.26	8.65	7.59	7.01	6.63	6.37	6.18	6.03	5.91	5.81	5.73	5.67
9	0.90	3.36	3.01	2.81	2.69	2.61	2.55	2.51	2.47	2.44	2.42	2.40	2.38
	0.95	5.12	4.26	3.86	3.63	3.48	3.37	3.29	3.23	3.18	3.14	3.10	3.07
	0.99	10.56	8.02	6.99	6.42	6.06	5.80	5.61	5.47	5.35	5.26	5.18	5.11
10	0.90	3.29	2.92	2.73	2.61	2.52	2.46	2.41	2.38	2.35	2.32	2.30	2.28
	0.95	4.96	4.10	3.71	3.48	3.33	3.22	3.14	3.07	3.02	2.98	2.94	2.91
	0.99	10.04	7.56	6.55	5.99	5.64	5.39	5.20	5.06	4.94	4.85	4.77	4.71
11	0.90	3.23	2.86	2.66	2.54	2.45	2.39	2.34	2.30	2.27	2.25	2.23	2.21
	0.95	4.84	3.98	3.59	3.36	3.20	3.09	3.01	2.95	2.90	2.85	2.82	2.79
	0.99	9.65	7.21	6.22	5.67	5.32	5.07	4.89	4.74	4.63	4.54	4.46	4.40
12	0.90	3.18	2.81	2.61	2.48	2.39	2.33	2.28	2.24	2.21	2.19	2.17	2.15
	0.95	4.75	3.89	3.49	3.26	3.11	3.00	2.91	2.85	2.80	2.75	2.72	2.69
	0.99	9.33	6.93	5.95	5.41	5.06	4.82	4.64	4.50	4.39	4.30	4.22	4.16
13	0.90	3.14	2.76	2.56	2.43	2.35	2.28	2.23	2.20	2.16	2.14	2.12	2.10
	0.95	4.67	3.81	3.41	3.18	3.03	2.92	2.83	2.77	2.71	2.67	2.63	2.60
	0.99	9.07	6.70	5.74	5.21	4.86	4.62	4.44	4.30	4.19	4.10	4.02	3.96
14	0.90	3.10	2.73	2.52	2.39	2.31	2.24	2.19	2.15	2.12	2.10	2.07	2.05
	0.95	4.60	3.74	3.34	3.11	2.96	2.85	2.76	2.70	2.65	2.60	2.57	2.53
	0.99	8.86	6.51	5.56	5.04	4.69	4.46	4.28	4.14	4.03	3.94	3.86	3.80

(*Continued*)

APPENDIX 9.3 *(Continues)*

		n_1											
		1	2	3	4	5	6	7	8	9	10	11	12
15	0.90	3.07	2.70	2.49	2.36	2.27	2.21	2.16	2.12	2.09	2.06	2.04	2.02
	0.95	4.54	3.68	3.29	3.06	2.90	2.79	2.71	2.64	2.59	2.54	2.51	2.48
	0.99	8.68	6.36	5.42	4.89	4.56	4.32	4.14	4.00	3.89	3.80	3.73	3.67
16	0.90	3.05	2.67	2.46	2.33	2.24	2.18	2.13	2.09	2.06	2.03	2.01	1.99
	0.95	4.49	3.63	3.24	3.01	2.85	2.74	2.66	2.59	2.54	2.49	2.46	2.42
	0.99	8.53	6.23	5.29	4.77	4.44	4.20	4.03	3.89	3.78	3.69	3.62	3.55
17	0.90	3.03	2.64	2.44	2.31	2.22	2.15	2.10	2.06	2.03	2.00	1.98	1.96
	0.95	4.45	3.59	3.20	2.96	2.81	2.70	2.61	2.55	2.49	2.45	2.41	2.38
	0.99	8.40	6.11	5.19	4.67	4.34	4.10	3.93	3.79	3.68	3.59	3.52	3.46
18	0.90	3.01	2.62	2.42	2.29	2.20	2.13	2.08	2.04	2.00	1.98	1.95	1.93
	0.95	4.41	3.55	3.16	2.93	2.77	2.66	2.58	2.51	2.46	2.41	2.37	2.34
	0.99	8.29	6.01	5.09	4.58	4.25	4.01	3.84	3.71	3.60	3.51	3.43	3.37
19	0.90	2.99	2.61	2.40	2.27	2.18	2.11	2.06	2.02	1.98	1.96	1.93	1.91
	0.95	4.38	3.52	3.13	2.90	2.74	2.63	2.54	2.48	2.42	2.38	2.34	2.31
	0.99	8.18	5.93	5.01	4.50	4.17	3.94	3.77	3.63	3.52	3.43	3.36	3.30
20	0.90	2.97	2.59	2.38	2.25	2.16	2.09	2.04	2.00	1.96	1.94	1.91	1.89
	0.95	4.35	3.49	3.10	2.87	2.71	2.60	2.51	2.45	2.39	2.35	2.31	2.28
	0.99	8.10	5.85	4.94	4.43	4.10	3.87	3.70	3.56	3.46	3.37	3.29	3.23
22	0.90	2.95	2.56	2.35	2.22	2.13	2.06	2.01	1.97	1.93	1.90	1.88	1.86
	0.95	4.30	3.44	3.05	2.82	2.66	2.55	2.46	2.40	2.34	2.30	2.26	2.23
	0.99	7.95	5.72	4.82	4.31	3.99	3.76	3.59	3.45	3.35	3.26	3.18	3.12
24	0.90	2.93	2.54	2.33	2.19	2.10	2.04	1.98	1.94	1.91	1.88	1.85	1.83
	0.95	4.26	3.40	3.01	2.78	2.62	2.51	2.42	2.36	2.30	2.25	2.22	2.18
	0.99	7.82	5.61	4.72	4.22	3.90	3.67	3.50	3.36	3.26	3.17	3.09	3.03
26	0.90	2.91	2.52	2.31	2.17	2.08	2.01	1.96	1.92	1.88	1.86	1.83	1.81
	0.95	4.23	3.37	2.98	2.74	2.59	2.47	2.39	2.32	2.27	2.22	2.18	2.15
	0.99	7.72	5.53	4.64	4.14	3.82	3.59	3.42	3.29	3.18	3.09	3.02	2.96
28	0.90	2.89	2.50	2.29	2.16	2.06	2.00	1.94	1.90	1.87	1.84	1.81	1.79
	0.95	4.20	3.34	2.95	2.71	2.56	2.45	2.36	2.29	2.24	2.19	2.15	2.12
	0.99	7.64	5.45	4.57	4.07	3.75	3.53	3.36	3.23	3.12	3.03	2.96	2.90
30	0.90	2.88	2.49	2.28	2.14	2.05	1.98	1.93	1.88	1.85	1.82	1.79	1.77
	0.95	4.17	3.32	2.92	2.69	2.53	2.42	2.33	2.27	2.21	2.16	2.13	2.09
	0.99	7.56	5.39	4.51	4.02	3.70	3.47	3.30	3.17	3.07	2.98	2.91	2.84
40	0.90	2.84	2.44	2.23	2.09	2.00	1.93	1.87	1.83	1.79	1.76	1.74	1.71
	0.95	4.08	3.23	2.84	2.61	2.45	2.34	2.25	2.18	2.12	2.08	2.04	2.00
	0.99	7.31	5.18	4.31	3.83	3.51	3.29	3.12	2.99	2.89	2.80	2.73	2.66
60	0.90	2.79	2.39	2.18	2.04	1.95	1.87	1.82	1.77	1.74	1.71	1.68	1.66
	0.95	4.00	3.15	2.76	2.53	2.37	2.25	2.17	2.10	2.04	1.99	1.95	1.92
	0.99	7.08	4.98	4.13	3.65	3.34	3.12	2.95	2.82	2.72	2.63	2.56	2.50

Note: Entries in the table are values of F_p, where $p = P\{F \leq F_p\}$, F has the F distribution with n_1 and n_2 degrees of freedom.

APPENDIX 9.4 Median, 90%, 95% Ranks

Median Ranks

Rank order	Sample size														
	1	2	3	4	5	6	7	8	9	10	11	12	13	14	15
1	0.5000	0.2929	0.2063	0.1591	0.1294	0.1091	0.0943	0.0830	0.0741	0.0670	0.0611	0.0561	0.0519	0.0483	0.0452
2		0.7071	0.5000	0.3864	0.3147	0.2655	0.2295	0.2021	0.1806	0.1632	0.1489	0.1368	0.1266	0.1178	0.1101
3			0.7937	0.6136	0.5000	0.4218	0.3648	0.3213	0.2871	0.2594	0.2366	0.2175	0.2013	0.1873	0.1751
4				0.8409	0.6853	0.5782	0.5000	0.4404	0.3935	0.3557	0.3244	0.2982	0.2760	0.2568	0.2401
5					0.8706	0.7345	0.6352	0.5596	0.5000	0.4519	0.4122	0.3789	0.3506	0.3263	0.3051
6						0.8909	0.7705	0.6787	0.6065	0.5481	0.5000	0.4596	0.4253	0.3958	0.3700
7							0.9057	0.7979	0.7129	0.6443	0.5878	0.5404	0.5000	0.4653	0.4350
8								0.9170	0.8194	0.7406	0.6756	0.6211	0.5747	0.5347	0.5000
9									0.9259	0.8368	0.7634	0.7018	0.6494	0.6042	0.5650
10										0.9330	0.8511	0.7825	0.7240	0.6737	0.6300
11											0.9389	0.8632	0.7987	0.7432	0.6949
12												0.9436	0.8734	0.8127	0.7599
13													0.9481	0.8822	0.8249
14														0.9517	0.8899
15															0.9548

90% Ranks

Rank Order	Sample size														
	1	2	3	4	5	6	7	8	9	10	11	12	13	14	15
1	0.9001	0.6838	0.5359	0.4377	0.3691	0.3188	0.2803	0.2501	0.2259	0.2057	0.1889	0.1746	0.1624	0.1517	0.1424
2		0.9487	0.8042	0.6796	0.5839	0.5104	0.4526	0.4062	0.3684	0.3369	0.3103	0.2875	0.2679	0.2507	0.2357
3			0.9655	0.8575	0.7516	0.6669	0.5962	0.5382	0.4901	0.4497	0.4152	0.3855	0.3599	0.3373	0.3174
4				0.9741	0.8878	0.7992	0.7214	0.6554	0.5995	0.5518	0.5108	0.4758	0.4443	0.4170	0.3929

Rank Order	1	2	3	4	5	6	7	8	9	10	11	12	13	14	15
5					0.9792	0.9075	0.8304	0.7603	0.6991	0.6458	0.5995	0.5590	0.5235	0.4920	0.4641
6						0.9827	0.9212	0.8581	0.7869	0.7327	0.6823	0.6377	0.5983	0.5631	0.5318
7							0.9851	0.9314	0.9705	0.8125	0.7595	0.7118	0.6692	0.6309	0.5966
8								0.9869	0.9393	0.8842	0.8308	0.7818	0.7364	0.6995	0.6586
9									0.9884	0.9455	0.8953	0.8458	0.7896	0.7569	0.7179
10										0.9896	0.9506	0.9044	0.8585	0.8149	0.7745
11											0.9905	0.9548	0.9121	0.8691	0.8291
12												0.9913	0.9584	0.9186	0.8783
13													0.9920	0.9614	0.9242
14														0.9925	0.9641
15															0.9931

95% Ranks

	Sample size														
Rank Order	1	2	3	4	5	6	7	8	9	10	11	12	13	14	15
1	0.9500	0.7764	0.6316	0.5271	0.4507	0.3930	0.3482	0.3123	0.2831	0.2589	0.2384	0.2209	0.2058	0.1926	0.1810
2		0.9747	0.8646	0.7514	0.6574	0.5818	0.5207	0.4707	0.4291	0.3942	0.3644	0.3387	0.3163	0.2967	0.2794
3			0.9830	0.9024	0.8107	0.7287	0.6587	0.5997	0.5496	0.5069	0.4701	0.4381	0.4101	0.3854	0.3634
4				0.9873	0.9326	0.8468	0.7747	0.7108	0.6551	0.6076	0.5644	0.5273	0.4946	0.4657	0.4398
5					0.9898	0.9371	0.8713	0.8071	0.7436	0.6965	0.6502	0.6091	0.5726	0.5400	0.5107
6						0.9915	0.9466	0.8889	0.8312	0.7776	0.7287	0.6848	0.6452	0.6096	0.5774
7							0.9926	0.9532	0.9032	0.8500	0.7993	0.7535	0.7117	0.6737	0.6392
8								0.9935	0.9590	0.9127	0.8637	0.8176	0.7745	0.7348	0.6984
9									0.9943	0.9632	0.9200	0.8755	0.8329	0.7918	0.7541
10										0.9949	0.9667	0.9281	0.8873	0.8473	0.8091
11											0.9953	0.9693	0.9335	0.8953	0.8576
12												0.9957	0.9719	0.9389	0.9033
13													0.9960	0.9737	0.9426
14														0.9963	0.9755
15															0.9966

10

Fatigue Analysis in the Frequency Domain

Yung-Li Lee
DaimlerChrysler

10.1 Introduction

Fatigue analysis is often performed in the time domain, in which all input loading and output stress or strain response are time-based signals. In this case, the response time history can be calculated either by the static stress analysis (or inertia relief method) (Lee et al., 1995) or by the modal transient analysis (Crescimanno and Cavallo, 1999; Vellaichamy, 2002). In the static stress analysis, stress can be obtained by superimposing all stress influences from the applied loads at every time step. In the modal transient analysis, stress can be calculated by using the modal stress and the modal coordinates. The modal transient analysis is recommended if resonant fatigue is of primary concern; that is, the loading frequencies appear closer to the system natural frequencies.

In some situations, however, response stress and input loading are preferably expressed as frequency-based signals, usually in the form of a power spectral density (PSD) plot. In this case, a system function (a characteristic of the structural system) is required to relate an input PSD of loading to the output PSD of response. The PSD techniques for dynamic analysis have been widely used in the offshore engineering field. PSD represents the energy of the time signal at different frequencies and is another way of denoting the loading signal in time domain. The Fast Fourier Transform (FFT) of a time signal can be used to obtain the PSD of the loading, whereas the Inverse Fourier Transform (IFT) can be used to transform the frequency-based signal to the time-based loading. The transform of loading history between the time domain and frequency domain is subject to certain requirements, as per which the signal must be stationary, random, and Gaussian (normal).

This chapter has two objectives. First, the fundamental theories of random vibration, which can be found elsewhere (Newland, 1984; Bendat and Piersol, 1986; Wirsching et al., 1995), are summarized in Sections 10.1–10.5 for the benefit of readers. Second, in Sections 10.6 and 10.7, the methods of predicting fatigue damage from an output PSD of stress response are introduced.

10.2 A RANDOM SAMPLE TIME HISTORY

A system produces a certain response under excitation. If the excitation or the response motion, $X(t)$, is unpredictable, the system is in random vibration because the exact value of $X(t)$ cannot be precisely predicted in advance. It can only be described probabilistically.

The probability density function (PDF) of a time history $X(t)$ can be obtained by calculating its statistical properties. Figure 10.1 shows a sample time history for a random process, $x(t)$, during the time interval T, where $X(t)$ exits between the values of x and $x+dx$ for a total time of $(dt_1+dt_2+dt_3+dt_4)$. The probability that $x \leq X(t) \leq x+dx$ is therefore given by

$$P[x \leq X(t) \leq x+dx] = \frac{dt_1+dt_2+dt_3+dt_4}{T} \tag{10.2.1}$$

If the duration T is long enough, the PDF $f_X(x)$ is given by

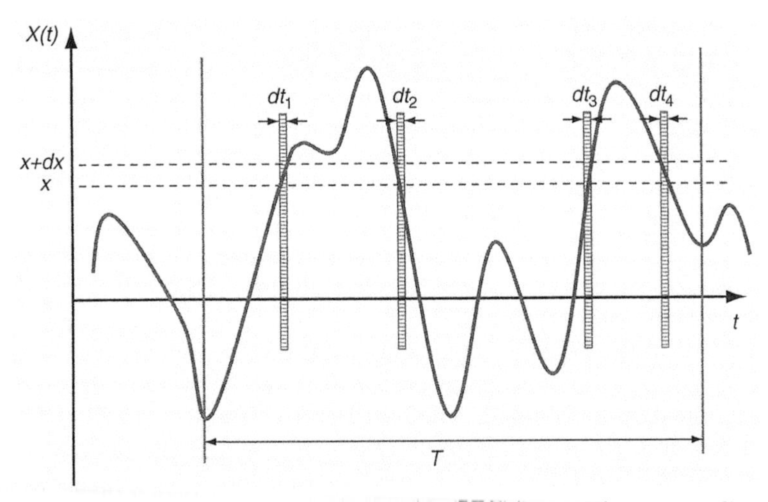

FIGURE 10.1 Probability density function (PDF) for a random process $X(t)$.

$$f_X(x) = P[x \leq X(t) \leq x + dx] = \frac{\sum_{i=1}^{k} dt_i}{T} \tag{10.2.2}$$

Equation 10.2.2 is correct if the time duration T goes to infinity, and implies that the sample time history continues forever. Measurement of the time segments, $\sum dt_i$, for the PDF $f_X(x)$ in Equation 10.2.2 is very cumbersome.

Alternatively, the PDF can be determined by the fraction of the total number of samples in the band between x and $x + dx$. This can be done by digitizing the time history at a certain sampling rate in the time interval T. As shown in Figure 10.2, $f_X(x)$ is then given by

$$f_X(x) = P[x \leq X(t) \leq x + dx] = \frac{\#\ \text{sample}_{\text{band}}}{\#\ \text{sample}_{\text{T}}} = \frac{6}{69} \tag{10.2.3}$$

With a given PDF $f_X(x)$, some statistical properties of the random process $X(t)$ can be obtained. The mean, μ_X, and the variance, σ_X^2, of the process can be calculated as

$$\mu_X = \int_{-\infty}^{+\infty} x f_X(x) dx \cong \frac{1}{T}\int_0^T X(t) dt \tag{10.2.4}$$

$$\sigma_X^2 = \int_{-\infty}^{+\infty} [x - \mu_X]^2 f_X(x) dx \cong \frac{1}{T}\int_0^T [X(t) - \mu_X]^2 dt \tag{10.2.5}$$

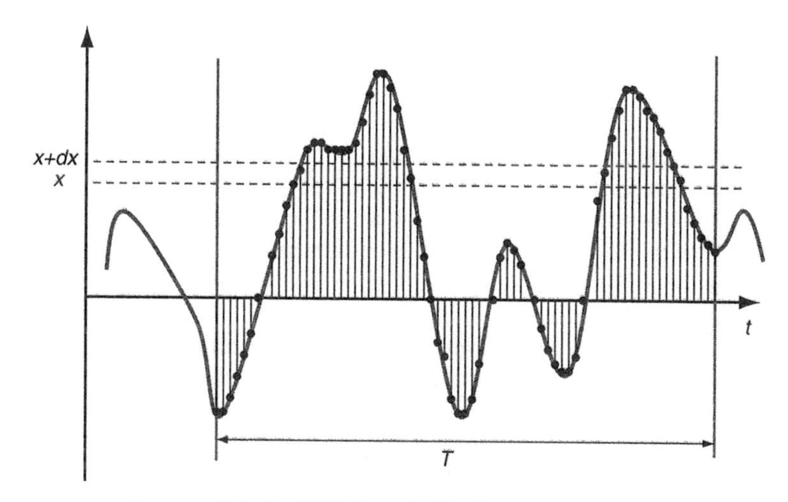

FIGURE 10.2 PDF for a digitized random process $X(t)$.

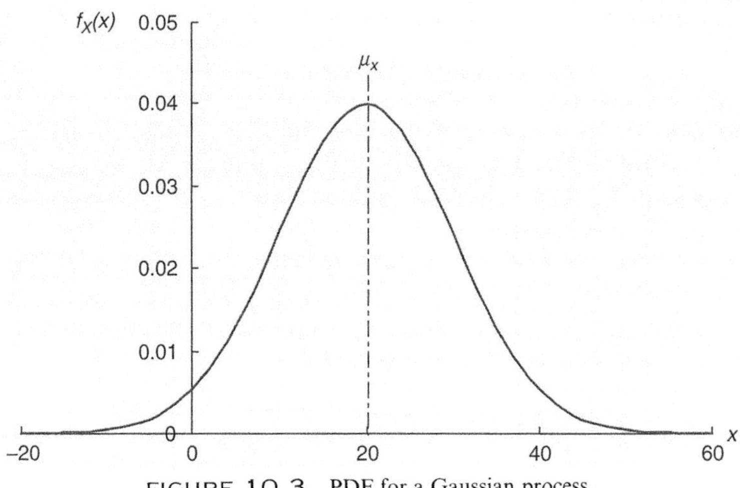

FIGURE 10.3 PDF for a Gaussian process.

where T is the record length. When $\mu_X = 0, \sigma_X$ is the root mean square (RMS) of the random process $X(t)$. The RMS is a measure of the amplitude of the process. The process $X(t)$ is called Gaussian if its PDF $f_X(x)$ follows the bell-shape distribution of Figure 10.3. The PDF is given by

$$f_X(x) = \frac{1}{\sqrt{2\pi}\sigma_X}\exp\left[-\frac{1}{2}\left(\frac{x-\mu_X}{\sigma_X}\right)^2\right] \quad -\infty < x < +\infty \tag{10.2.6}$$

where μ_X and σ_X are the mean and standard deviation of the process, respectively.

10.3 A RANDOM PROCESS

A collection of an infinite number of sample time histories, such as $X_1(t)$, $X_2(t)$, and $X_3(t)$ makes up the random process $X(t)$ as shown in Figure 10.4. Each time history can be generated from a separate experiment. For example, the front axle shaft torque of a vehicle is of interest. It is necessary to learn how the torque varies during long-distance driving. A large number of time histories for various road and weather conditions must be recorded. In engineering, the ensemble of a sufficiently large number of sample time histories approximates the infinite collection representing a random process.

Instead of being measured along a single sample, the ensemble statistical properties are determined across the ensemble, as shown in Figure 10.5. For a Gaussian random process, the ensemble probability density at each time instant and at any two time units must be Gaussian. A random process is said to be stationary if the probability distributions for the ensemble remain

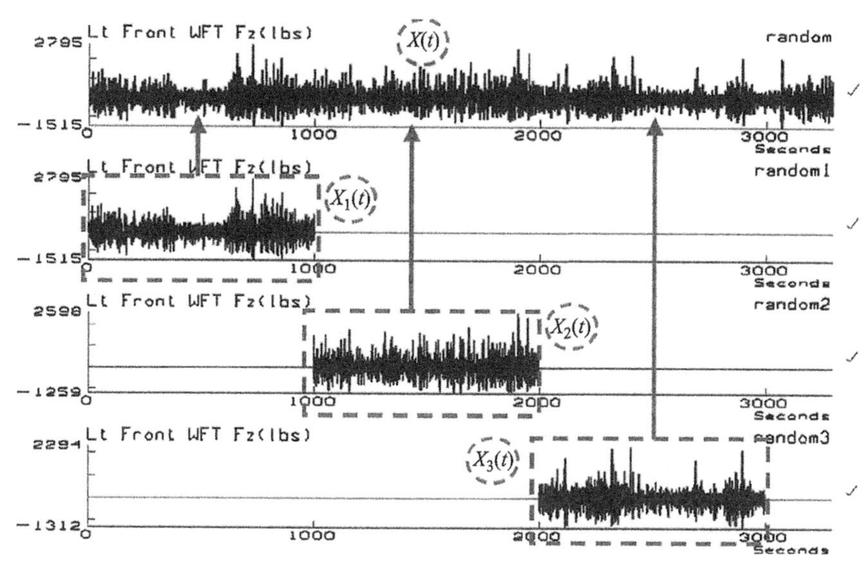

FIGURE 10.4 Random process—ensemble of random sample time histories.

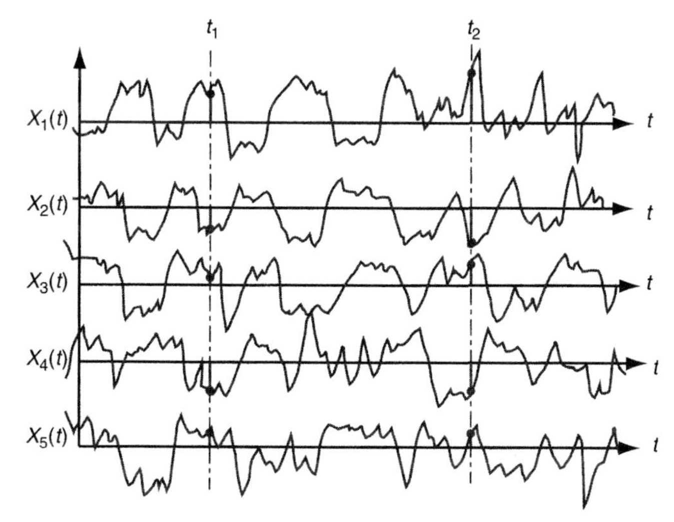

FIGURE 10.5 Illustration of ensemble statistical data.

the same (stationary) for each time instant. This implies that the ensemble mean, standard deviation, variance, and mean square are time invariant. A stationary process is called ergodic if the statistical properties along any single sample are the same as the properties taken across the ensemble. That means each sample time history completely represents the ensemble. Note that if a random process is ergodic, it must be stationary. However, the converse is not true: a stationary process is not necessary ergodic.

The autocorrelation function of a random process is the mean value of the product $X(t_1)X(t_2)$, and is denoted by $E[X(t_1)X(t_2)]$. It can be approximated by the average value of the product $X(t_1)X(t_2)$, which can be obtained by sampling the random variable x at times t_1 and t_2.

For a stationary random process, the value of $E[X(t_1)X(t_2)]$ is time invariant. However, it depends only on the time difference $\tau = |t_2 - t_1|$. Therefore, the autocorrelation function of $X(t)$, denoted by $R(\tau)$, is actually a function of τ. It is expressed as

$$R(\tau) = E[X(t_1)X(t_2)] \tag{10.3.1}$$

Also, because $X(t)$ is stationary, its mean and standard deviation are independent of t. Thus,

$$E[X(t_1)] = E[X(t_2)] = \mu_X \tag{10.3.2}$$

$$\sigma_X(t_1) = \sigma_X(t_2) = \sigma_X \tag{10.3.3}$$

The correlation coefficient, ρ, for $X(t_1)$ and $X(t_2)$ is given by

$$\rho = \frac{R_X(\tau) - \mu_X^2}{\sigma_X^2} \tag{10.3.4}$$

If $\rho = \pm 1$, there is perfect correlation between $X(t_1)$ and $X(t_2)$, and if $\rho = 0$, there is no correlation. Because the value of ρ lies between -1 and 1, it has

$$\mu_X^2 - \sigma_X^2 \leq R_X(\tau) \leq \mu_X^2 + \sigma_X^2 \tag{10.3.5}$$

Figure 10.6 shows the properties of the autocorrelation function $R(\tau)$ of a stationary random process $X(t)$. When the time interval approaches infinity, the random variables at times t_1 and t_2 are not correlated. That means

$$R_X(\tau \to \infty) \cong \mu_X \quad \text{and} \quad \rho \to 0 \tag{10.3.6}$$

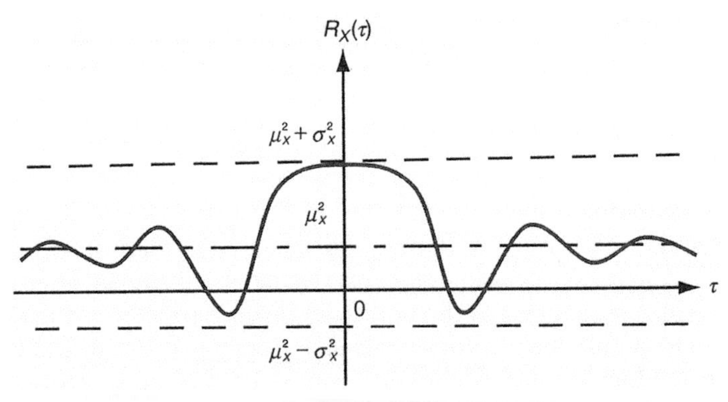

FIGURE 10.6 Autocorrelation function $R_x(\tau)$ of a stationary random process.

When the time interval is zero,

$$R_X(\tau = 0) = \mu_X^2 + \sigma_X^2 = E[X^2] \quad (10.3.7)$$

and $R(0)$ becomes the mean square value of the process. Also, for a stationary random process, $R(\tau)$ is an even function, i.e.,

$$R_X(\tau) = R_X(-\tau) \quad (10.3.8)$$

10.4 FOURIER ANALYSIS

Any periodic time history can be represented by the summation of a series of sinusoidal waves of various amplitude, frequency, and phase. If $X(t)$ is a periodic function of time with a period T, $X(t)$ can be expressed by an infinite trigonometric series of the following form:

$$X(t) = A_0 + \sum_{k=1}^{\infty}\left(A_k \cos\frac{2\pi kt}{T} + B_k \sin\frac{2\pi kt}{T}\right) \quad (10.4.1)$$

where

$$A_0 = \frac{1}{T}\int_{-T/2}^{T/2} X(t)dt$$

$$A_k = \frac{2}{T}\int_{-T/2}^{T/2} X(t)\cos\frac{2\pi kt}{T}dt$$

$$B_k = \frac{2}{T}\int_{-T/2}^{T/2} X(t)\sin\frac{2\pi kt}{T}dt$$

The Fourier series can be also expressed by using complex coefficients as

$$X(t) = \sum_{k=1}^{\infty} C_k e^{i2\pi kt/T} \quad (10.4.2)$$

where the complex coefficients C_k are given by

$$C_k = \frac{1}{T}\int_{-T/2}^{T/2} X(t)e^{-i2\pi kt/T}dt \quad (10.4.3)$$

The Fourier transform can be considered as the limit of the Fourier series of $X(t)$ as T approaches infinity. This can be illustrated as follows by rewriting Equation 10.4.2 with infinite T:

$$X(t) = \lim_{T\to\infty} \sum_{k=-\infty}^{\infty} \left(\frac{1}{T} \int_{-T/2}^{T/2} X(t) e^{-i2\pi kt/T} dt \right) e^{i2\pi kt/T} \tag{10.4.4}$$

If the frequency of the kth harmonic, ω_k, in radians per second, is

$$\omega_k = \frac{2\pi k}{T} \tag{10.4.5}$$

and the spacing between adjacent periodic functions, $\Delta\omega$, is

$$\Delta\omega = \frac{2\pi}{T} \tag{10.4.6}$$

Equation 10.4.4 becomes

$$X(t) = \lim_{T\to\infty} \sum_{k=-\infty}^{\infty} \left(\frac{\Delta\omega}{2\pi} \int_{-T/2}^{T/2} X(t) e^{-ik\Delta\omega t} dt \right) e^{ik\Delta\omega} \tag{10.4.7}$$

As T goes to infinity, the frequency spacing, $\Delta\omega$, becomes infinitesimally small, denoted by $d\omega$, and the sum becomes an integral. As a result, Equation 10.4.7 can be expressed by the well-known Fourier transform pair $X(t)$ and $X(\omega)$:

$$X(\omega) = \frac{1}{2\pi} \int_{-\infty}^{\infty} X(t) e^{-i\omega t} dt \tag{10.4.8}$$

$$X(t) = \int_{-\infty}^{\infty} X(\omega) e^{i\omega t} d\omega \tag{10.4.9}$$

The function $X(\omega)$ is the forward Fourier transform of $X(t)$, and $X(t)$ is the inverse Fourier transform of $X(\omega)$. The Fourier transform exists if the following conditions are met:

1. The integral of the absolute function exists, i.e., $\int_{-\infty}^{\infty} |X(t)| dt < \infty$.
2. Any discontinuities are finite.

10.5 SPECTRAL DENSITY

The Fourier transform of a stationary random process $X(t)$ usually does not exist because the condition

$$\int_{-\infty}^{\infty} |X(t)| dt < \infty \tag{10.5.1}$$

is not met. However, the Fourier transform of the autocorrelation function $R_X(\tau)$ always exists. If the stationary random process $X(t)$ is adjusted (or normalized) to a zero-mean value, i.e.,

$$R_X(\tau \to \infty) = 0 \tag{10.5.2}$$

the condition

$$\int_{-\infty}^{\infty} |R_X(\tau)| dt < \infty \tag{10.5.3}$$

is met. In this case, the forward and inverse Fourier transforms of $R_X(\tau)$ are given by

$$S_X(\omega) = \frac{1}{2\pi} \int_{-\infty}^{\infty} R_X(\tau) e^{-i\omega\tau} d\tau \tag{10.5.4}$$

$$R_X(\tau) = \int_{-\infty}^{\infty} S_X(\omega) e^{i\omega\tau} d\omega \tag{10.5.5}$$

where $S_X(\omega)$ is the spectral density of the normalized stationary random process $X(t)$. If $\tau = 0$, Equation 10.5.5 reduces to

$$E[X^2] = R_X(0) = \int_{-\infty}^{\infty} S_X(\omega) d\omega = \sigma_X^2 \tag{10.5.6}$$

This means that the square root of the area under a spectral density plot $S_X(\omega)$ is the RMS of a normalized stationary random process. $S_X(\omega)$ is also called mean square spectral density and is illustrated in Figure 10.7.

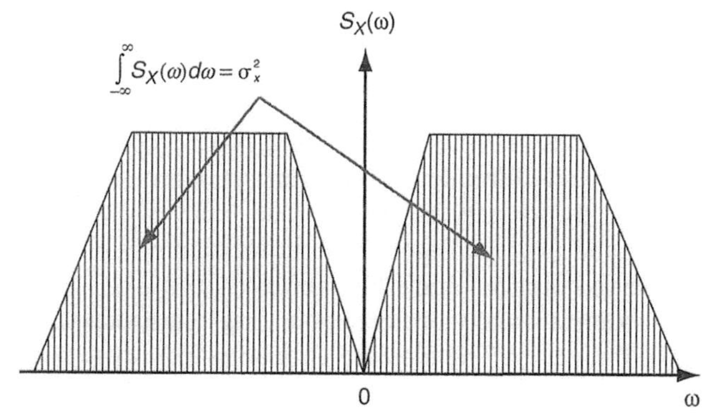

FIGURE 10.7 Relationship between the spectral density and root mean square of a normalized stationary random process.

The idea of a negative frequency has been introduced for mathematical completeness. However, it has no physical meaning. It is common practice to consider the frequency from zero to infinity and to have the frequency expressed in hertz (cycle/second) rather than radians/second. Therefore, the two-sided spectral density, $S_X(\omega)$, can be transformed into an equivalent one-sided spectral density $W_X(f)$ as follows:

$$E[X^2] = \sigma_X^2 = \int_0^\infty W_X(f)df \tag{10.5.7}$$

where

$$W_X(f) = 4\pi S_X(\omega) \tag{10.5.8}$$

is the PSD and

$$f = \frac{\omega}{2\pi} \tag{10.5.9}$$

The following spectral density relationships exist for first and second derivatives of a stationary random process $X(t)$:

$$S_{\dot{X}}(\omega) = \omega^2 S_X(\omega) \tag{10.5.10}$$

$$W_{\dot{X}}(f) = (2\pi)^2 f^2 W_X(f) \tag{10.5.11}$$

$$\sigma_{\dot{X}}^2 = \int_{-\infty}^{\infty} S_{\dot{X}}(\omega)d\omega = \int_{-\infty}^{\infty} \omega^2 S_X(\omega)d\omega = (2\pi)^2 \int_0^\infty f^2 W_X(f)df \tag{10.5.12}$$

$$S_{\ddot{X}}(\omega) = \omega^4 S_X(\omega) \tag{10.5.13}$$

$$W_{\ddot{X}}(f) = (2\pi)^4 f^4 W_X(f) \tag{10.5.14}$$

$$\sigma_{\ddot{X}}^2 = \int_{-\infty}^{\infty} S_{\ddot{X}}(\omega)d\omega = \int_{-\infty}^{\infty} \omega^4 S_X(\omega)d\omega = (2\pi)^4 \int_0^\infty f^4 W_X(f)df \tag{10.5.15}$$

A random process is called a narrow-band process if its spectral density has only a narrow band of frequencies. In contrast, a broad-band process is a process whose spectral density covers a broad band of frequencies. White noise is a broad-band process. Figure 10.8 shows examples of a narrow-band process, a wide-band process, and white noise.

The PSD function is usually presented on a log–log scale. An octave (oct) is a doubling of frequency. The increase in octaves from f_1 to f_2 is

$$oct = \frac{1}{\log_{10} 2} \log \frac{f_2}{f_1} \tag{10.5.16}$$

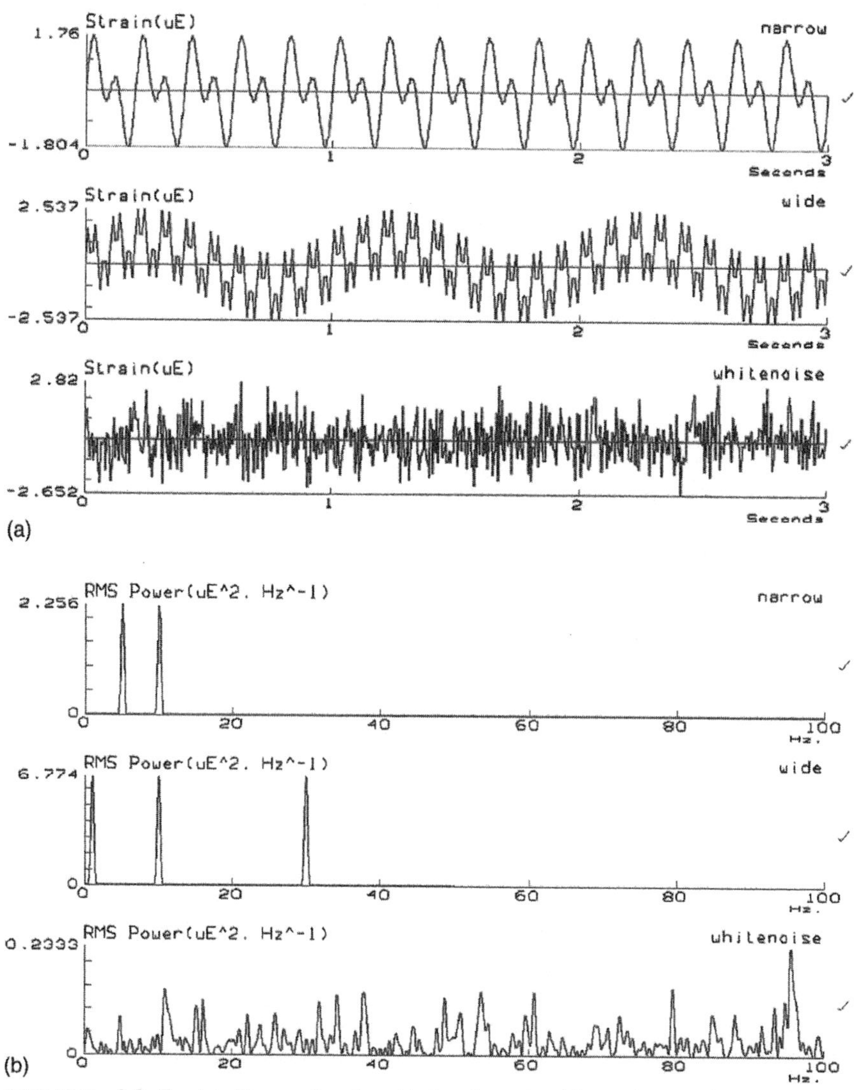

FIGURE 10.8 (a) Narrow-band, wide-band, and white noise random processes, (b) power spectral densities (PSDs) of narrow-band, wide-band, and white noise random processes.

A bel is the common logarithm of the ratio of two measurements of power. A decibel (db) is one-tenth of a bel and is defined by

$$\text{db} = 10\log_{10}\frac{W_X(f_2)}{W_X(f_1)} \tag{10.5.17}$$

A doubling of PSD corresponds to an increase of approximately 3 db. If the PSD doubles for each doubling of frequency, the spectrum increases at 3 db/octave or the spectrum has a positive roll-off rate of 3 db/octave. Figure 10.9 shows a random vibration test profile with a constant positive and negative roll-off rate of 6 db/octave. It is noted that on a log–log scale the constant decibel per octave lines are straight. The negative roll-off rate of 6 db/octave indicates a power decrease of 4:1 in one octave from 1000 to 2000 Hz.

Example 10.1 Calculate the RMS acceleration for the $W_X(f)$ profile shown in Figure 10.9.

Solution. The calculation of the RMS acceleration may appear relatively simple, at first, as being the RMS sum of the three areas under the PSD curve in Figure 10.9. But because the profile is drawn on a log scale, the slopes that appear to be straight lines are actually exponential curves. Thus, we need to replot using linear graph paper as in Figure 10.10 and subsequently calculate the area under the exponential curves. Mathematically, the exponential curve can be expressed in the following form:

$$W_X(f) = A \times (f)^B \tag{10.5.18}$$

where A and B are the exponential parameters. B is the slope of the roll-off line on log–log scales and is defined as

$$B = \frac{\log \dfrac{W_X(f_2)}{W_X(f_1)}}{\log \dfrac{f_2}{f_1}} = \frac{\dfrac{1}{10} \times \left(10 \log \dfrac{W_X(f_2)}{W_X(f_1)}\right)}{\log 2 \times \left(\dfrac{1}{\log 2} \log \dfrac{f_2}{f_1}\right)} \tag{10.5.19}$$
$$= \frac{1}{3}(\text{the roll-off rate in db/octave})$$

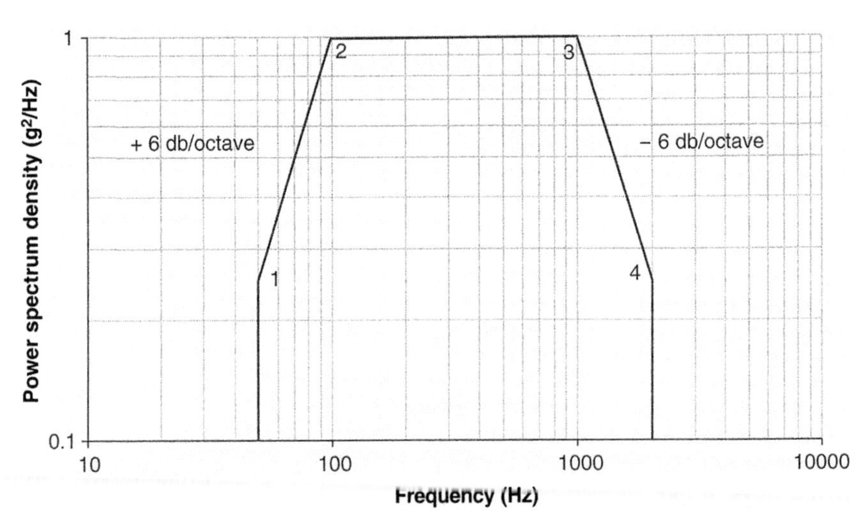

FIGURE 10.9 Example of a random vibration test profile on log–log scales.

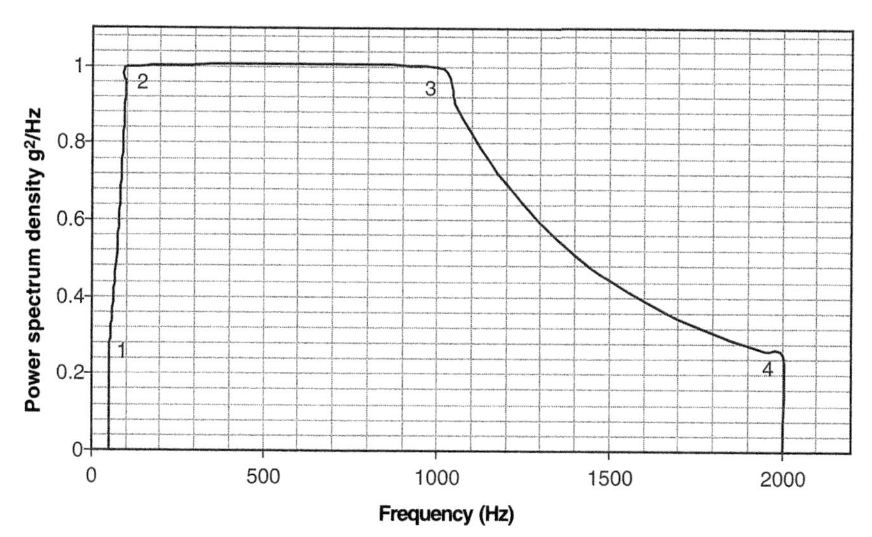

FIGURE 10.10 Example of a random vibration test profile on linear scales.

A is the exponential coefficient and is obtained as

$$A = \frac{W_X(f_1)}{(f_1)^B} = \frac{W_X(f_2)}{(f_2)^B} \qquad (10.5.20)$$

The area under the roll-off curve is therefore

$$\text{Area} = \int_{f_1}^{f_2} W_X(f)df = \int_{f_1}^{f_2} A \times (f)^B df = \frac{A}{B+1}\left[(f_2)^{B+1} - (f_1)^{B+1}\right] \qquad (10.5.21)$$

For the first segment 1–2, that the following is known:

$$f_1 = 50\,\text{Hz}, \quad W_X(f_1) = 0.25\,\text{g}^2/\text{Hz}$$

and

$$f_2 = 100\,\text{Hz}, \quad W_X(f_2) = 1.00\,\text{g}^2/\text{Hz}.$$

B and A are therefore equal to

$$B = \frac{\log\dfrac{1.0}{0.25}}{\log\dfrac{100}{50}} = 2$$

and

$$A = \frac{W_X(f_1)}{(f_1)^B} = \frac{0.25}{50^2} = 10^{-4}$$

The area under the segment 1–2 is then equal to

$$\text{Area}_{1-2} = \frac{A}{B+1}\left[(f_2)^{B+1} - (f_1)^{B+1}\right] = \frac{10^{-4}}{2+1}[100^{2+1} - 50^{2+1}] = 29.2\,\text{g}^2$$

For the second segment 2–3, the calculation of the rectangular area is

$$\text{Area}_{2-3} = 1.0 \times (1000 - 100) = 900\,\text{g}^2$$

For the third segment 3–4, the following is known:

$$f_1 = 1000\,\text{Hz}, \; W_X(f_1) = 1.00\,\text{g}^2/\text{Hz}$$

and

$$f_2 = 2000\,\text{Hz}, \; W_X(f_2) = 0.25\,\text{g}^2/\text{Hz}.$$

By using Equations 10.5.19 and 10.5.20, B and A are equal to -2 and 10^6, respectively. The area under the segment 3–4 is

$$\text{Area}_{3-4} = \frac{A}{B+1}\left[(f_2)^{B+1} - (f_1)^{B+1}\right] = \frac{10^6}{-2+1}\left[2000^{-2+1} - 1000^{-2+1}\right] = 500\,\text{g}^2$$

Thus, the total area is equal to

$$\text{Area}_{1-2} + \text{Area}_{2-3} + \text{Area}_{3-4} = 29.2 + 900 + 500 = 1429.2\,\text{g}^2$$

and the RMS acceleration is equal to $\sqrt{1429.2} = 37.8\,\text{g}$.

10.6 LEVEL CROSSING RATE OF NARROW-BAND RANDOM PROCESSES

For a continuous and differentiable stationary process, $X(t)$, the expected number of positively sloped crossing (upcrossing) in an infinitesimal interval is only dependent on dt. We have

$$E[N_{a^+}(dt)] = \upsilon_{a^+}dt \tag{10.6.1}$$

where υ_{a^+} is the expected rate of upcrossing per time unit. If A denotes the event that any random sample from $X(t)$ has an upcrossing $x = a$ in an infinitesimal time interval dt, the probability of Event A, $P(A)$, is given

$$P(A) = \upsilon_{a^+}dt \tag{10.6.2}$$

Equation 10.6.2 allows us to express υ_{a^+} in terms of P(A). For Event A to exist, we must have

$$a - \dot{X}(t) < X(t) < a \quad \text{and} \quad \dot{X}(t) > 0$$

Combining these two conditions, $P(\text{A})$ can be written as

$$P(\text{A}) = P(a - X(t) < X(t) < a \cap \dot{X}(t) > 0) \tag{10.6.3}$$

These conditons define a triangle area in the $X(t) - \dot{X}(t)$ plane, as shown in Figure 10.11.

The probability of Event A is calculated by integrating the joint PDF of $X(t)$ and $\dot{X}(t)$ over this region, i.e.,

$$P(\text{A}) = \int_0^{\infty} \int_{a-vdt}^{a} f_{X\dot{X}}(u, v)dudv \tag{10.6.4}$$

Substituting Equation 10.6.4 into Equation 10.6.2 leads to the following expression of the level upcrossing rate for a stationary random process:

$$\upsilon_{a^+} = \int_0^{\infty} \upsilon f_{X\dot{X}}(a, \upsilon)d\upsilon \tag{10.6.5}$$

If $X(t)$ is Gaussian, the expected upcrossing rate of $x = a$ is

$$\upsilon_{a^+} = \frac{1}{2\pi}\frac{\sigma_{\dot{X}}}{\sigma_X}\exp\left(\frac{-a^2}{2\sigma^2}\right) \tag{10.6.6}$$

The expected rate of zero upcrossings $E[0^+]$ is found by letting $a = 0$ in Equation 10.6.6:

$$E[0^+] = \frac{1}{2\pi}\frac{\sigma_{\dot{X}}}{\sigma_X} \tag{10.6.7}$$

By using Equations 10.5.7 and 10.5.12, the expected rate of zero upcrossing is

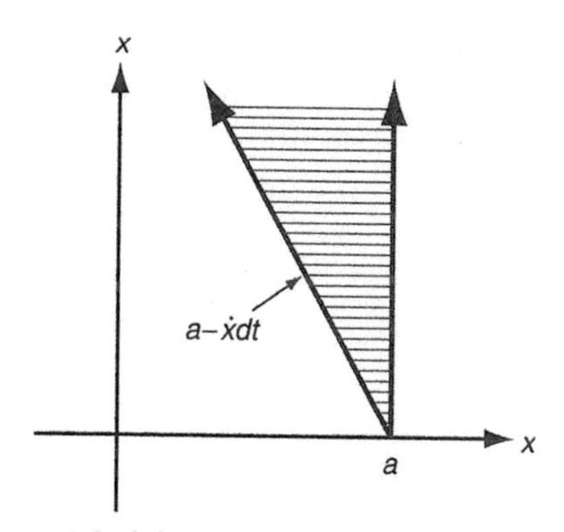

FIGURE 10.11 The region where Event A occurs.

$$E[0^+] = \sqrt{\frac{\int_0^\infty f^2 W_X(f)df}{\int_0^\infty W_X(f)df}} \tag{10.6.8}$$

The expected rate of peak crossing, $E[P]$, is found from a similar analysis of the velocity process $\dot{X}(t)$. The rate of zero downcrossing of the velocity process corresponds to the occurrence of a peak in $\dot{X}(t)$. The result for a Gaussian process is

$$E[P] = \frac{1}{2\pi}\frac{\sigma_{\ddot{X}}}{\sigma_X} \tag{10.6.9}$$

By using Equations 10.5.12 and 10.5.15, we have

$$E[P] = \sqrt{\frac{\int_0^\infty f^4 W_X(f)df}{\int_0^\infty f^2 W_X(f)df}} \tag{10.6.10}$$

A narrow-band process is smooth and harmonic. For every peak there is a corresponding zero upcrossing, meaning $E[0^+]$ is equal to $E[P]$. However, the wide-band process is more irregular. A measure of this irregularity is the ratio of the zero upcrossing rate to the peak crossing rate. The ratio is known as the irregularity factor, γ, i.e.,

$$\gamma = \frac{E[0^+]}{E[P]} \tag{10.6.11}$$

Figure 10.12 shows a simple example of how to calculate the irregularity factor. Note that when $\gamma \to 0$, there is an infinite number of peaks for every zero upcrossing. This is considered a wide band random process. The vaue of $\gamma = 1$ corresponds to one peak per one zero upcrossing and it represents a narrow-band random process.

Alternatively, a narrow- or wide-band process can be judged by the width of its spectrum. For this reason, the spectral width parameter, λ, is introduced as

$$\lambda = \sqrt{1 - \gamma^2} \tag{10.6.12}$$

Note that $\lambda \to 0$ represents a narrow-band random process.

If M_j is the jth moment of a one-sided PSD function (Figure 10.13) defined as

$$M_j = \int_0^\infty f^j W_{S_a}(f)df \tag{10.6.13}$$

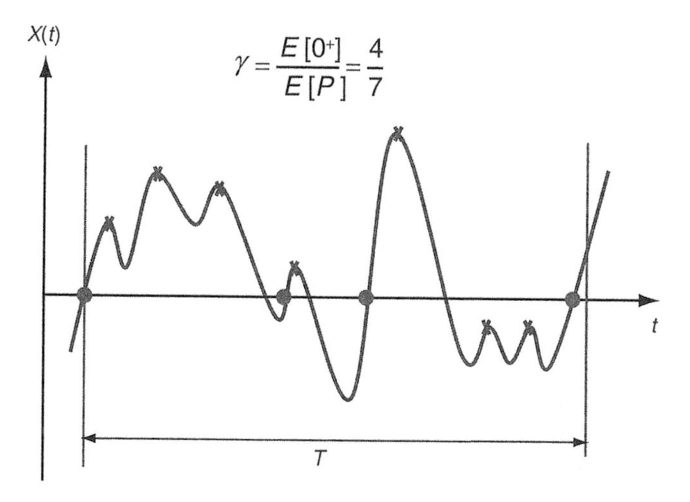

FIGURE 10.12 Calculation of the irregularity factor γ.

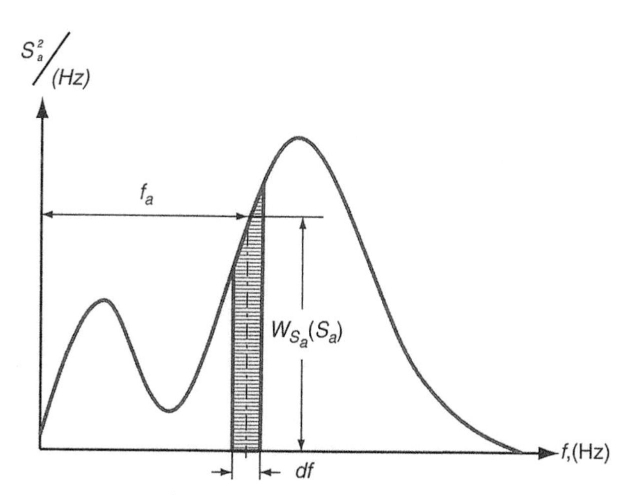

FIGURE 10.13 Moments from a one-side PSD.

the rate of zero crossings $E[0^+]$ and rate of peaks $E[P]$ are given by

$$E[0^+] = \sqrt{\frac{M_2}{M_0}} \tag{10.6.14}$$

$$E[P] = \sqrt{\frac{M_4}{M_2}} \tag{10.6.15}$$

$$\gamma = \sqrt{\frac{M_2^2}{M_0 M_4}} \tag{10.6.16}$$

$$\lambda = \sqrt{1 - \frac{M_2^2}{M_0 M_4}} \tag{10.6.17}$$

10.7 MODELS OF FATIGUE DAMAGE UNDER NARROW-BAND RANDOM PROCESSES

In this section, a linear S–N curve in log–log coordinates and the linear damage accumulation rule are employed. It is assumed that all stress cycles have the same mean value. If a mean stress exists, it can be accounted for by various mean stress correction models (see Chapter 4).

Variable amplitude loading is simulated by a sequence of blocks of constant amplitudes. Damage D is defined as

$$D = \sum_{i=1}^{k} \frac{n_i}{N_{f,i}} \tag{10.7.1}$$

where n_i is the total number of cycles in the ith block of constant-stress amplitude $S_{a,i}$, $N_{f,i}$ is the number of cycles to failure under $S_{a,i}$, and k is the total number of blocks. Failure occurs when $D \geq 1$. Recall that the relationship between constant stress amplitude $S_{a,i}$ (or constant stress range $S_{r,i}$) and the fatigue life $N_{f,i}$ (see Chapter 4) has the following expression:

$$S_{a,i} = S_f'(2N_{f,i})^b \tag{10.7.2}$$

or

$$N_{f,i} = \frac{1}{2}\left(\frac{S_{a,i}}{S_f'}\right)^{1/b} \tag{10.7.3}$$

where S_f' is the fatigue strength coefficient and b is the fatigue strength exponent. If $m = -1/b$ and $A = 0.5 \times \left(S_f'\right)^m$, an alternative form of the S–N curve is given by

$$N_{f,i} = A \times S_{a,i}^{-m} \tag{10.7.4}$$

This is a very convenient formula to predict $N_{f,i}$ for a given $S_{a,i}$.

The cycle counting histogram (also called discrete stress spectrum) for a narrow-band stress process $S(t)$ can be established by either performing the rainflow cycle counting technique (see Chapter 3) or by the counting the number of peaks n_i in the window ΔS around a stress level. Suppose that the total number of peaks counted in the $S(t)$ process is denoted by $\sum_{i=1}^{k} n_i$. The probability f_i that a stress amplitude $S_a = S_{a,i}$ will occur is

$$f_i = \frac{n_i}{\sum_{i=1}^{k} n_i} \tag{10.7.5}$$

Thus, f_i is the PDF of the random variable S_a. In this case, the total fatigue damage can be written as

$$D = \sum_{i=1}^{k} \frac{n_i}{N_{f,i}} = \sum_{i=1}^{k} \frac{f_i \sum_{i=1}^{k} n_i}{N_{f,i}} \tag{10.7.6}$$

By using the linear S–N model, the expression for fatigue damage is

$$D = \frac{\sum_{i=1}^{k} n_i}{A} \sum_{i=1}^{k} f_i S_{a,i}^m \tag{10.7.7}$$

Also, the expected value of S_a^m is

$$E(S_a^m) = \sum_{i=1}^{k} f_i S_i^m \tag{10.7.8}$$

Therefore,

$$D = \frac{\sum_{i=1}^{k} n_i}{A} E(S_a^m) \tag{10.7.9}$$

where the total count of cycles $\sum_{i=1}^{k} n_i$ is equal to the rate of upcrossing multiplying the total time period $T\,(= E[0^+] \times T)$.

Assume that the PDF of stress amplitude S_a can be treated as a continuous random variable as shown in Figure 10.14. The expected value of S_a^m is

$$E(S_a^m) = \int_0^{\infty} s_a^m f_{S_a}(s_a) ds_a \tag{10.7.10}$$

Even though any statistical model of S_a can be employed, it is common to use the Weibull distribution with the following CDF:

$$F_{S_a}(s_a) = 1 - \exp\left[-\left(\frac{s_a}{\alpha}\right)^{\beta}\right] \tag{10.7.11}$$

where α and β are the scale parameter (characteristic life) and shape parameter (Weibull slope), respectively. For the Weibull distribution,

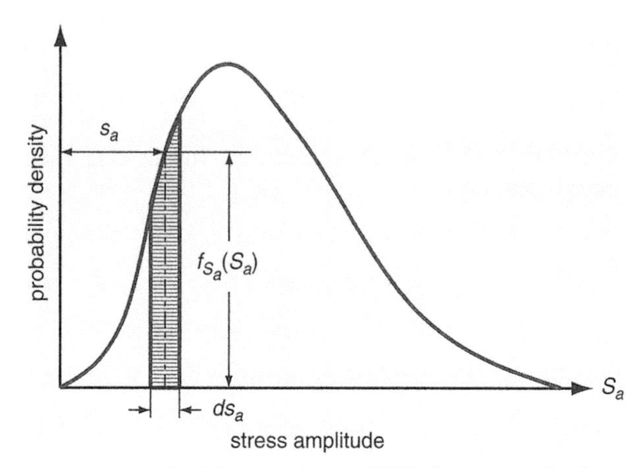

FIGURE 10.14 Continuous PDF of stress amplitude.

$$E(S_a^m) = \alpha^m \Gamma\left(\frac{m}{\beta} + 1\right) \tag{10.7.12}$$

where $\Gamma(.)$ is the gamma function.

In the special case in which $\beta = 2$, the Weibull distribution reduces to the Rayleigh distribution. This is an important case, because Rayleigh is the distribution of peaks or ranges or amplitude in a stationary narrow-band Gaussian process that has an RMS value of σ_S. Also, it can be shown that

$$\alpha = \sqrt{2}\sigma_S \tag{10.7.13}$$

Therefore, if $S(t)$ is stationary narrow-band Gaussian and the stress amplitudes follow the Rayleigh distribution,

$$E(S_a^m) = \left(\sqrt{2}\sigma_S\right)^m \Gamma\left(\frac{m}{2} + 1\right) \tag{10.7.14}$$

where $\sigma_S = \sqrt{M_0}$.

In such a case, therefore, the expected total fatigue damage D_{NB} of a zero-mean stationary narrow-band Gaussian stress process over a time interval τ can be written as

$$D_{\text{NB}} = \frac{\sum_{i:1}^{k} n_i}{A} E(S_a^m) = \frac{E[0^+] \times T}{A} \left(\sqrt{2M_0}\right)^m \Gamma\left(\frac{m}{2} + 1\right) \tag{10.7.15}$$

10.8 MODELS OF FATIGUE DAMAGE UNDER WIDE-BAND RANDOM PROCESSES

Based on the rainflow counting method, a model for predicting fatigue damage under stationary wide band Gaussian stress process has been proposed by Wirsching and Light (1980). By using the narrow-band approach as a starting point, the general expression for the damage $D_{\text{WB, Wirsching}}$ over a time interval τ is

$$D_{\text{WB, Wirsching}} = \zeta_W D_{\text{NB}} \tag{10.8.1}$$

where D_{NB} is the fatigue damage under a narrow-band random process and ζ_W is the rainflow correction factor. ζ_W is an empirical factor derived from extensive Monte Carlo simulations that include a variety of spectral density functions. It is expressed as follows:

$$\zeta_S = a_S + [1 - a_S](1 - \lambda)^{b_W} \tag{10.8.2}$$

where

$$a_S = 0.926 - 0.033m \quad b_S = 1.587m - 2.323$$

Note that m is the slope of the S–N curve defined in Equation 10.7.4, and λ is the spectral width parameter defined in Equations 10.6.12 or 10.6.17.

Oritz and Chen (1987) also derived another similar expression for fatigue damage, $D_{\text{WB, Ortiz}}$, under wide-band stresses as

$$D_{\text{WB, Oritz}} = \zeta_O D_{\text{NB}} \tag{10.8.3}$$

where

$$\zeta_O = \frac{1}{\gamma}\sqrt{\frac{M_2 M_k}{M_0 M_{k+2}}} \quad \text{and} \quad k = \frac{2.0}{m} \tag{10.8.4}$$

The irregularity factor γ is defined in Equation 10.6.11 or 10.6.16.

Instead of using the damage correction factor from the narrow-band stresses to the wide-band stresses, Dirlik (1985) developed an empirical closed-form expression for the PDF of rainflow amplitude, $f_{S_a}(s_a)$, based on extensive Monte Carlo simulations of the stress amplitudes. Dirlik's solutions were sucessfully verified by Bishop (1988) in theory. Dirlik's damage model for a time period of τ is as follows:

$$D_{\text{WB, Dirlik}} = \frac{E[P]\tau}{A} \int_0^\infty s_a^m f_{S_a}(S_a) dS_a \tag{10.8.5}$$

$$f_{S_a}(s_a) = \frac{D_1}{2\sqrt{M_0}Q} e^{\frac{-Z}{Q} \times s_a} + \frac{D_2 \times Z}{2\sqrt{M_0}R^2} e^{\frac{-Z^2}{2R^2} \times s_a^2} + \frac{D_3 \times Z}{2\sqrt{M_0}} \times e^{\frac{-Z^2}{2} \times s_a^2} \tag{10.8.6}$$

where

$$Z = \frac{1}{2\sqrt{M_0}} \quad \gamma = \frac{M_2}{\sqrt{M_0 M_4}} \quad X_m = \frac{M_1}{M_0}\sqrt{\frac{M_2}{M_4}}$$

$$D_1 = \frac{2(X_m - \gamma^2)}{1+\gamma^2} \quad R = \frac{\gamma - X_m - D_1^2}{1 - \gamma - D_1 + D_1^2} \quad D_2 = \frac{1 - \gamma - D_1 + D_1^2}{1 - R}$$

$$D_3 = 1 - D_1 - D_2 \quad Q = \frac{1.25(\gamma - D_3 - D_2 \times R)}{D_1}$$

Bishop and Sherratt (1989) and Bishop (1994) concluded that the Dirlik solution gives the better results as compared with the corresponding time domain results.

Example 10.7.1. A hot-rolled component made of SAE 1008 steel is subjected to random loading process. The fatigue strength coefficient (S_f') and the fatigue strength exponent, (b) are 1297 MPa and −0.18, respectively The stress response at a critical location is calculated in terms of the PSD in Figure 10.15. The PSD has two frequencies of 1 and 10 Hz, corresponding to 10,000 and 2500 MPa^2/Hz, respectively. Determine the fatigue damage of this component, using the equations for wide-band stresses. Note that a crest factor is defined as the ratio of peak (maximum) to RMS values. A sine wave has a crest factor of $\sqrt{2} = 1.414$.

Solution. We will calculate the fatigue damage of the component in time domain first and use it as baseline information to compare with the predicted damage based on the PSD. The original stress time history can be obtained by adding two sine waves, one for each block in the PSD. For a sine wave, the amplitude of each is calculated from 1.414 times the RMS value, i.e., the area of each PSD block. The stress amplitude $S_{a,1}$ of the sine wave at 1 Hz is

$$S_{a,1} = \sqrt{10000 \times 1} \times 1.414 = 141.4\,\text{MPa}$$

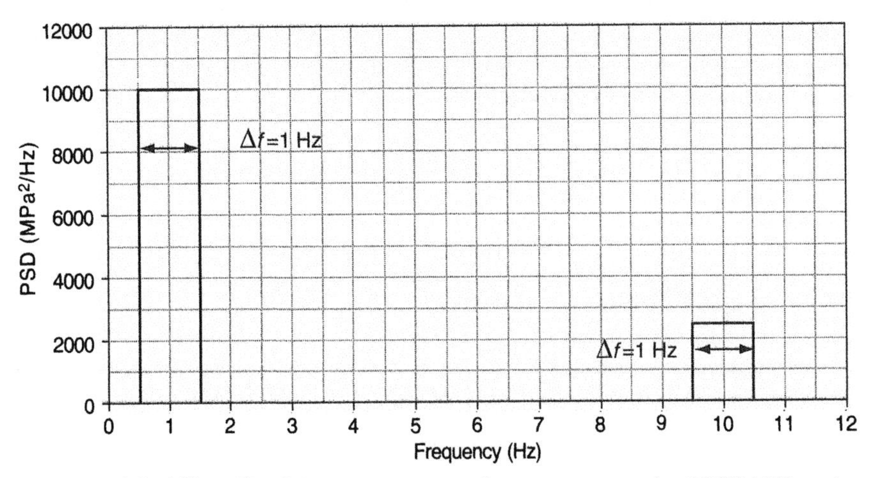

FIGURE 10.15 PSD of the stress response of a component made of SAE 1008 steel.

The stress amplitude $S_{a,2}$ of the second sine wave at 10 Hz is

$$S_{a,2} = \sqrt{2500 \times 1} \times 1.414 = 70.7\,\text{MPa}$$

The material S–N curve for vibration fatigue usually follows the following expression:

$$N_{f,i} = A \times S_{a,i}^{-m}$$

where

$$m = -1/b = -1/(-0.18) = 5.56$$

$$A = 0.5 \times \left(S_f'\right)^m = 0.5 \times 1297^{5.56} = 1.02 \times 10^{17}\,\text{MPa}$$

With this S–N curve, we determine the fatigue life for each sine wave as follows:

$$N_{f,1} = 1.02 \times 10^{17} \times 141.4^{-5.56} = 1.13 \times 10^5 \text{ cycles}$$

$$N_{f,2} = 1.02 \times 10^{17} \times 70.7^{-5.56} = 5.32 \times 10^6 \text{ cycles}$$

In a 1-sec time interval, the sine waves at 1 and 10 Hz represent 1 cycle (i.e., $n_1 = 1$) and 10 cycles (i.e., $n_2 = 10$) of reversed loading, respectively. The linear damage calculation for this time interval gives

$$D_{\text{time}} = \frac{n_1}{N_{f,1}} + \frac{n_2}{N_{f,2}} = \frac{1}{1.13 \times 10^5} + \frac{10}{5.32 \times 10^6} = 1.08 \times 10^{-5}$$

This corresponds to a fatigue life of 93,000 seconds ($= 1/1.08 \times 10^{-5}$).

We then calculate the fatigue damage to the component based on the given PSD. The aforementioned frequency domain methods will be used for the damage estimation. The jth moment of the PSD can be easily calculated as

$$M_j = \int_0^\infty f^j W_{S_a}(f)df = 1^j \times 10000 \times 1 + 10^j \times 2500 \times 1$$

$$M_0 = 1^0 \times 10000 \times 1 + 10^0 \times 2500 \times 1 = 12500$$

$$M_2 = 1^2 \times 10000 \times 1 + 10^2 \times 2500 \times 1 = 260000$$

$$M_4 = 1^4 \times 10000 \times 1 + 10^4 \times 2500 \times 1 = 25010000$$

From this we can compute

$$E[0^+] = \sqrt{\frac{M_2}{M_0}} = \sqrt{\frac{260000}{12500}} = 4.56$$

$$E[P] = \sqrt{\frac{M_4}{M_2}} = \sqrt{\frac{25010000}{260000}} = 9.81$$

$$\gamma = \frac{E[0^+]}{E[P]} = \frac{4.56}{9.81} = 0.465$$

$$\lambda = \sqrt{1-\gamma^2} = \sqrt{1-0.465^2} = 0.885$$

1. Wirsching and Light Method

Fatigue damage D_{NB} of a zero-mean stationary narrow-band Gaussian stress process over a time interval $\tau = 1$ sec can be written as

$$D_{NB} = \frac{E[0^+] \times \tau}{A}\left(\sqrt{2M_0}\right)^m \Gamma\left(\frac{m}{2}+1\right)$$

$$D_{NB} = \frac{4.56 \times 1}{1.02 \times 10^{17}}\left(\sqrt{2 \times 12500}\right)^{5.56} \Gamma\left(\frac{5.56}{2}+1\right)$$

$$D_{NB} = 0.000345$$

The rainflow correction factor is calculated as

$$a_w = 0.926 - 0.033m = 0.926 - 0.033 \times 5.56 = 0.743$$

$$b_w = 1.587m - 2.323 = 1.587 \times 5.56 - 2.323 = 6.501$$

$$\zeta_W = a_W + [1-a_W](1-\lambda)^{b_W} = 0.743 + (1-0.743) \times (1-0.885)^{6.501} = 0.743$$

Finally, the fatigue damage $D_{WB,\,\text{Wirsching}}$ is computed as

$$D_{WB,\,\text{Wirsching}} = \zeta_W D_{NB} = 0.743 \times 0.000345 = 0.000256/\text{second}$$

This corresponds to a fatigue life of 3900 seconds (= 1/0.000256), which is very conservative as compared with the baseline fatigue life (93,000 seconds).

2. Oritz and Chen Method

$D_{NB} = 0.000345$ is the same as that calculated previously. Calculation of the rainflow correction factor ζ_O is required. Given the slope of the S–N curve, $m = 5.56$,

$$k = \frac{2.0}{m} = \frac{2.0}{5.56} = 0.3597$$

$$M_k = 1^{0.3597} \times 10000 \times 1 + 10^{0.3597} \times 2500 \times 1 = 15723$$

$$M_{k+2} = 1^{0.3597+2} \times 10000 \times 1 + 10^{0.3597+2} \times 2500 \times 1 = 582338$$

$$\zeta_O = \frac{1}{\gamma}\sqrt{\frac{M_2 M_k}{M_0 M_{k+2}}} = \frac{1}{0.465}\sqrt{\frac{260000 \times 15723}{12500 \times 582338}} = 0.432$$

The fatigue damage $D_{WB,\,\text{Lutes}}$ is computed as

$$D_{\text{WB, Oritz}} = \zeta_O D_{\text{NB}} = 0.432 \times 0.000345 = 0.00015/\text{second}$$

This corresponds to a fatigue life of 6700 seconds (= 1/0.00015), which is very conservative as compared with the baseline fatigue life (93,000 seconds).

3. Dirlik Method

It is required to determine the following parameters for the PDF of stress amplitudes that have been rainflow cycle counted.

$$Z = \frac{1}{2\sqrt{M_0}} = \frac{1}{2\sqrt{12500}} = 0.004472$$

$$X_m = \frac{M_1}{M_0}\sqrt{\frac{M_2}{M_4}} = \frac{35000}{12500}\sqrt{\frac{260000}{25010000}} = 0.2855$$

$$D_1 = \frac{2(X_m - \gamma^2)}{1+\gamma^2} = \frac{2(0.2855 - 0.465^2)}{1+0.465^2} = 0.1074$$

$$R = \frac{\gamma - X_m - D_1^2}{1 - \gamma - D_1 + D_1^2} = \frac{0.465 - 0.2855 - 0.1074^2}{1 - 0.465 - 0.1074 + 0.1074^2} = 0.3825$$

$$D_2 = \frac{1 - \gamma - D_1 + D_1^2}{1 - R} = \frac{1 - 0.465 - 0.1074 + 0.1074^2}{1 - 0.3825} = 0.7111$$

$$D_3 = 1 - D_1 - D_2 = 1 - 0.1074 - 0.7111 = 0.1815$$

$$Q = \frac{1.25(\gamma - D_3 - D_2 \times R)}{D_1} = \frac{1.25(0.465 - 0.1815 - 0.7111 \times 0.3825)}{0.1074} = 0.1339$$

Substituting these values into Equation 10.8.6 gives the Dirlik's PSD of stress amplitudes as follows:

$$f_{S_a}(s_a) = 0.003587e^{-0.0334 s_a} + 9.72 \times 10^{-5} e^{-6.84\times 10^{-5} s_a^2} + 3.63 \times 10^{-6} e^{-1.0\times 10^{-5} s_a^2}$$

The fatigue damage in a given time period of 1 second is

$$D_{\text{WB, Dirlik}} = \frac{E[P]\tau}{A}\int_0^\infty S_a^m f_{S_a}(s_a)ds_a = \frac{9.82 \times 1}{1.02 \times 10^{17}}\int_0^\infty s_a^{5.56} f_{S_a}(s_a)ds_a$$

$$= 9.627 \times 10^{-17}\int_0^\infty s_a^{5.56} f_{S_a}(s_a)ds_a$$

The previous equation can be approximated by the discrete intergration scheme listed next:

$$D_{\text{WB, Dirlik}} \cong 9.627 \times 10^{-17} \sum_{i=1}^{\infty} (0.003587 s_{a,i}^{5.56} e^{-0.0334 s_{a,i}} +$$
$$9.72 \times 10^{-5} s_{a,i}^{5.56} e^{-6.84\times10^{-5} s_{a,i}^2}) \Delta s_a +$$
$$9.627 \times 10^{-17} (\sum_{i=1}^{\infty} 3.63 \times 10^{-6} s_{a,i}^{5.56} e^{-1.0\times10^{-5} s_{a,i}^2}) \Delta s_a$$

where

$$s_{a,i} = \frac{\Delta s_a}{2} + (j-1)\Delta s_a$$

The numerical integration technique leads to $D_{\text{WB, Dirlik}} = 1.268 \times 10^{-5}$/second. This corresponds to a fatigue life of 79,000 seconds, which correlates better to the baseline fatigue life (93,000 seconds). Therefore, the Dirlik method is the preferred method for fatigue damage calculation based on the PSD and has been widely adopted by many commercial fatigue software packages.

REFERENCES

Bendat, J. S., and Piersol, A. G., *Random Data, Analysis and Measurement Procedures*, 2nd ed., Wiley Interscience, New York, 1986.

Bishop, N. W. M., Spectral methods for estimating the integrity of structural components subjected to random loading. In *Handbook of Fatigue Crack Propagation in Metallic Structures*, Vol. 2, Carpinteri, A. (Eds.), Elsevier, Dordrecht, 1994, pp. 1685–1720.

Bishop, N. W. M., *The Use of Frequency Domain Parameters to Predict Structural Fatigue*, Ph.D. Thesis, Warwick University, 1988.

Bishop, N. W. M., and Sherratt, F., Fatigue life prediction from power spectral density data. Part 2: Recent Development, *Environmental Engineering*, Vol. 2, Nos. 1 and 2, 1989, pp. 5–10.

Crescimanno, M. R., and Cavallo, P., On duty simulation of a trimmed body under dynamics loads: modal superposition approach to evaluate fatigue life, SAE Paper 1999-01-3150, 1999.

Dirlik, T., *Application of Computers in Fatigue Analysis*, Ph.D. Thesis, Warwick University, 1985.

Lee, Y., Raymond, M. N., and Villaire, M. A., Durability design process of a vehicle suspension component, *Journal of Testing and Evaluation*, Vol. 23, No. 5, 1995, pp. 354–363.

Newland, D. E., *An Introduction to Random Vibrations and Spectral Analysis*, 2nd ed., Longman, New York, 1984.

Oritz, K., and Chen, N. K., Fatigue damage prediction for stationary wide-band stresses. Presented at the *5th International Conference on the Applications of Statistics and Probability in Civil Engineering*, Vancouver, Canada, 1987.

Vellaichamy, S., Transient dynamic fatigue analysis using inertia relief approach with modal resonance augmentation, SAE Paper 2002-01-3119, 2002.

Wirsching, P. H., and Light, M. C., Fatigue under wide band random stresses, *ASCE Journal of Structural Division*, Vol. 106, 1980, pp. 1593–1607.

Wirsching, P. H., Paez, T. L., and Oritiz, K., *Random Vibrations: Theory and Practice*, Wiley, New York, 1995.

INDEX

A
Accelerated life test, 313, 314
A-D conversion, 55
Aliasing, 58
Aliasing error, 63
Amplifying, 54
Amplitude ratio, 104
Analog filters, 58
Analog-digital converter, 59
Anti-aliasing filters, 62
ASTM standard
  ASTM E1012, 183
  ASTM E606, 182–184
  ASTM E739, 186
  ASTM E8M, 182, 183
Asymptotic crack tip stress, 240
  HRR crack tip stress, 259
  mode I, 242
  mode II, 242
  mode III, 243
  non-singular stress, 247
  T stress, 247
Attenuating, 54
Attribute analysis method, 349
Autocorrelation function, 374

B
Basquin relation, 105
Basquin equation, 192, 198
Bayesian approach, 350, 351
Bel
  decibel, 379, 380
Bias sampling approach, 347–350
Binomial test method
  for a finite population, 340–342
Block load cycle test, 326–329, 334–336
Bogey, 316, 329, 335
B_q life
  B_{10} life, 343–346,356,357
  B_{50} life, 357
Bridgman correction factor, 186
Brittle coating, 41
Button, 286

C
Calibration resistor, 31
Cantilever beam, 34
Characteristic life, 343–345, 353
Coefficient of variation, 109
Coffin-Manson equation, 192, 198
Cold forming, 182
Comparison testing analysis method,
    356, 357
Component life test, 314, 326–336
Confidence, 111, 196–198, 338, 339
Confidence interval
  double-sided, 112
Constant-amplitude fatigue test, 326–327
Constant Current Wheatstone Bridge, 17
Constant Voltage Wheatstone Bridge, 21

Correlation coefficient, 374
Corrosion, 127
Covariance matrix, 333
Crack, 104, 105, 187
  growth, 191, 201
  initiation, 124, 181, 201, 210
  macrocrack, 127
  microcrack, 127, 134, 141
  nucleation, 128, 181
  propagation, 124, 141
Crack geometry
  center crack, 243
  center cracked panel, 245
  circular crack, 248
  crack emanating from hole, 247
  edge crack, 244
  edge cracked plate, 246
  elliptical crack, 250
  kinked crack, 267
  penny shaped crack, 245, 248
  semi-circular crack, 249
  semi-elliptical crack, 251
  surface crack, 249
Critical plane approach, 295
Cross-talk, 33
Cumulative damage concept, 60
Cumulative damage model, 60–66
  damage curve approach, 60–67, 75
  double-linear (bilinear) damage rule, 72–74
  linear damage model (rule), 66–68
Cumulative exceedance histogram, 327–329, 335
Cumulative frequency distribution, 327
Cutoff frequency, 66
Cycle counting techniques
  one-parameter, 78–82
    level crossing, 78–79
    peak-valley, 79–80
    range counting, 80–82
  two-parameter, 82–96
    four-point cycle counting, 89–96
    rainflow cycle, 82–84
    range-pair method, 84, 85, 88
    three-point cycle counting, 84–89
Cycle extraction, 79, 85, 97
Cycle extrapolation, 314, 318, 319, 326, 327–332, 334
Cycle ratio, 61–62, 66–69, 77
Cyclic hardening, 187
Cyclic softening, 187
Cyclic strain hardening exponent, 209
Cyclic stress coefficient, 209

D

Damage accumulation
  Miner-Haibach model, 127
  Miner's rule, 127
Damage parameter, 289, 294, 301
Damage rule, 316, 320
Data acquisition, 1, 58
Data reduction techniques
  Dixon and Mood method, 117
  Zhang and Kececioglu method, 117
Deformation
  cyclic, 181
  cyclic plastic, 181
Digital filters, 59
Digitizing, 54
Discrete Fourier transform, 73
Discrete stress spectrum, 376
Dislocation, 187
Distortion energy theory, 155
Distribution
  chi-square, 363, 364
  F, 340, 365, 366
  hypergeometric, 341
  lognormal, 108
  mean, 118
  normal, 118, 354, 356, 362
  standard deviation, 118
  Weibull, 342–346
Drift, 70
Driver variability, 314, 317, 319–320, 334
Dropout, 70, 72
Dummy gages, 27, 30
Dynamometer
  differential, 315–317, 321–322
  transmission, 315, 323–324, 326

E

Effective stress, 154
Elastic modulus, 12
Electrical noise, 70
Endurance limit. *See* fatigue limit.
Energy release rate, 239, 258, 259
  J-integral, 259, 302
Equivalent damage cycles, 65
Equivalent force, 290
Excitation voltage, 5, 16
Expulsion, 286

Extended test method, 345, 346
Extensometer, 182

F

Fatigue
high cycle, 124, 141, 182, 193, 196
life, 103
low cycle, 124, 141, 182, 193, 195
Fatigue crack growth, 260
crack growth rate, 261
Forman equation. *See* load ratio
life estimation. *See* load ratio
load ratio, 261
Paris law, 261
stress intensity factor range, 260
stress intensity factor range threshold, 261
Walker equation. *See* load ratio
Fatigue damage, 77, 78, 386–394
Fatigue damage mechanism, 57–59
beach markings, 59
clamshell, 59
discontinuities, 58
inclusions, 58
persistent slip bands, 58
porosity, 58
shear lip, 59
striations, 59–60
Fatigue ductility coefficient, 192, 209
Fatigue ductility exponent, 192, 209
Fatigue limit, 59, 63, 104, 117, 127
bending, 124, 127, 127
estimated, 126
Fatigue notch factor. *See* fatigue strength reduction factor
Fatigue process, 58–59
crack nucleation, 58–59
final fracture, 57–58, 60
long crack growth, 58–59
short crack growth, 58–59
stage I crack, 58
stage II crack, 58
Fatigue strength
estimated at 1000 cycles, 125
estimated at 1000 cycles for various loading coefficient, 192, 202, 208
conditions, 125
exponent, 192, 208
reduction factor, 211
testing, 117
test methods, 117
Fatigue strength coefficient, 105, 109, 321, 329, 386, 390
Fatigue strength exponent, 105, 109, 321, 390
Fatigue strength reduction factor, 124
Fatigue testing
alignment, 183
design allowable testing, 107
preliminary testing, 107
reliability testing, 107
replicate data, 107
research and development testing, 107
strain-controlled, 181
strain-controlled testing system, 183
strain-life, 182
stress control, 184
F-distribution, 112
Filtering, 54
Filters
active filters, 57
band pass filters, 56
high pass filters, 56
low pass filters, 56
notch pass filters, 56
passive filters, 56
Finite element analysis, 289
Finite element method, 301
Fourier analysis
fast Fourier transformation (FFT), 369
forward Fourier transform, 376
Fourier series, 375
inverse Fourier transformation (IFT), 369
trigonometric series, 375
Fourier transform, 73
Fracture ductility, 184, 186, 209
Fracture mechanics, 302
Fracture mode
mode I, 240
mode II, 240
mode III, 240
Fracture strength, 184, 208

G

Gamma function, 388
Grain, 128, 135
barrier, 127, 128

H

Hardness
Brinell, 125, 208

Hole
  circular hole, 238
  elliptical hole, 237
Holographic interferometry, 43
Hysteresis loop
  cycle(s), 78, 81, 97–100
  hanging cycle, 90, 95, 97
  reversal(s), 78, 81, 97–100
  stabilized, 192
  standing cycle, 90, 95, 97

I
Inertia relief method, 369
Intensity function, 359, 360
Interface forces, 305
Irregularity factor, 384, 385

J
J-integral, 259, 302

K
Kernel estimator, 332
Knee. *See* fatigue limit
K-type thermocouple, 48

L
Lead wire signal attenuation, 20
Least squares analysis, 186, 192, 196
LEFM approach; see *linear elastic fracture mechanics*
Level crossing rate, 382–385
Limited load factor, 216
Linear elastic fracture mechanics (LEFM), 77
  fracture toughness, 257
  fracture toughness testing, 256
  plane strain requirement. *See* plastic zone
  requirement. *See* plastic zone
Loading
  axial, 105, 125
  bending, 105, 125
  constant amplitude, 77, 78, 82, 87
  fully-reversed, 104
  proportional, 154
  torque, 105
  torsional, 125
  variable amplitude, 78, 82
Load history
  irregular, 181
Load-life, 290
Load parameters, 78
  acceleration, 78
  deflection, 78
  force, 78
  strain, 78
stress, 78
torque, 78
Load sequence effects, 181

M
Massing equation, 188, 197, 198
Matrix based load transducer, 26–27
Mean deviation, 69
Mean stress, 141, 181
  multiaxial, 155
  shear, 144
Mean stress correction, 201
  Gerber model, 141, 142
  Goodman model, 141, 142, 155
  Haigh model, 141
  mean stress sensitivity factor, 143
  Morrow model, 143
  Morrow's method, 201
  Smith, Watson, and Topper (SWT) model, 143, 202
  Soderberg model, 142
Mean stress effect, 141, 182
Mean stress sensitivity, 296
Mechanical properties
  monotonic, 184
Memory effect, 206
Modal transient analysis
  modal stress, 369
  modal coordinates, 369
Modifying factors, 124, 129
  Collin' notch effect method, 138
  Heywood's notch effect method, 138
  loading, 129
  reliability, 125, 126
  specimen size, 131
  surface finish, 130
Modulus of elasticity, 184
Mohr's circle, 10
Monte Carlo simulations, 389
Morrow proposal, 192
Multiaxial, 288

N
e-N approach
  strain-life, 77
Natural frequency, 74
Necking, 185, 186

Neuber rule, 211, 215
modified, 212
Non-homogeneous Poisson process (NHPP), 358, 359
Nonlinear damage curve, 61–62
Nonparametric density estimator, 332
Nonproportional, 292, 296
Notch, 181
analysis, 210
blunting, 135
effect, 133
maximum elastic stress, 133
nominal stress, 133
nominally elastic behavior, 212
nominally gross yielding of a net section, 214
plasticity, 182
strain, 211
stress, 211
Notch effect
at intermediate and short lives, 137
Notch root, 213
plasticity, 211
Notch sensitivity, 135
Heywood model, 136
Neuber model, 136
Peterson model, 135
relative stress gradient, 136
Nyquist frequency, 64

O

Octave, 378–380
One-side power spectral density function, 384
Optical pyrometers, 51
Optical Strain Measuring, 43
Over-constraint, 303
Overload, 104, 127

P

Peak
highest, 85, 86
maximum, 85
Peak crossing, 384
Percent elongation, 184, 185
Percent reduction area, 184, 185
Percent replication, 107
Photo-stress coating, 42
Piezoresistance, 8
Plastic deformation, 141
Plastic limit load, 215
Plastic zone
cyclic loading, 248
monotonic loading, 257
non-singular stress, 255
plane strain, 254, 255
plane stress, 252, 255
Plasticity, 144
Poisson's ratio, 8, 12
Posterior distribution, 351
Power law damage equation, 61, 63
Power spectral density (PSD), 73, 369, 370, 378–380, 390, 391
Power spectrum, 67
Power spectrum density, 67
Powertrain endurance (PTE), 316
Principle strains, 10
Principle stresses, 12
Prior distribution, 350
Probability
mean, 371–374, 386
of occurrence, 314, 317, 331
probability density function, 330–332, 370–372
standard deviation, 372–374
variance, 371, 373
Proving grounds (PG), 313

Q

Quantile extrapolation, 314, 319–320, 321, 326–327, 332–333
Quenching, 128

R

R^2 regression statistic, 196
Rainflow correction factor, 389, 392
Rainflow matrix
from-to, 87, 90, 95
range-mean, 93, 94
Ramberg-Osgood equation, 186
Random process
ensemble, 372, 373
ergodic, 373
Gaussian, 370, 372
narrow band, 378, 379, 382–388
stationary, 372, 373
white noise, 378, 379
wide band, 378, 379, 384, 388–394
Rank
median, 344, 354, 355, 367
90%, 354, 367
tables, 355, 362–368
Rate of occurrence of failures (ROCOF), 359

Rayleigh distribution, 388
Reconstruction, 96–101
Reference life level, 62
Reference load level, 79, 80
Reference stress amplitude, 63
Regression, 112, 189, 196–198
  least squares, 108
Regression analysis
  least squares method, 108
Reliability, 111, 125, 196–198, 337–339
Repairable system reliability prediction, 358, 359
Residual strain, 213
Residual stress, 130, 181, 182, 213
Resistance, 2
Resistivity, 2
Reversal, 103
Root mean square (RMS), 71, 372, 377, 380
Rosette, 10
Rotating bending testing, 123
Rotating moment histogram (RMH), 316–319, 321

S

SAE Fatigue Design Handbook, 79, 80
Safe design stress region, 142
Sampling frequency, 62
Sampling rates, 67
Semiconductor strain gages, 17
Sensitivity coefficient matrix, 37, 38
Shannon's sampling theorem, 63
Shear force load cell, 35
Shear modulus, 12
Shear strain, 10
Shearography, 45
Signal conditioner, 58
Signal-to-noise ratio, 8
Skewness, 69
Slip ring, 75
S-N approach
  stress-life, 77
S-N curve, 106, 155. *See* Wöhler curve
  baseline, 124, 125, 154
  component, 123
  for design, 111
  notch effects, 138
  test methods, 106
Specimen fabrication, 182
Speckle-shearing interferometry, 44
Spectral density
  mean square spectral density, 377
  spectral width, 384, 389
Spike, 70, 72
Spot weld
  cup specimen, 267
Spot-weld fatigue, 285
Stacked rectangular strain rosettes, 10
Staircase testing method, 117
Standard deviation, 69
Static stress analysis
  stress influences, 369
Step-stress accelerated test (SSAT) method, 352–355
Step-stress testing, 184
Strain
  elastic, 186
  engineering, 184
  nominal, 211
  plastic, 186
  true, 185
Strain amplitude
  plastic, 188, 189
Strain concentration, 211
Strain energy density, 214
Strain gages, 1, 2, 9, 59
Strain gage
  bridge calibration, 30
  gage factor, 8, 31
  gage length, 6
  gage material, 7
Strain hardening exponent, 189
  cyclic, 187, 188, 197
  monotonic, 184
Strain-life
  relationship, 181
Strain-life approach, 181, 201
Strain-life curve
  elastic, 197
  plastic, 197
  total, 197
Strain-life equation, 192
Strain-life method, 181
Strain range, 187
Strength coefficient
  cyclic, 187, 188, 197
  monotonic, 184
Stress
  amplitude, 103, 184, 188
  engineering, 184

equivalent fully-reversed bending, 155
maximum, 103
maximum principal, 154
mean, 104
minimum, 103
multiaxial, 154
nominal, 211
range, 103
ratio, 104
true, 185
Stress amplitude
equivalent bending, 155
Stress concentration, 133, 181
elastic, 211
Stress concentration factor, 238
elastic, 133, 211
Stress field intensity, 134
Stress gradient, 211
Stress intensity, 290, 301
Stress intensity factor
mode I, 241
mode II, 242
mode III, 242
Stress-life approach. *See* S-N
Stress range, 187
Stress-strain curve, 185
cyclic, 188, 189, 212
monotonic, 189
Stress-strain relationship, 183
Stress-strain response
steady state cyclic, 187
Structural stress, 290–292, 302, 306
Surface finish, 181, 182

T

Telemetry systems, 75
Temperature effects, 183
Tensile-shear, 288
specimen, 287, 288, 306
Tension load cell, 34
Tension test
monotonic, 183
Test bogey, 338, 346
Test equipment, 182
Test equpment, 182
Testing
monotonic tensile, 182
Thermistors, 49
Three lead wire configuration, 28
Tolerance limit
approximate Owen, 114
double-sided confidence interval, 112
lower Owen single-sided, 112, 193
one-side lower bound, 111
single-sided, 118, 119, 120
Torsion load cell, 35
Transient cyclic response, 186
Transition fatigue life, 192, 193, 199, 209
Transmission gear ratio, 323, 324
Transmission gear efficiency, 323
Two lead wire configuration, 27
Two-phase process, 68
Phase I damage, 68–69, 71–74
Phase I fatigue life, 72
Phase II damage, 68, 69, 71, 72, 74, 75
Phase II fatigue life, 72
Two-step high–low sequence loading, 61

U

Ultimate strength, 124, 125
estimated, 125
shear, 125
tensile, 184, 185, 208
Uniform Material Law, 209
Up-and-down testing method. *See* staircase testing method

V

Valley
Minimum, 85
Variable amplitude loading, 386
Variance, 69, 128
homoscedasticity, 108
Voltage divider circuit, 15
von Mises stress, 154

W

Weibull analysis method, 342–345
Weibull analysis of reliability data with few or no failures, 346, 347
Weibull distribution
characteristic life, 387
scale parameter, 387
shape parameter, 387
Weibull slope, 343–345, 387
Weld
spot, 104, 144
Welding, 182, 213
Weld schedule, 286

Welds, 144, 181
Wheatstone bridges, 14, 59
  full bridge, 22
  half bridge, 22
  quarter bridge, 22, 23
Wöhler curve. *See* S-N curve
Wöhler, August, 106
Work, 187

Y
Yield strength, 184, 185
Yield stress, 211
  cyclic, 187,188

Z
Zero upcrossings 383, 384

Printed and bound by CPI Group (UK) Ltd, Croydon, CR0 4YY
08/05/2025
01864816-0002